10 0517134 4
D0303023

IET POWER AND ENERGY SERIES 50

Wind Power Integration

Other volumes in this series:

Volume 1	**Power circuit breaker theory and design** C.H. Flurscheim (Editor)
Volume 4	**Industrial microwave heating** A.C. Metaxas and R.J. Meredith
Volume 7	**Insulators for high voltages** J.S.T. Looms
Volume 8	**Variable frequency AC-motor drive systems** D. Finney
Volume 10	**SF6 switchgear** H.M. Ryan and G.R. Jones
Volume 11	**Conduction and induction heating** E.J. Davies
Volume 13	**Statistical technjiques for high voltage engineering** W. Hauschild and W. Mosch
Volume 14	**Uninterruptible power supplies** J. Platts and J.D. St Aubyn (Editors)
Volume 15	**Digital protection for power systems** A.T. Johns and S.K. Salman
Volume 16	**Electricity economics and planning** T.W. Berrie
Volume 18	**Vacuum switchgear** A. Greenwood
Volume 19	**Electrical safety: a guide to causes and prevention of hazards** J. Maxwell Adams
Volume 21	**Electricity distribution network design, 2nd edition** E. Lakervi and E.J. Holmes
Volume 22	**Artificial intelligence techniques in power systems** K. Warwick, A.O. Ekwue and R. Aggarwal (Editors)
Volume 24	**Power system commissioning and maintenance practice** K. Harker
Volume 25	**Engineers' handbook of industrial microwave heating** R.J. Meredith
Volume 26	**Small electric motors** H. Moczala *et al.*
Volume 27	**AC-DC power system analysis** J. Arrillaga and B.C. Smith
Volume 29	**High voltage direct current transmission, 2nd edition** J. Arrillaga
Volume 30	**Flexible AC Transmission Systems (FACTS)** Y-H. Song (Editor)
Volume 31	**Embedded generation** N. Jenkins *et al.*
Volume 32	**High voltage engineering and testing, 2nd edition** H.M. Ryan (Editor)
Volume 33	**Overvoltage protection of low-voltage systems, revised edition** P. Hasse
Volume 34	**The lighting flash** V. Cooray
Volume 35	**Control techniques drives and controls handbook** W. Drury (Editor)
Volume 36	**Voltage quality in electrical power systems** J. Schlabbach *et al.*
Volume 37	**Electrical steels for rotating machines** P. Beckley
Volume 38	**The electric car: development and future of battery, hybrid and fuel-cell cars** M. Westbrook
Volume 39	**Power systems of electromagnetic transients simulation** J. Arrillaga and N. Watson
Volume 40	**Advances in high voltage engineering** M. Haddad and D. Warne
Volume 41	**Electrical operation of electrostatic precipitators** K. Parker
Volume 43	**Thermal power plant simulation and control** D. Flynn
Volume 44	**Economic evaluation of projects in the electricity supply industry** H. Khatib
Volume 45	**Propulsion systems for hybrid vehicles** J. Miller
Volume 46	**Distribution switchgear** S. Stewart
Volume 47	**Protection of electricity distribution networks, 2nd edition** J. Gers and E. Holmes
Volume 48	**Wood pole overhead lines** B. Wareing
Volume 49	**Electric fuses, 3rd edition** A. Wright and G. Newbery
Volume 50	**Wind power integration** B. Fox *et al.*
Volume 51	**Short circuit currents** J. Schlabbach
Volume 52	**Nuclear power** J. Wood
Volume 905	**Power system protection, 4 volumes**

Wind Power Integration

Connection and system operational aspects

Brendan Fox, Damian Flynn, Leslie Bryans,
Nick Jenkins, David Milborrow, Mark O'Malley,
Richard Watson and Olimpo Anaya-Lara

The Institution of Engineering and Technology

Published by The Institution of Engineering and Technology, London, United Kingdom

First published 2007

The Institution of Engineering and Technology
Michael Faraday House
Six Hills Way, Stevenage
Herts, SG1 2AY, United Kingdom

www.theiet.org

British Library Cataloguing in Publication Data
Wind power integration: connection and system operational aspects
1. Wind power
I. Fox, Bendan II. Institution of Electrical Engineers
621.3'12136

ISBN 978-0-86341-449-7

Typeset in India by Newgen Imaging Systems (P) Ltd, Chennai
Printed in the UK by MPG Books Ltd, Bodmin, Cornwall

Contents

Preface

The impetus for the book is the rapid growth of wind power and the implications of this for future power system planning, operation and control. This would have been a considerable challenge for the vertically integrated power companies pre-1990. It has become an even greater challenge in today's liberalised electricity market conditions. The aim of the book is to examine the main problems of wind power integration on a significant scale. The authors then draw on their knowledge and expertise to help guide the reader through a number of solutions based on current research and on operational experience of wind power integration to date.

The book's backdrop was the commitment of the UK government (and European governments generally) to a target of 10 per cent of electrical energy from renewable energy sources by 2010, and an 'aspirational goal' of 20 per cent by 2020. There has also been a significant reduction in the cost of wind power plant, and hence energy cost. Where average wind speeds are 8 m/s or more, as is the case for much of Great Britain and Ireland, the basic production cost of wind energy is nearly competitive with electricity from combined cycle gas turbine (CCGT) plant, without the concern about long-term availability and cost. The downside is that the supply over the system operational time-scale is difficult to predict. In any case, wind power cannot provide 'firm capacity', and therefore suffers commercially in markets such as BETTA (British electricity trading and transmission arrangements). On the other hand, green incentives in the form of renewable obligation certificates (ROCs) provide wind generators with a significant extra income. This is encouraging developers to come forward in numbers which suggest that the 10 per cent target may be attained. Indeed, Germany, Spain and Ireland are already experiencing wind energy penetration levels in the region of 5 per cent, while Denmark reached a level of 20 per cent some years ago.

The book attempts to provide a solid grounding in all significant aspects of wind power integration for engineers in a variety of disciplines. Thus a mechanical engineer will learn sufficient electrical power engineering to understand wind farm voltage regulation and fault ride-through problems; while an electrical engineer will benefit from the treatment of wind turbine aerodynamics. They will both wish to understand electricity markets, and in particular how wind energy is likely to fare.

The introductory chapter charts the remarkable growth of wind energy since 1990. The various technical options for wind power extraction are outlined. This chapter then goes on to describe the potential problems of large-scale wind integration, and outlines some possible solutions. The second chapter is essentially an electrical power engineering primer, which will enable non-electrical engineers to cope with the concepts presented in Chapters 3 and 4. Chapter 3 deals with wind turbine generator technology, with particular attention being paid to current variable-speed designs. Chapter 4 is concerned with wind farm connection, and the implications for network design - an area lacking an established methodology to deal with variable generation.

Chapter 5 addresses the key issue of power system operation in the presence of largely unpredictable wind power with limited scope for control. Energy storage provides a tempting solution: the treatment here concentrates on realistic, low-cost options and imaginative use of existing pumped storage plant. The importance of wind power forecasting is emphasised, and forms the main theme of Chapter 6. The encouraging progress in the last decade is described. Ensemble forecasting offers a useful operational tool, not least by providing the system operator with an indication of forecast reliability. Finally, Chapter 7 summarises the main types of electricity market, and discusses the prospects for wind power trading. The main renewable energy support schemes are explained and discussed.

The book arose largely from a number of workshops organised as part of an EPSRC (Engineering and Physical Sciences Research Council) network on 'Operation of power systems with significant wind power import'. This was later known simply as the BLOWING (Bringing Large-scale Operation of Wind power Into Networks and Grids) network. The book reflects many lively discussions involving the authors and members of the network, especially Graeme Bathurst, Richard Brownsword, Edward Clarke, Ruairi Costello, Lewis Dale, Michael Farrell, Colin Foote, Paul Gardner, Sean Giblin, Nick Goodall, Jim Halliday, Brian Hurley, Michael Jackson, Daniel Kirschen, Lars Landberg, Derek Lumb, Andy McCrea, Philip O'Donnell, Thales Papazoglou, Andrew Power and Jennie Weatherill. Janaka Ekanayake, Gnanasamnbandapillai Ramtharan and Nolan Caliao helped with Chapter 3. We should also mention Dr Shashi Persaud, whose PhD studies at Queen's University in the late 1990s were instrumental in drawing up the network proposal, and who later helped with its ongoing administration.

B Fox, Belfast
November 2006

Acronyms

AC	alternating current
AWPT	Advanced Wind Prediction Tool
BETTA	British Electricity Trading and Transmission Arrangements
CCGT	combined cycle gas turbine
DC	direct current
DFIG	doubly-fed induction generator
DSO	Distribution System Operator
DTI	Department of Trade and Industry (UK)
EMS	energy management system
FACTS	flexible alternating current transmission system
GTS	Global Telecommunication System
HIRLAM	High Resolution Limited Area Model
HV	high voltage
IGBT	insulated-gate bipolar transistor
ISO	Independent System Operator
LDC	line drop compensation
MAE	mean absolute error
MOS	model output statistics
NETA	New Electricity Trading Arrangements (BETTA's predecessor)
NFFO	Non-Fossil Fuel Obligation
NWP	numerical weather prediction
OCGT	open cycle gas turbine
OFGEM	Office of Gas and Electricity Markets (UK)
PFCC	power factor correction capacitance
PI	proportional + integral
PWM	pulse width modulation
RAS	remedial action scheme
RMSE	root mean square error
ROC	Renewable Obligation Certificate
ROCOA	rate of change of angle
ROCOF	rate of change of frequency
SCADA	supervisory, control and data acquisition (system)

SDE	standard deviation of errors
SEM	Single Electricity Market (Ireland)
STATCOM	static compensator
SVC	static VAr compensator
TSO	Transmission System Operator
UTC	Co-ordinated Universal Time
VIU	vertically integrated utility
VSC	voltage source converter
WAsP	Wind Atlas Analysis and Applications Program
WPPT	Wind Power Prediction Tool
WTG	wind turbine generator

Chapter 1

Introduction

1.1 Overview

The main impetus for renewable energy growth has been increasing concern over global warming, and a range of policy instruments have been used to promote carbon-free technologies; these are briefly reviewed. Unsurprisingly, most growth has taken place in locations with generous subsidies, such as California (in the early days), Denmark, Germany and Spain.

The attributes of the renewable energy technologies are summarised, including those which may develop commercially in the future. However, wind energy has sustained a 25 per cent compound growth rate for well over a decade, and total capacity now exceeds 60,000 MW. With the growth of the technology has come increased reliability and lower generation costs, which are set alongside those of the other thermal and renewable sources.

'What happens when the wind stops blowing?' is an intuitive response to the growth of wind energy for electricity generation, but it is simplistic. In an integrated electricity system what matters to the system operators is the additional uncertainty introduced by wind. Several studies have now quantified the *cost of intermittency* – which is modest – and also established that the wind can displace conventional thermal plant.

While wind turbines for central generation now approach 100-m in diameter and up to 5 MW in rating, with even larger machines under development, off-grid applications are generally much smaller and the criteria for successful commercial exploitation are different, as generation costs are frequently high, due to the use of imported fuels.

In the future, offshore wind is likely to play an increasing role, partly on account of the reduced environmental impact. It is being actively deployed in Denmark, Germany, Ireland, the United Kingdom and elsewhere, and is likely to contribute to the continuing strong growth of wind capacity, which may reach 200,000 MW by 2013.

1.2 World energy and climate change

World primary energy demand almost doubled between 1971 and 2003 and is expected to increase by another 40 per cent by 2020. During the last 30 years there has been a significant shift away from oil and towards natural gas. The latter accounted for 21 per cent of primary energy and 19 per cent of electricity generation, worldwide, in 2003 (International Energy Agency, 2005).

When used for heating or electricity, natural gas generates lower carbon dioxide emissions than coal or oil, and so the rise in carbon dioxide emissions for the past 30 years did not match the growth in energy demand, but was about 70 per cent. However, increasing concerns over global warming have led world governments to discuss ways of slowing down the increase in carbon dioxide emissions. International climate change negotiations are proceeding under the auspices of the United Nations, and – at a key meeting at Kyoto – in December 1997, an overall target for global reduction of greenhouse gases by 5 per cent was agreed (between 1990 and the target date window of 2008–12) and national targets were then set. The Kyoto protocol finally became legally binding on 16 February 2005, following Russia's ratification in November 2004.

The European Union has undertaken to reduce emissions by 8 per cent and has agreed a share-out of targets among its members. The United Kingdom is committed to a 12.5 per cent reduction during the same period. It may be noted that the government is also committed, in its 1997 manifesto, to a reduction of 20 per cent in carbon dioxide emissions during the same period.

1.2.1 Renewable energy

In 2003, renewable energy contributed 13.5 per cent of world total primary energy (2.2 per cent hydro, 10.8 per cent combustible renewables and waste, and 0.5 per cent geothermal, solar and wind). As much of the combustible renewables are used for heat, the contributions for electricity generation were somewhat different: hydro contributed 15.9 per cent and geothermal, solar, wind and combustible renewables contributed 1.9 per cent. The capacity of world hydro plant is over 800 GW and wind energy – which is developing very rapidly – accounted for about 60 GW by the end of 2005. Although significant amounts of hydro capacity are being constructed in the developing world, most renewable energy activity in the developed world is centred on wind, solar and biomass technologies. In 2003, world electricity production from hydro accounted for 2,654 TWh and all other renewables delivered 310 TWh.

There are no technical or economic reasons constraining the further development of hydroelectric energy, but large-scale developments need substantial areas of land for reservoirs and such sites tend to be difficult to identify. For this reason, large-scale hydro generally does not come under the umbrella of most renewable energy support mechanisms. Small-scale developments, including *run of river* schemes, are generally supported, but these tend to have higher generation costs – typically around 5p/kWh and above. Future developments of tidal barrage schemes are likely to be constrained for similar reasons, although the technology is well understood and proven. Tidal stream technology, on the other hand, is a relatively new technology, which involves harnessing tidal currents using underwater turbines that are similar in concept to wind turbines. It is the focus of considerable research activity within the European Union and prototype devices are currently being tested. Wave energy is also at a similar stage of development, with research activity right across the Organisation for Economic Co-operation and Development (OECD) and a number of prototype devices currently under test.

1.2.1.1 Support mechanisms

Although some renewable energy sources can – and do – compete commercially with fossil sources of energy, their emerging status is generally recognised by various methods of support. Over the last decade or so, several types of system have appeared:

- *Capital subsidies*: These were at least partially responsible for a very rapid expansion of wind energy activity in California in the early 1980s. (Generous production subsidies were also a contributory factor.) Capital subsidies also appeared in Europe but are now rare.
- Several European countries have supported renewables through a system of standard payments per unit of electricity generated – often a percentage of the consumer price. Of these mechanisms, the German and Danish mechanisms have stimulated the markets extremely effectively. The German support mechanism for wind is now more sophisticated as it is tailored to the wind speed at the specific sites.
- *Competitive bidding*: This is exemplified by the non-fossil fuel obligation (NFFO) in the United Kingdom. Developers bid for contracts, specifying a price at which they are able to generate. Those bidding underneath a hurdle set by government in the light of the capacity that is required are then guaranteed long-term contracts. This mechanism underwent various changes in the United Kingdom and has been emulated in France and Ireland.
- *Partial subsidies of the energy price*: The US production tax credit is a good example of this; successful renewable energy projects qualify for a premium of $0.018 for each unit of electricity generated.
- *Obligations*: These are typified in the Renewables Portfolio Standard in the United States and the Renewables Obligation in Britain. In essence, electricity suppliers are mandated to source specified percentages of their electricity from renewable sources by specified dates. Failure to meet the obligation is penalised by *buy-out* payments.

Standard payments in general have been relatively successful in encouraging deployment, although they do not provide strong incentives to reduce prices.

(The current German system attempts to overcome this difficulty by stepping down the payment, year-by-year.) Competitive schemes, such as the NFFO, are sometimes less successful in terms of deployment, although the UK NFFO was very successful in bringing prices down.

1.2.1.2 UK renewable energy: capacity and targets

The UK Government has introduced a range of measures aimed at reducing greenhouse gas emissions, including the Climate Change Levy, energy efficiency measures and initiatives in other sectors, such as transport and agriculture. The renewable energy target is for 10.4 per cent of electricity to be provided by renewable sources by 2010; it builds up year-by-year to this target. A draft EU directive to promote electricity from renewable sources proposes the same (non-binding) target for the United Kingdom.

Electricity suppliers meet their obligation by any combination of the following means:

- Physically supplying power from renewable generating stations.
- Purchasing Renewable Obligation Certificates (ROCs) independently of the power that gave rise to their issue.
- Paying a *buy-out* price to the regulator. This was initially set at 3p/kWh by the Office of Gas and Electricity Markets (OFGEM) in the United Kingdom.

The UK Department of Trade and Industry (DTI) expects the buy-out price to become the market price (due to a lack of liquidity) and estimates the eventual cost of the obligation will rise to about £1,000 million per annum, adding about 5 per cent to electricity bills. However, there is considerable scepticism as to whether the target will be reached. It means that the average rate at which renewables are deployed needs to increase by a factor of around 5. Generation from renewable energy sources is expected to build towards 38 TWh by that date.

In 2002, total generation amounted to 12.6 TWh, of which 4.7 TWh came from hydro, 2.7 TWh from landfill gas and 1.3 TWh from wind. Total wind capacity was just under 600 MW.

1.3 Wind energy

1.3.1 Background

World wind energy capacity doubled every three years from 1990 to 2005, as shown in Figure 1.1 (*Windpower Monthly*, 'The Windicator', 1997–2005). It is doubtful whether any other energy technology is growing, or has grown, at such a remarkable rate. Installed capacity in 2005 grew by 26 per cent so that total world wind turbine capacity at the end of the year was around 60,000 MW. Germany, with over 18,000 MW, has the highest capacity but Denmark, with over 3,000 MW, has the highest per capita level and wind production in western Denmark accounts for about

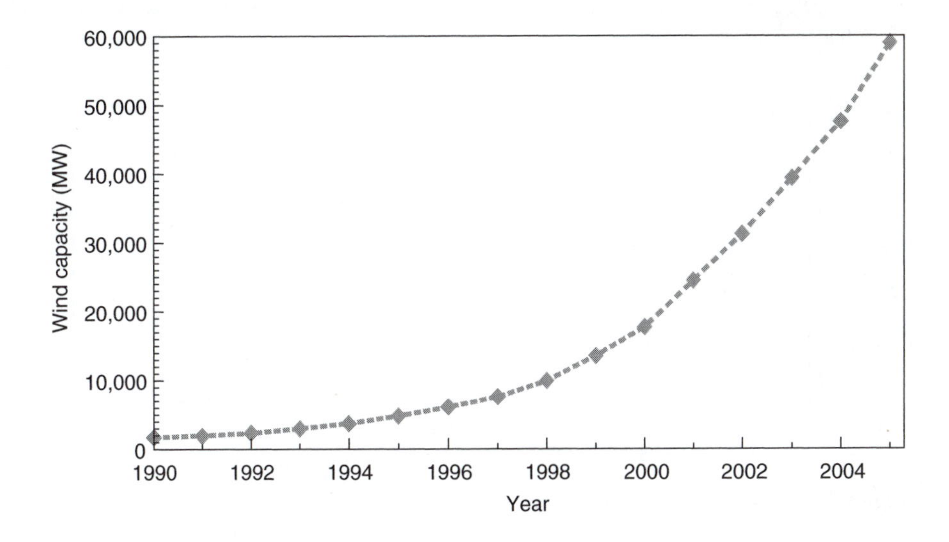

Figure 1.1 Development of world wind capacity, 1990–2005

25 per cent of electricity consumed. At times, the power output from the wind turbines matches the total consumption in Jutland.

This rapid growth has not only been stimulated by financial support mechanisms of various kinds, but also by a very rapid maturing of the technology. Energy outputs have improved, partly due to better reliability, partly due to the development of larger machines. Economies of scale produce quite modest increases of efficiency, but larger machines mean taller towers, so that the rotors intercept higher wind speeds. Technical improvements have run parallel with cost reductions and the latter, in turn, have been partly due to economies of scale, partly to better production techniques. Finally, wind's success has also been due to the growing awareness that the resources are substantial – especially offshore – and that energy costs are converging with those of the *conventional* thermal sources of electricity generation. In some instances, the price of wind-generated electricity is lower than prices from the thermal sources.

More recently, there have been several developments of offshore wind farms and many more are planned. Although offshore wind-generated electricity is generally more expensive than onshore, the resource is very large and there are few environmental impacts.

1.3.2 Changes in size and output

Early machines, less than 20 years ago, were fairly small (50–100 kW, 15–20-m diameter), but the size of commercial wind turbines has steadily increased. Figure 1.2 tracks the average size of machine installed in Germany from 1990 to 2004 (Ender, 2005); during that period, average machine capacity increased by a factor of ten – from 160 to 1,680 kW.

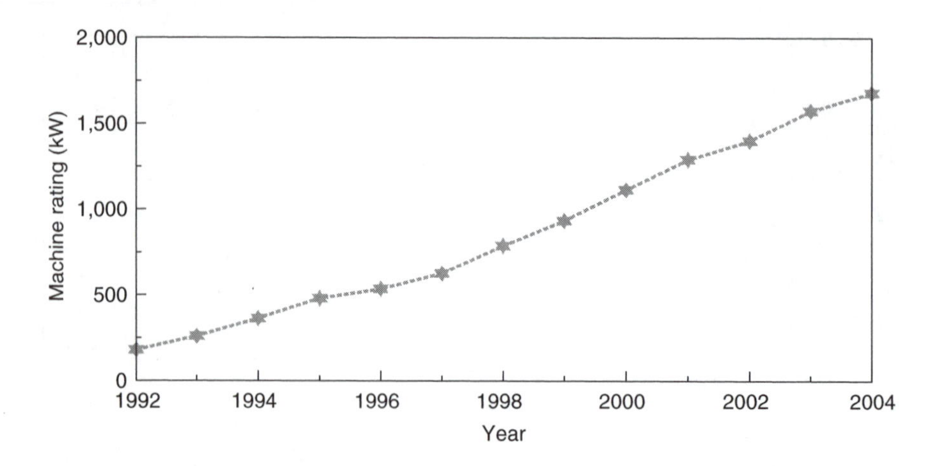

Figure 1.2 Average machine ratings – Germany

As machine ratings have increased, so have hub heights. As a rule of thumb, hub height is usually similar to rotor diameter, but an analysis of data from Germany (Milborrow, 2001a) suggests that many manufacturers are offering the option of taller towers to achieve even higher output. Several machines with diameters around 70 m have hub heights of 100 m.

Larger machines, on taller towers, intercept higher velocity winds, which improves energy productivity. Another way of increasing energy yields is to increase the rating of the generator. However, the higher the rating, the lesser the time that maximum output is achieved. It does not make economic sense to install generators with very high ratings, as the high winds needed to reach maximum output will only be encountered for a few hours per year. As wind turbine manufacturers are subject to similar economic pressures, most have settled for rated outputs corresponding to about 400 W/m^2 of rotor area. A 40-m diameter machine, marketed in the early 1990s, would therefore have an output around 500 kW. With an average wind speed of, say, 7.5 m/s and a typical wind speed distribution pattern, this meant that maximum output would be achieved for around 8 per cent of the year. Ratings have increased, however, as the markets have become increasingly competitive, and those of the very largest machines now approach 600 W/m^2.

1.3.3 Energy productivity

The increase in size and specific rotor outputs, coupled with a small contribution from aerodynamic scale effects and from design improvements, have all combined to bring about marked increases in energy production. To illustrate this point, Figure 1.3 shows that annual yields, per unit area, increase by over 50 per cent as rotor diameters increase from 20 to 80 m. This figure was constructed using actual performance data from manufacturers' specifications and taking a reference wind speed of 7 m/s at 30 m. It was assumed that wind speed increases with height according to a 1/7-power law.

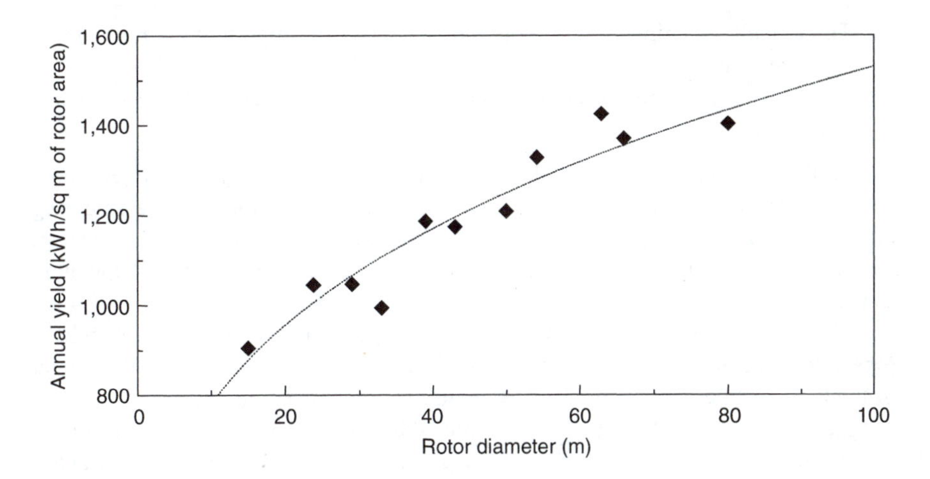

Figure 1.3 Annual yields from turbine rotors

1.4 Design options

Although energy yield is important, so is price, and there are various ways in which manufacturers seek to achieve the best balance between high yield and low price. At one end of the spectrum are the simple and rugged *stall-regulated* wind turbines, running at a fixed speed and using induction generators. At the other end are the variable-speed, direct-drive machines with power conditioning equipment that is able to deliver either leading or lagging power factor.

Numerous permutations of the design options are possible. There has been some reduction in the types of machine on the market, but no real move towards uniformity. The majority of the world's wind turbines now have three blades. Most of the machinery is mounted in the nacelle, which is yawed into the wind under power and mounted on a steel tower.

1.4.1 Blades

A wide range of blade materials has been used for blade manufacture, including aluminium, steel (for the spar, with a light fairing), wood epoxy and glass-reinforced plastic. The two latter materials are now most common as they have the best combination of strength, weight and cost. It is essential to keep weight to the minimum, as the weight of a wind turbine has a strong influence on its overall cost. The cost of wind turbines usually accounts for 65–75 per cent of the total cost of a wind farm and capital repayments typically account for around 75 per cent of electricity-generating costs.

The preference for three blades probably arises as there are some distinct advantages:

1. The moment of inertia about the yaw axis, defined by the wind direction, does not vary substantially, whatever the disposition of the blades; this reduces the cyclic gyroscopic forces encountered when yawing two-blade machines.

2. Three-blade machines rotate more slowly, which is important as this reduces noise generation.
3. The rotation of a three-blade rotor is easier on the eye – a consideration which is important to planners.

Perhaps contrary to intuition, three-blade rotors are only slightly heavier (about 15 per cent) than two-blade rotors. There has been speculation that the emerging offshore market might reawaken interest in two-blade rotors, as two of the onshore constraints – noise and visual – are less important. There is little sign of this happening as yet.

Wind turbines are large structures and so weight is important. Blade weight is especially important, as savings in rotor weight allows related reductions in the weight of the hub, nacelle and tower structure. Fairly simple reasoning suggests that blade weight increases with the cube of the rotor diameter and this is borne out by an examination of rotor weights in the size range from 20 to 100 m. However, the better understanding of rotor aerodynamics and blade loads, which has been acquired over the years, means that substantial reductions in weight have been achieved.

1.4.2 Control and the power train

Roughly half of the world's wind turbines have fixed blades and are of the stall-regulated type and the remainder have variable-pitch blades to limit the power in high winds. Stall-regulated machines can dispense with potentially troublesome controls for changing the pitch of the blades, but still need to have some form of movable surface to regulate rotational speed before synchronisation and in the event of disconnection. It must be emphasised, however, that the reliability of both these types of machine is now very high.

The majority of machines run at a fixed speed and use induction generators, but increasing numbers run at variable speed, using power conditioning equipment (Chapter 3), or run at two speeds, with pole switching on the induction generator. The advantage of using slower speeds in low winds is that noise levels are reduced and, in addition, aerodynamic efficiency – and hence energy yield – is slightly increased. Increasing numbers of machines dispense with a gearbox and use a direct drive to a multi-pole generator.

A further advantage of using power conditioning equipment is that it avoids the need to draw reactive power from the electricity network. This benefits the utility and may mean savings for the wind turbine operator as reactive power is often subject to a charge. A further benefit accrues if the wind turbine is able to sell energy with a leading power factor to the utility at a premium.

The most recent developments in power train technology involve the use of direct-drive electrical generators. The German manufacturer, Enercon, has sold a large number of machines of this type, worldwide, and currently offers machines in a range of sizes from 200 to 5,000 kW. The diameter of the latter is 112 m.

1.4.3 Summary of principal design options

A summary of the principal options is given in Table 1.1.

Table 1.1 Features of typical electricity-generating wind turbin

Rotor diameter	Up to 110 m (6 MW)
Number of blades	Most have three, some have two, a few have one
Blade material	Most in glass-reinforced plastic, increasingly wood epoxy
Rotor orientation	Usually upwind of tower; some downwind machines
Rotational speed	Usually constant, about 25 revolutions/minute at 52-m diameter, faster at smaller sizes Some machines use variable speed, or two speed
Power control	The most common methods are: 'Stall control': blades are fixed, but stall in high winds 'Pitch control': all or part of the blade rotated to limit power
Power train	Step-up gearboxes most common, but direct drives (no gearbox) with multi-pole generators now used in some commercial designs – see next entry
Generator	Induction usual, four or six pole; doubly-fed induction generators becoming established Variable speed machines use AC/DC/AC systems Also direct-drive machines with large diameters (up to 6 m)
Yaw control	Sensors monitor wind direction; rotor moved under power to line up with wind A few machines respond passively
Towers	Cylindrical steel construction most common Lattice towers used in early machines A few (large) machines have concrete towers

AC: alternating current; DC: direct current.

1.5 Wind farms

The nature of the support mechanisms has influenced the way in which wind energy has developed. Early developments in California and subsequently in the United Kingdom, for example, were mainly in the form of wind farms, with tens of machines, but up to 100 or more in some cases. In Germany and Denmark, the arrangements favoured investments by individuals or small cooperatives, and so there are many single machines and clusters of two or three. Economies of scale can be realised by building wind farms, particularly in the civil engineering and grid connection costs, and possibly by securing *quantity discounts* from the turbine manufacturers. Economies of scale deliver more significant savings in the case of offshore wind farms and many of the proposed developments involve large numbers of machines.

able 1.2 *Key features of an onshore and an offshore wind farm*

	Onshore	Offshore
Project name	Hagshaw Hill	Middelgrunden
Project location	50 km south of Glasgow in the Southern Highlands of Scotland	Near Copenhagen, Denmark
Site features	High moorland surrounded by deep valleys	Water depth of 2–6 m
Turbines	26, each 600 kW	20, each 2 MW
Project rating (MW)	15.6	40
Turbine size	35-m hub height, 41-m rotor diameter	60-m hub height, 76-m rotor diameter
Special features of turbines	Turbine structure modified for high extreme gust wind speed; special low-noise features of blades	Modified corrosion protection, internal climate control, built-in service cranes
Turbine siting	Irregular pattern with two main groups, typical spacing three rotor diameters	180 m apart in a curve and a total wind farm length of 3.4 km
Energy production (MWh) (annual)	57,000	85,000 (3% of Copenhagen's needs)
Construction period	August to November 1995	March 2000 to March 2001

Source: Bonus Energy A/S, Denmark, www.bonus.dk, 30 June 2005.

Table 1.2 gives an indication of parameters for offshore and onshore wind farms.

Wind speed is the primary determinant of electricity cost, on account of the way it influences the energy yield and, roughly speaking, developments on sites with wind speeds of 8 m/s will yield electricity at one third of the cost for a 5 m/s site. Wind speeds around 5 m/s can be found, typically, away from the coastal zones in all five continents, but developers generally aim to find higher wind speeds. Levels around 7 m/s are to be found in many coastal regions and over much of Denmark; higher levels are to be found on many of the Greek Islands, in the Californian passes – the scene of many early wind developments – and on upland and coastal sites in the Caribbean, Ireland, Sweden, the United Kingdom, Spain, New Zealand and Antarctica.

1.5.1 *Offshore wind*

Offshore wind energy has several attractions, including huge resources and minimal environmental impacts. In Europe, the resources are reasonably well located relative to the centres of electricity demand.

Wind speeds are generally higher offshore than on land, although the upland regions of the British Isles, Italy and Greece do yield higher speeds than offshore.

Ten kilometres from the shore, speeds are typically around 1 m/s higher than at the coast. There are large areas of the North Sea and Baltic with wind speeds above 8 m/s (at 50 m) (Hartnell and Milborrow, 2000). Turbulence is lower offshore, which reduces the fatigue loads. However, wind/wave interactions must be taken into account during design. Wind speeds are inevitably less well characterised than onshore; this is unimportant as far as resource assessments are concerned, but accurate estimates are needed to establish generation costs. Potential offshore operators are currently making measurements and further studies are also underway.

Denmark, Sweden, Belgium, the Netherlands, Germany, the United Kingdom and Ireland have already built wind turbines in marine environments, either in the sea or on harbour breakwaters. Further activity is planned in most of these countries, including Italy. Responding to this interest, several manufacturers are now offering machines specifically for the offshore market. Most are in the range 1.5–3.5 MW in size and with design modifications such as sealed nacelles and special access platforms for maintenance. Larger machines tend to be more cost effective, because more expensive foundations are at least partially justified by a higher energy yield.

The construction, delivery to site and assembly of the megawatt-size machines demands specialist equipment, suitable ports and careful timetabling to maximise the possibilities of calm weather windows. Although it was anticipated that access for maintenance might be a problem, early experience from Danish installations is encouraging, although, again, experience from some of the more hostile seas is still lacking.

1.6 Economics

1.6.1 Wind turbine prices

Large machines have been developed as they offer lower costs, as well as better energy productivity, as discussed earlier. Machine costs (per unit area of rotor) fall with increasing size, and the use of larger machines means that fewer machines are needed for a given amount of energy, and so the costs of transport, erection and cabling are all reduced.

1.6.2 Electricity-generating costs

There is no single answer to the question, 'is wind power economic?' The answer depends on the local wind regime, the price of competing fuels and institutional factors. None of these can be considered in isolation and the importance of institutional factors cannot be overlooked.

To provide an indication of how onshore and offshore energy prices compare, Figure 1.4 (Milborrow, 2006) shows data for a range of wind speeds, with the following assumptions.

- *Installed cost*: €1000–€1400/kW onshore, €1300–€1900/kW offshore;
- 15-year depreciation, 6 per cent real discount rate;
- *O&M costs*: €15 (low onshore estimates) to €80/kW/year (high offshore estimates) + €3/MWh.

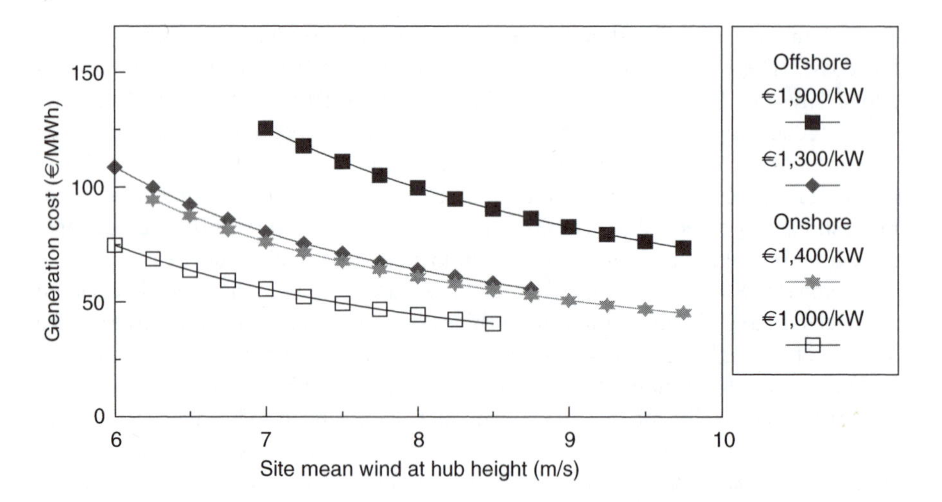

Figure 1.4 Indicative comparisons between offshore and onshore wind energy prices

Comparable electricity prices for gas and coal plant are around €40–€55/MWh, and for nuclear in the range €30–€75/MWh, depending on how the plant is financed. So, from Figure 1.4, the best onshore wind sites produce electricity at competitive prices.

This comparison, however, is simplistic. It ignores three important issues:

1 Wind energy, in common with most of the other renewable energy sources, feeds into low voltage distribution networks, closer to the point of use and therefore perhaps has a higher value.
2 The *external costs* of wind energy are much lower than those of the thermal sources of electricity. These are costs which are not accounted for, such as those due to acid rain damage and coal subsidies.
3 These above points enhance the value of wind. On the debit side, the variable nature of wind energy means that it has to carry the costs of extra reserve, as it increases the uncertainty in matching supply and demand.

The first and last of these issues will be addressed in the next section. The European Commission funded an authoritative study of external costs (European Commission, 1995), which provided estimates that are included in Table 1.3. It may be noted that many governments tacitly accept the principle by providing the support mechanisms for the renewable energy sources discussed earlier.

1.6.3 Carbon dioxide savings

Although there have been proposals for the *carbon tax* that would, at least qualitatively, reflect the external costs of generation from fossil fuels, the EU's Emissions Trading Scheme aims to establish a market price for carbon dioxide emissions that depends on the levels of the ceilings that are set on emissions. In early 2006,

Table 1.3 Estimates of external costs (in €/kWh)

Category	Coal	Oil	CCGT	Nuclear	Wind
Human health and accidents	0.7–4	0.7–4.8	0.1–0.2	0.030	0.040
Crops/forestry	0.07–1.5	1.600	0.080	Small	0.080
Buildings	0.15–5	0.2–5	0.05–0.18	Small	0.1–0.33
Disasters				0.11–2.5	
TOTAL, damage	0.7–6	0.7–6	0.3–0.7	0.2–2.5	0.2–0.5
Global warming estimates	0.05–24	0.5–1.3	0.3–0.7	0.020	0.018
Indicative totals	1.7–40	3.7–18.7	0.83–1.86	0.36–5	0.4–1.0

CCGT: combined cycle gas turbine.

carbon dioxide was trading at around €25/tonne, so imposing a penalty of around €20/MWh on coal-fired generation and up to around €6 – €8/MWh on gas-fired generation. This improves the competitive position of wind energy in the European Union significantly.

The EUs Emissions Trading Scheme short circuits the need to establish the exact level of carbon dioxide savings achieved by the introduction of wind energy. There is some debate on this point. The carbon dioxide savings that result from the introduction of renewable energy depend on which fuel is displaced. As wind energy is invariably a *must run* technology, this implies that changes in output from wind plant are reflected in changes in output from the *load-following* plant. In most of Europe, coal- or oil-fired plant tend to be used for load following, which implies that each unit of renewable electricity displaces between 650 and 1,000 g of carbon dioxide (the lower figure applies to oil). However, as the amount of wind energy on the system rises, the extra carbon dioxide emissions from the additional reserve plant need to be taken into account, although these are very modest.

Until recently, there was a reasonable consensus on the method of estimating carbon dioxide savings, based on the above reasoning. More recently, however, alternative approaches have been suggested, perhaps exemplified by the supporting analysis to the UK DTI Renewable Energy consultation document (ETSU, 1999), which proposed three possible scenarios for assessing the range of carbon savings from the UK renewables programme:

1. Renewables displace combined cycle gas turbines.
2. Renewables displace modern coal plant.
3. Renewables displace the current generating mix.

The document suggested that the first scenario was unlikely in practice, but did not comment further. Despite this, the DTI has since adopted *renewables displace gas* as its preferred approach, on the basis that renewables will inhibit the construction of new gas plant. Given that the construction of new gas plant in the United Kingdom has slowed to a trickle, this argument now seems more difficult to sustain. The alternative argument is that wind, like gas, forces the closure of old coal plant.

There appears little technical justification for *renewables displace current generating mix*, although published studies of carbon dioxide savings subscribe to each of the three options in roughly equal numbers.

1.7 Integration and variability: key issues

There is a widespread, but incorrect, perception that the introduction of wind energy into an electricity network will cause problems – and financial penalties. That perception was reinforced in the United Kingdom with the introduction of the New Electricity Trading Arrangements (NETA, later British Electricity Trading and Transmission Arrangements [BETTA]), under which variable sources of energy are likely to be accorded less value. BETTA muddies the waters, as the contracts at the heart of the system are generally based on matching the needs of electricity suppliers with the capabilities of generators. Supply and demand are both disaggregated, to a degree. This puts variable renewables at a disadvantage.

It is important, therefore, to draw a distinction between technical reality and accounting convenience. There is scant evidence, as yet, that the new trading arrangements will improve the overall technical efficiency of the system. In fact, there are indications that they may result in higher provisions of spinning reserve (OFGEM, 2000).

The technical criteria for absorbing renewables on the transmission network operated by the National Grid Company have been restated and are set out in Table 1.4 (National Grid Company, 1999). Of the three criteria, the third is probably the most severe as far as wind is concerned and is discussed in the next section. It should be noted that the thresholds are not barriers and that higher levels of wind can be absorbed – at a cost.

1.7.1 Wind fluctuations

Just as consumer demands are smoothed by aggregation, so is the output from wind plant, and geographic dispersion dramatically reduces the wind fluctuations. This is clearly illustrated by data from the German wind programme (ISET, 1999/2000), and by an analysis of wind fluctuations in Western Denmark (Milborrow, 2001b). The data cover 350 MW of plant in Germany and 1,900 MW in Western Denmark.

In each case, there were no recorded upward power excursions in an hour greater than 20 per cent of rated power. However, from individual machines, or a single wind farm, power excursions up to about 96 per cent of rated power were recorded, albeit rarely. With a single machine, excursions of 10 per cent of rated power were measured for about 7 per cent of the time, whereas these fluctuations only occurred about 1 per cent of the time from the aggregated output. Figure 1.5 compares the output from all the wind plant in Western Denmark with data from a single wind farm. The latter is based on power measurements from Bessy Bell wind farm in Northern Ireland, by kind permission of Northern Ireland Electricity. These data were analysed in the same way as the Danish data to show the power swings.

Table 1.4 Criteria for absorption of renewable energy

Impact	Threshold	Mitigation options
Change in renewable generation output	Generation subject to fluctuation >20% of peak demand	Purchase additional controllable output
Unpredictable instantaneous reduction in generation output	Potential instantaneous loss >2% of peak demand	Purchase additional frequency control measures
Unpredictable short-notice reduction in output	Potential loss >3% of peak demand in an hour	Purchase additional reserve services

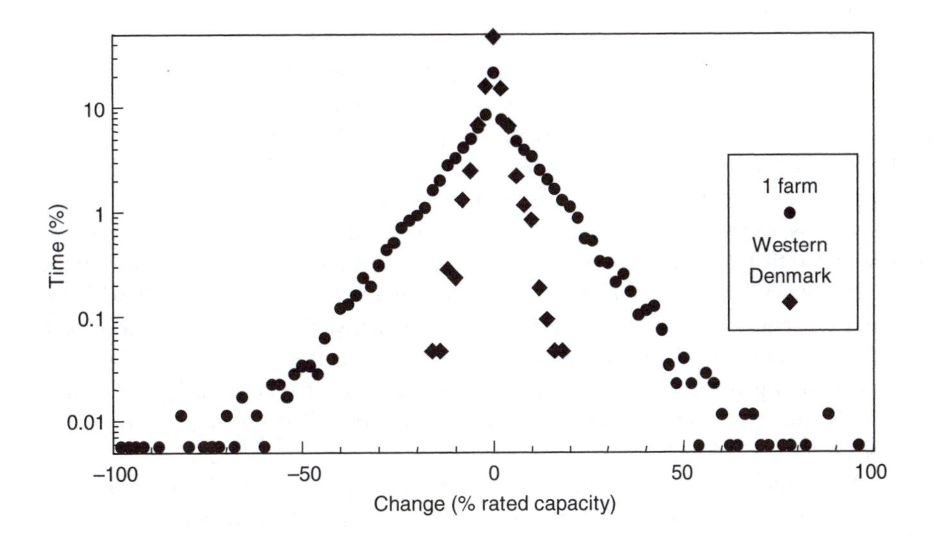

Figure 1.5 Changes in average generated output from wind plant, based on hourly averages

The measurements from the German wind programme also indicate that the power fluctuations are lower than suggested by previous modelling and analyses. This, in turn, means that the operational penalties associated with running networks with wind may be lower than early estimates. The extra cost of running more thermal plant as spinning reserve is modest – below 0.1p/kWh with 2 per cent of wind on the system, rising to around 0.2p/kWh with 10 per cent wind energy (Dale, 2002). Other studies have yielded similar results, as the key issue is the additional uncertainty that the introduction of wind energy imposes on a network. Demand and generation levels cannot be predicted with total precision on any network and typical *demand prediction errors* are around 1–2 per cent. This means that some spinning reserve must always be scheduled and it must also be able to cope with the loss of the largest single unit on the system.

The precise level at which wind becomes difficult to absorb on a system depends on the characteristics of other plant. Systems with hydro- or pumped storage (which can be used to respond to changes in wind output) can absorb more variable renewables. Systems with high proportions of nuclear or combined heat and power tend to be less flexible. What is clear is that the vagaries of the wind or the sun, at modest levels of penetration, are not a problem for most centralised systems. Despite the steadily increasing penetration of wind on the Danish grid, energy from the plant continues to be absorbed. Greater threats to stability abound, particularly the loss of interconnectors or of large thermal plant.

1.7.2 Capacity credits

Few topics generate more controversy than capacity credits for wind plant. The capacity credit of any power plant may be defined as a measure of the ability of the plant to contribute to the peak demands of a power system. Capacity credit is often defined as the ratio (firm power capability)/(rated output). As thermal plant is not 100 per cent reliable, values for all plant are less than unity. To a first order, 1,000 MW of nuclear plant corresponds to about 850 MW of firm power and hence has a capacity credit of 0.85; coal plant has a capacity credit of about 0.75. These figures are, roughly, the statistical probability of the plant being available at times of peak demand.

Almost every authoritative utility study of wind energy in large networks has concluded that wind can provide firm capacity – roughly equal, in northern Europe, to the capacity factor in the winter quarter. This implies that if, say, 1,000 MW of wind plant was operating on the mainland UK network, it might be expected to displace around 300 MW of thermal plant. It is difficult to say whether this would happen in a privatised and fragmented electricity industry but, ideally, the market signals should encourage it to happen, as this would lead to a technically optimised system.

With smaller networks, which are not large enough to benefit from geographical dispersion, the capacity credit may be smaller, or non-existent. A recent study of the Irish network, for example, assumed that wind had no capacity credit, although it acknowledged that the evidence was conflicting and advocated further work to clarify the position (Gardner *et al.*, 2003). More recently, however, the network operator in Ireland has carried out an analysis of the impact of wind on the system (ESBNG, 2004a) and suggested that the capacity credit, with small amounts of wind, is about 30 per cent of the rated capacity of the wind plant, which is in line with most other European studies.

1.7.2.1 Power available at times of peak demand

As the risk of a generation deficiency is highest at times of peak demand, values of capacity credit are strongly influenced by the availability of variable renewable energy sources at times of peak demand. Several authors have examined this issue, in the context of wind availability. Palutikof *et al.* (1990), for example, observed that '…peak demand times occur when cold weather is compounded by a wind chill factor, and low temperature alone appears insufficient to produce the highest demand of the year'. Using four sites, they showed that the summed average wind turbine outputs during eight winter peak demands was about 32 per cent of rated output. National Wind Power, similarly, have found that 'wind farm capacity factors during periods of peak demand are typically 50 per cent higher than average all-year capacity factors' (Warren, 1995).

1.7.3 Embedded generation benefits

The principle that small-scale generation may save transmission and distribution losses (and charges) is now established and small additional payments may be made for voltage support, reactive power and other ancillary services. It is important to recognise, however, that concentrations of embedded generation can increase distribution losses in rural areas where demand is low and so should be avoided. These are complex issues, which vary both regionally and locally, but studies are under way to identify the issues and suggest ways of removing institutional barriers (DTI, 2000). It is important to do this, as problems at the distribution level may be more of an inhibiting factor than at the transmission level (Laughton, 2000).

A study of a ten-machine, 4 MW wind farm connected to an 11 kV system in Cornwall, England has provided valuable information on the impact on a distribution system (South Western Electricity Plc, 1994). The study examined a range of issues and concluded that 'the wind farm caused surprisingly little disturbance to the network or its consumers'. In particular:

- Voltage dips on start-up were well within the limits prescribed. There were no problems with flicker during any operating conditions.
- During periods of low local load, the output from the farm was fed *backwards* through the distribution network, but no problems were reported.
- Reduced activity of the automatic tap changers at the adjacent 33/11 kV transformers was significant and would lead to lower maintenance costs.

1.7.4 Storage

Energy storage, it is often claimed, can enhance the value of electricity produced by variable sources such as wind. In some circumstances this may be true, but only if the extra value that storage gives to the electricity system is greater than the cost of providing it. Electricity from storage devices actually costs about the same as generation from conventional thermal plant, although it has a variable *fuel* cost – that of the electricity used to charge the store.

Although electricity systems with storage enable variable renewables to be assimilated more easily, it does not necessarily follow that it is worthwhile to build storage specifically to increase the value of renewable energy. However, if storage is available it is quite possible that its output can substitute for a shortfall in the output from the renewable capacity on some occasions. It may also be sensible to use surplus output from wind plant during, say, windy nights to charge a storage device.

Apart from pumped storage, large-scale storage of electricity has generally been too expensive or too demanding of specific site requirements to be worth serious consideration. A recent development is the reversible fuel cell concept (see Section 5.4.1). This could provide affordable and flexible storage – just the type needed to make wind power that much more attractive, especially in new competitive markets dependent on bilateral contracts.

The role of energy storage in the operational integration of wind power is considered further in Chapter 5.

The intermittent sources of renewable energy, such as wind, are more likely to benefit from load management, although there are potential benefits for all technologies. The potential benefits for wind have been examined in one study (Econnect Ltd, 1996), but such arrangements will be dependent on the agreement of the supplier.

1.8 Future developments

1.8.1 Technology

Wind turbine technology has now come *full circle* since the days of the early government-funded programmes, which spawned a number of megawatt-size machines around 20 years ago. Although few of those machines were economic, the industry learned a lot and is now, once again, building machines of similar size. As this progression towards larger sizes shows little sign of slowing, the question is often asked whether there is any technical or economic limit. Roughly speaking, rotor weight increases as the cube of the size, whereas energy yield increases as the square. Perhaps more importantly, there will eventually come a point at which gravitational bending forces start to dominate the design process and beyond this point weights increase with the fifth power of the diameter. However, a simplified analysis of the crucial design drivers has suggested that this crossover point is unlikely to be reached until rotor diameters of about 200 m (Milborrow, 1986). Even then, the use of strong, lightweight materials such as carbon fibre-reinforced plastic (CFRP) may raise the limit higher.

The use of these lightweight materials is also likely to bring about further weight and cost savings in the large machines now being developed. Further sophistication in control techniques and in electrical innovation is also likely to improve energy yields and further improve the attractions of wind turbines to electricity network operators.

Several manufacturers are now marketing machines specifically for the offshore market, with ratings up to 5 MW. In addition, it is possible that the removal of some of the onshore constraints may lead to significant changes in design. The use of faster rotational speeds and of two-bladed machines are two options which would result in significant weight reductions. The use of CFRP for the blades would also result in significant weight reduction. Although CFRP is presently too expensive, it is possible that an increased demand for the product would lead to cost reductions.

1.8.2 Future price trends

World wind energy capacity seems likely to continue doubling every three years or so, accompanied each time, assuming recent trends continue, by a 15 per cent reduction in production costs. During 2005, however, there was a marked upturn in the installed costs of wind plant, primarily due to increases in the price of wind turbines. This, in turn, was due to increases in the prices of steel, copper and blade materials. Nevertheless, the trends that were responsible for the earlier cost reductions are still at work. Manufacturers are developing more cost-effective production techniques, so bringing down the price of machines. Machine sizes are increasing, which means that fewer are needed for a given capacity, and so installed costs of wind farms are decreasing. In addition, larger wind farms are being built, which spreads the costs of overheads, roads, electrical connections and financing over greater capacities. The use of larger wind turbines means that they intercept higher wind speeds, and this impacts on the energy generation and so on the generation costs.

1.8.3 Market growth

The rate at which the wind energy market develops and the rate at which prices fall are linked. Strong market growth has led to a steady fall in prices and, assuming these trends continue, the market growth will be sustained as wind energy becomes steadily more competitive in comparison with gas-fired generation. One recent projection of future trends (BTM Consult ApS, 2003) suggests that the annual increase of capacity will rise from the present level of about 8,000 to around 14,000 MW by 2008. Total global capacity may then reach 100,000 MW shortly after 2008 and exceed 200,000 MW by 2014. The European market will continue to account for most capacity for some years yet, and offshore wind is likely to increase gradually in significance. By 2013, wind energy worldwide may account for 2 per cent of all generated electricity.

1.8.4 Integration issues

Increasing numbers of electricity networks are coping with, or considering, the issues associated with increasing amounts of wind energy, and so there is an increasing understanding of the possible problems and solutions. Broadly speaking, there is a consensus on the key question of the additional costs associated with additional reserve, although there are some variations due to the differing costs of the reserve itself. With the benefit of more operational experience, the additional costs associated with variability may be expected to be quantified more precisely and may possibly fall. In addition, new techniques of demand-side management, currently being investigated in America, may mean that the need for extra physical frequency response plant may be reduced, with consequential reductions in cost (Kirby, 2003).

Chapter 2
Power system fundamentals

2.1 Introduction

Most wind power capacity is connected to electricity supply networks, and this is likely to continue for the foreseeable future. The advantages of connection to a grid include:

- the ability to locate wind farms where the wind resource is plentiful, irrespective of demand;
- the ability of an interconnected grid to absorb variations in wind generation unrelated to overall demand variation;
- provision of excitation, enabling simple induction machines to be used as generators.

These advantages are qualified by various limitations of the power supply system, and Chapters 4 and 5 will consider how the resulting problems can be overcome. However, the fact remains that grid connection has provided a major impetus to the growth of wind energy. Hence it is important to understand the fundamentals of electrical power engineering presented in this chapter.

The chapter will start with the basic principles of electrical engineering. The discussion will lead naturally to the transformer, found in all wind farms as well as throughout power supply systems. We then consider alternating current (AC) systems, with particular emphasis on active and reactive power and the use of phasors. Power supply systems are then considered. The chapter will close with an introduction to AC power transmission.

2.2 Basic principles

2.2.1 Electromagnetism

Öersted observed in 1820 that a magnetic compass needle is deflected by the flow of electric current. He was able to show that the magnetic field may be

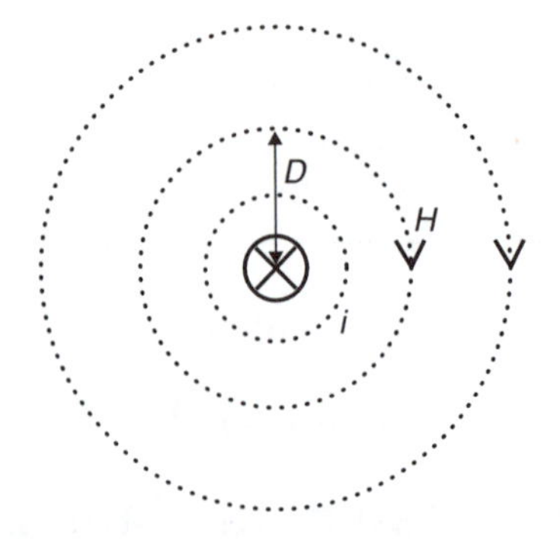

Figure 2.1 Magnetic field due to current

represented by concentric circles round the conductor axis. This is shown in Figure 2.1.

Note that current flowing away – represented by an '×' – produces a clockwise field. Current flowing towards the observer – represented by a '.' – produces an anti-clockwise field. This relationship between field and current direction is known as the *right-hand screw* rule: a right-handed screw will move away (the current) when the screw is rotated clockwise (the field).

The magnitude of the magnetic field due to current may be quantified by Ampère's Law (Christopoulos, 1990). When the field due to current i in an infinitely long, straight conductor is considered, application of Ampère's Law leads to a simple and very useful result. The *magnetic field strength* H of the field at a distance D from the conductor, as illustrated in Figure 2.1, is given by

$$H = \frac{i}{2\pi D} \tag{2.1}$$

The magnetic field strength describes the magnetic stress on the medium due to current. The actual magnetic field reflects the *permeability* of the medium. In the case of a vacuum, or a non-magnetic medium such as air, the field is described by *flux density* B. Flux density is related to magnetic field strength by

$$B = \mu_0 H \tag{2.2}$$

where μ_0 is the *permeability of free space*. When the medium is magnetic, the flux density for a particular magnetic field strength will be much greater, and is given by

$$B = \mu_r \mu_0 H \tag{2.3}$$

where μ_r is the *relative permeability* of the magnetic material. Obviously, non-magnetic materials have a relative permeability of 1. Ferromagnetic materials have relative permeabilities ranging from 100 to 100,000.

It may be noted from Equation (2.1) that

$$H \times 2\pi D = i \tag{2.4}$$

This is a particular case of an important general principle of electromagnetism. The line integral of H around a closed path is equal to the total current enclosed or *magnetomotive force* (m.m.f.) F:

$$\oint H\,\mathrm{d}l = F \tag{2.5}$$

Öersted's discovery implies that a current exerts a force on a magnet, and in particular on the magnet's field. It may be inferred from Newton's third law of motion – every force has an equal and opposite reaction – that a magnetic field will exert a force on a current-carrying conductor. This is readily confirmed by experiment. If a conductor carrying current i is placed in a plane normal to the direction of a magnetic field of flux density B, it is found that the force on the conductor is given by

$$f \propto Bli \tag{2.6}$$

where l is the length of conductor in the magnetic field. The force is normal to the field and to the current. Its direction is given by Fleming's left-hand rule, which may be applied as follows:

first finger	**f**ield
m**i**ddle finger	current (***i***)
thu**m**b	**m**otion

It follows from Equation (2.6) that

$$f = \mathrm{k}Bli$$

where k is a constant of proportionality. The unit of flux density, the Tesla (T), is chosen such that k is unity, giving

$$f = Bli \tag{2.7}$$

Thus the Tesla (T) is the density of a magnetic field such that a conductor carrying 1 ampere normal to it experiences a force of 1 Newton/m.

Equation (2.7), combined with the definition of the unit of electric current, the ampere, may be used to determine the permeability of free space, μ_0. The ampere is defined as 'that current which, flowing in two long, parallel conductors 1 m apart in a vacuum, produces a force between the conductors of 2×10^{-7} Newton/m'. From Equations (2.1) and (2.2), the flux density at one conductor due to the current in the other will be

$$B = \frac{\mu_0}{2\pi}$$

The force on this conductor will therefore be, from Equation (2.7),

$$f = \frac{\mu_0}{2\pi} = 2 \times 10^{-7} \text{ Newton/m}$$

giving

$$\mu_0 = 4\pi \times 10^{-7} \text{ Henry/m}$$

2.2.2 Magnetic circuits

It is convenient to deal with electromagnetic systems in terms of *magnetic circuits*. A magnetic field may be visualised with the help of *flux lines* (see Fig. 2.1). A magnetic circuit may be defined as the 'complete closed path followed by any group of magnetic flux lines' (Hughes, 2005).

Total magnetic flux Φ through an area may be obtained from the flux density normal to it. In many cases of practical interest the flux density is indeed normal to the area of interest. If, in addition, the flux density is uniform over the area, the total flux is given simply by

$$\Phi = Ba \tag{2.8}$$

where a is the area.

We are particularly interested here in magnetic circuits containing ferromagnetic materials. It was noted above (Equation (2.3)) that these materials are characterised by high values of relative permeability μ_r. The behaviour of these materials is described by the linear relationship of Equation (2.3) for flux densities up to around 1 T. They are then subject to saturation, which effectively reduces the relative permeability as flux density increases further.

Consider the effect of applying a magnetic field of strength H to part of a magnetic circuit of length l. It is assumed that the flux density B is uniform and normal to a constant cross-sectional area a. From Equation (2.5) the m.m.f. will be simply

$$F = Hl$$

From Equations (2.3) and (2.8) we have

$$F = \frac{Bl}{\mu_r \mu_0} = \frac{l}{\mu_r \mu_0 a}\Phi = S\Phi \tag{2.9}$$

The quantity S is known as reluctance. It can be determined easily for the various sections of a magnetic circuit. Equation (2.9) has the same form as the familiar expression of Ohm's Law: $V = RI$. A magnetic circuit may therefore be analysed using electric circuit methods, with the following equivalences:

Electric circuits	Magnetic circuits
voltage V	m.m.f. F
current I	flux Φ
resistance R	reluctance S

2.2.3 Electromagnetic induction

The key to electric power generation was Faraday's discovery of electromagnetic induction in 1831. Faraday's Law is usually expressed in terms of the electromotive force (e.m.f.) e induced in a coil of N turns linked by a flux ϕ. Setting *flux linkage* $\lambda = N\Phi$ we have

$$e = \frac{d\lambda}{dt} = N\frac{d\Phi}{dt} \tag{2.10}$$

This equation encapsulates the key idea that it is the change of flux which creates the e.m.f., rather than flux *per se* as Faraday had expected. The polarity of the induced e.m.f. is governed by Lenz's Law, which states that 'the direction of an induced e.m.f. is such as to tend to set up a current opposing the motion or the change of flux responsible for inducing that e.m.f.' (Hughes, 2005).

It is useful for power system analysis to recast Equation (2.10) in terms of coil voltage drop v and coil current i. From Equations (2.5) and (2.9),

$$\phi = \frac{F}{S} = \frac{Ni}{S}$$

We then have

$$v = e = \frac{N^2}{S} \times \frac{di}{dt} = L\frac{di}{dt} = \frac{d(Li)}{dt} \tag{2.11}$$

where L is the *inductance* of the coil. It may be seen from comparison of Equations (2.10) and (2.11) that inductance may be defined as 'flux linkage per ampere causing it'.

2.2.4 Electricity supply

The discovery of electromagnetic induction paved the way for electricity supply on a useful scale. The first schemes appeared in Britain and the United States at about the same time (1882), supplying the newly developed electric lighting. The voltage was induced in stationary coils, linked by flux produced on a rotating member or *rotor*. The general arrangement is shown schematically in Figure 2.12. The constant rotor flux is seen as an alternating flux by the stationary coils as a result of the rotor's rotation. It follows from Faraday's Law that an e.m.f. will be induced in each stator coil.

The machine shown in Figure 2.12 is known as a synchronous generator or *alternator*. Alternators have provided almost all of the world's electricity to date. However, in spite of the alternator's extreme simplicity, the pioneers of electricity supply decided to provide a direct – rather than an alternating – voltage. This required brushes and a commutator to be fitted to the alternator to achieve rectification. The added complexity of the resulting *direct current* (DC) generator contributed to poor supply reliability. A further difficulty with DC supply arises when a switch is opened, perhaps to disconnect a load. The attempt to reduce the current to zero in a short time creates a large voltage; this is a direct consequence of Faraday's Law as expressed in Equations (2.10) and (2.11). This voltage appears across the switch, creating an arc. This problem can be overcome by suitable switch design, but adds to the difficulty and expense of DC supply. The much greater currents to be interrupted under short-circuit conditions create a very severe system protection problem.

Challenging as these problems were, the greatest limitation of the early DC supply systems was that power had to be supplied, distributed and consumed at the same voltage. This was typically 110 V, which was deemed to be the highest safe value for consumers. Extra load required cables of ever increasing cross-sectional area for a given power. It soon became clear that distribution at such a low voltage was untenable. However, a move to a higher distribution voltage – and lighter cables for the same power – would require voltage transformation. And this, in the late nineteenth century, meant the use of AC supply and transformers.

Fortunately, the transformer is a very simple device that was invented by Faraday in the course of his discovery of electromagnetic induction. The transformer can be used to transform a convenient alternator generation voltage to a much higher transmission voltage. The high transmission voltage can then be stepped down to an intermediate voltage for distribution, and finally to a suitable voltage for consumption, currently 230 V in the United Kingdom.

The rapid growth of electrical energy in the twentieth century can be attributed, *inter alia*, to effective AC generation and supply. Paradoxically, though, DC interconnectors are increasingly being deployed. This is possible as a result of developments in power electronics. Power electronic devices are finding applications at various levels of electricity supply and utilisation. This is certainly true of wind power generation, as will be seen in Chapter 3. However, we will focus on AC systems for the moment. Our discussion will start with a look at the basic principles of the transformer.

2.2.5 *The transformer*

A transformer consists essentially of an iron core on which are wound two coils – referred to here for convenience as the primary (coil 1) and the secondary (coil 2). This arrangement is shown in Figure 2.2. The primary and secondary have N_1 and N_2 turns.

Assuming that the same flux links both windings, that is, no leakage flux, and applying Faraday's Law (Equation (2.10)), we obtain

$$v_1 = N_1 \frac{\mathrm{d}\Phi}{\mathrm{d}t}$$

$$v_2 = N_2 \frac{\mathrm{d}\Phi}{\mathrm{d}t}$$

The voltages are related as follows:

$$\frac{v_2}{v_1} = \frac{N_2}{N_1} \tag{2.12}$$

It may be seen that the voltage is transformed in proportion to the turns ratio. Note that the transformer's operation depends on flux variation, as Faraday observed. In particular, an alternating primary voltage will ensure a varying flux, and hence an alternating secondary voltage in phase with it.

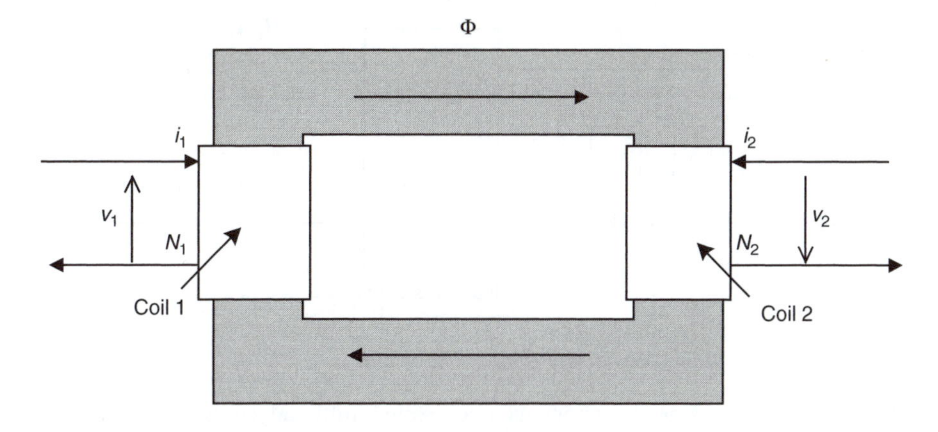

Figure 2.2 Simple transformer

The relation between the currents may be derived by considering the magnetic circuit formed by the iron core. The m.m.f. acting on the core is given by

$$F = N_1 i_1 - N_2 i_2 = S\Phi = \frac{l\Phi}{\mu a}$$

The sign of the secondary current reflects the fact that it opposes the assumed flux direction. l is the core length, a its cross-sectional area and μ its permeability. Assuming that the core permeability is infinite, the m.m.f. F must be zero for a finite flux. Hence

$$\frac{i_2}{i_1} = \frac{N_1}{N_2} \tag{2.13}$$

The power output from the secondary is given by

$$p_{\text{out}} = v_2 i_2 = \left(\frac{N_2}{N_1}\right) v_1 \times \left(\frac{N_1}{N_2}\right) i_1 = v_1 i_1 = p_{\text{in}}$$

The power delivered by the transformer equals the power supplied. This is unsurprising, given that we assumed an ideal core with no leakage, and took no account of copper losses in the windings.

A real transformer differs from the ideal in the following respects:

- Leakage flux – some primary flux does not link the secondary and *vice versa*.
- The windings have resistance, leading to a power loss under load.
- The core permeability is finite; hence a magnetising current is drawn.
- There are iron losses in the core due to magnetic hysteresis and eddy currents.

The most significant of these effects in the system context is flux leakage. The leakage flux links one coil only, and therefore manifests itself as an inductance in series with each coil, known as leakage inductance – see L_1, L_2 in Figure 2.3. The winding resistances (R_1, R_2) are easily included in series with the leakage inductances. These model transformer copper losses, which are proportional to the square of current.

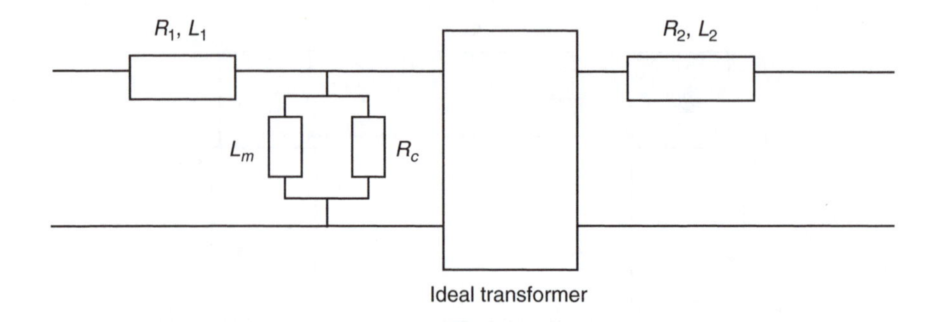

Figure 2.3 Practical transformer equivalent circuit

The effect of these series components is to create a full-load series voltage drop of up to 10 per cent of rated voltage at power frequencies. The predominance of the leakage inductance over winding resistance at power frequency ensures that the drop is largely at right angles to the line to neutral voltages. For this reason the change in the transformer voltage ratio over the loading range is minimal.

A relatively small current is needed to magnetise a practical transformer. This effect may be represented by a *magnetising inductance* placed across either winding of the ideal transformer – see L_m in Figure 2.3.

Finally, core losses may be represented by a resistance (R_c in Fig. 2.3) in parallel with the magnetising inductance. The core losses are independent of load. The overall efficiencies of power transformers tend to be in the range 95–97 per cent.

These various effects may be represented as shown in Figure 2.3. They are treated as external to essential transformer action, modelled by an ideal transformer embedded in the overall equivalent circuit.

2.3 AC power supply

The combination of alternators for generation and transformers to allow high-voltage power transmission has remained essentially unchanged for the past hundred years. It is important therefore to understand the behaviour of AC systems. The following sections will provide the necessary theoretical framework.

Initially we consider the meaning of power in AC systems. We will then introduce a very useful tool for analysing AC systems in the steady state, namely the phasor. This leads to a consideration of the behaviour of passive circuit components in AC circuits. We will then return to the representation of power, this time from the point of view of phasors. Finally we will consider the reasons for the adoption of three-phase systems for electricity supply.

2.3.1 Power in steady-state AC systems

Consider the following simple AC system, in which the instantaneous power flow from A to B may be positive or negative. It will be positive in the direction of A to B if v and i are positive (Fig. 2.4).

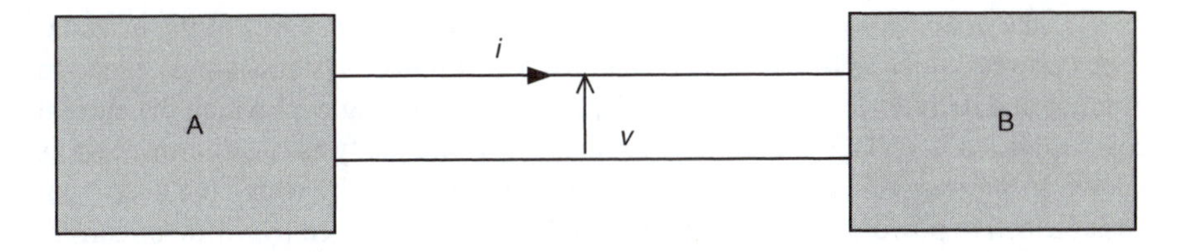

Figure 2.4 *Simple AC system*

The alternating voltages and currents may be represented as follows.

1. $v(t) = V_m \cos \omega t$
2. $i(t) = I_m \cos(\omega t - \phi)$

where $\omega =$ frequency in rad/s $= 2\pi f$ and $f =$ frequency in Hz.

The current is taken to lag the voltage by a time ϕ/ω, without any loss of generality. In electrical engineering, this time lag is often expressed as the corresponding angle, ϕ. This reflects the use of phasors, which will be introduced later.

The instantaneous power from A to B is given by the product of voltage and current as

$$p(t) = V_m I_m \cos \omega t \times \cos(\omega t - \phi)$$

Some rearrangement of the right-hand side of this equation will enable us to decouple the power into *active* and *reactive* components. First, noting that

$$\begin{aligned} \cos(A + B) &= \cos A \cos B - \sin A \sin B \\ \cos(A - B) &= \cos A \cos B + \sin A \sin B \\ \therefore \cos(A + B) + \cos(A - B) &= 2 \cos A \cos B \end{aligned}$$

and setting $A = \omega t$ and $B = \omega t - \phi$, we have

$$\begin{aligned} p(t) &= V_m I_m (\cos(2\omega t - \phi) + \cos \phi)/2 \\ &= \frac{V_m}{\sqrt{2}} \frac{I_m}{\sqrt{2}} (\cos 2\omega t \cos \phi + \sin 2\omega t \sin \phi + \cos \phi) \\ &= VI((1 + \cos 2\omega t) \cos \phi + \sin 2\omega t \sin \phi) \\ &= P(1 + \cos 2\omega t) + Q \sin 2\omega t \end{aligned}$$

Note that $V = V_m/\sqrt{2}$ and $I = I_m/\sqrt{2}$.

V and I are effective or *root mean square* (r.m.s.) values of voltage and current, respectively. From now on, all values for alternating voltage and current will be assumed to be r.m.s. unless it is stated otherwise.

The above analysis has introduced two very important quantities:

$$\begin{aligned} P &= VI \cos \phi \\ Q &= VI \sin \phi \end{aligned} \tag{2.14}$$

P is the *active power*, or simply *power*. This is the average value of the instantaneous power, which is of course the quantity which power systems are designed to produce and deliver. The active power depends on ϕ, the angle by which the current cosinusoid leads or lags the voltage cosinusoid in the analysis. The term $\cos\phi$ is known as the *power factor*. It can be thought of as the factor by which the maximum possible active power for a given voltage and current is reduced by virtue of current being out of phase with voltage.

Q is the *reactive power*. It is the amplitude of that component of instantaneous power which *oscillates*. It must be managed carefully by power system operators, not least because the associated current requires conductor capacity. Note that unity power factor implies zero reactive power.

The units (and unit symbols) of active and reactive power are as follows:

(active) power	watts (W)
reactive power	volt-amperes reactive (VAr)

2.3.2 *Phasors*

So far we have described time-varying cosinusoidal (or sinusoidal) quantities in the time domain: for example,

$$v(t) = V_m \cos(\omega t + \phi)$$

This equation contains a lot of information we do not normally require. Usually all we need are:

Magnitude	the peak or r.m.s. value is sufficient
Phase	the voltage (or current) leads a reference by angle ϕ
Frequency	often the same for all quantities

A shorthand version of the above equation is provided by a *phasor*. The phasor is a complex quantity, in this case voltage $\mathbf{V}$. This may be represented as a vector on an Argand diagram, as depicted in Figure 2.5. The magnitude of the phasor equals the r.m.s. value of the quantity being represented, $V_m/\sqrt{2}$ in this case. The angle of the phasor is the angle of the quantity relative to a reference, ϕ in our example.

The instantaneous value may be obtained from the phasor by reversing the above procedure. In most practical cases, the phasor quantity contains all the information we need.

2.3.2.1 Impedance: phasor representation

Phasors provide a convenient means of analysing circuits subject to alternating voltage and current. First, it is necessary to understand how the three passive circuit components – resistance, inductance and capacitance – respond to AC. The instantaneous voltage v resulting from an instantaneous cosinusoidal current i is considered in each case, based on the reference directions shown in Figure 2.6.

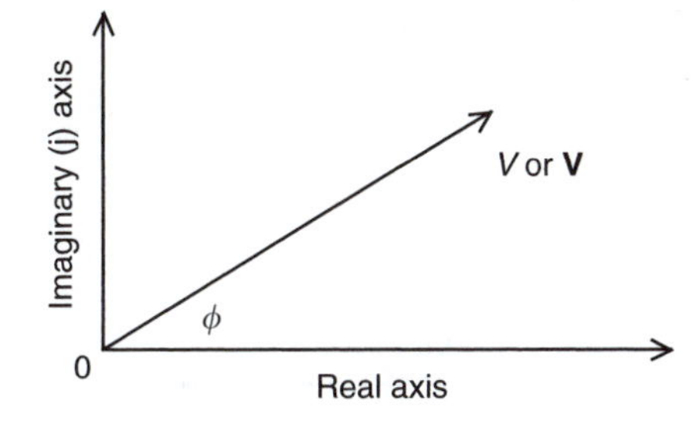

Figure 2.5 Representation of a phasor on an Argand diagram

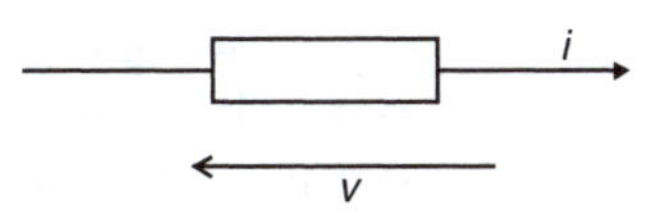

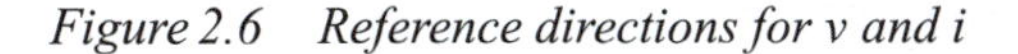
Figure 2.6 Reference directions for v and i

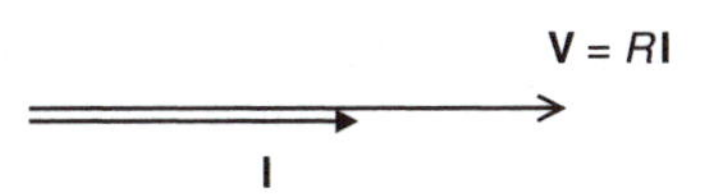

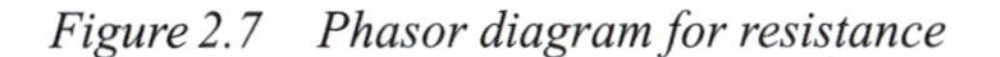
Figure 2.7 Phasor diagram for resistance

Resistance (R)
We know from Ohm's Law that $v = Ri$. Taking i as reference, we have

$$i = I_m \cos \omega t$$
$$\therefore \; v = RI_m \cos \omega t$$

The voltage drop across the resistor is in phase with the current through it. The corresponding phasor diagram is shown in Figure 2.7. Note that the complex phasor quantities are shown as bold letters.

The active and reactive powers for the resistance are given by Equation (2.14):

$$P = VI \cos \phi = VI = RI^2$$
$$Q = VI \sin \phi = 0$$

Inductance (L)
We may obtain inductor voltage for a cosinusoidal current from Equation (2.11):

$$v = L\frac{\mathrm{d}i}{\mathrm{d}t}$$
$$i = I_m \cos \omega t$$
$$\therefore \; v = \omega L I_m \cos(\omega t + \pi/2)$$

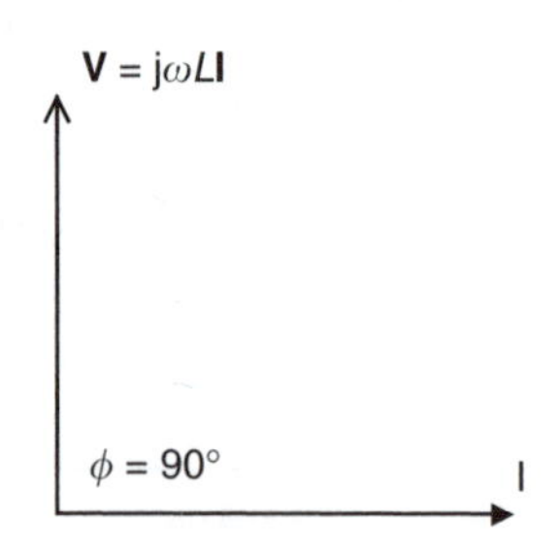

Figure 2.8 Phasor diagram for inductance

The corresponding phasor diagram is shown in Figure 2.8. The voltage phasor leads the current phasor by 90° ($\pi/2$ rad). This phase shift is achieved by use of the j operator. j is itself a phasor, with a magnitude of 1 and an angle of 90° measured anti-clockwise. If j is multiplied by itself, the product will be 1 with an angle of 180° anti-clockwise, or −1. From this it is seen that $\mathrm{j} = \sqrt{-1}$.

The impedance of the inductance is complex, and is given by $\mathbf{Z} = \mathrm{j}\omega L$ and the magnitude of the inductive impedance is referred to as its reactance, given by $X = \omega L$.

We may obtain the power and reactive power in the inductance from Equation (2.14):

$$P = VI\cos\phi = 0$$

$$Q = VI\sin\phi = VI = XI^2$$

An inductive load absorbs positive reactive power.

Capacitance (C)

Capacitive current may be related to the voltage drop across it by consideration of rate of change of charge:

$$q = Cv$$

$$i = \frac{\mathrm{d}q}{\mathrm{d}t} = C\frac{\mathrm{d}v}{\mathrm{d}t}$$

$$\therefore\ v = \frac{1}{\omega C} I_m \cos(\omega t - \pi/2)$$

In this case the voltage phasor lags the current phasor by 90° ($\pi/2$ rad). This corresponds to a phase shift of −j. The corresponding phasor diagram is shown in Figure 2.9.

The complex impedance of the capacitance is $\mathbf{Z} = 1/\mathrm{j}\omega C$ and the capacitive reactance or impedance magnitude is given by $X = 1/\omega C$.

The power and reactive power taken by the capacitor are given by Equation (2.14):

$$P = VI\cos\phi = 0$$

$$Q = VI\sin\phi = -VI$$

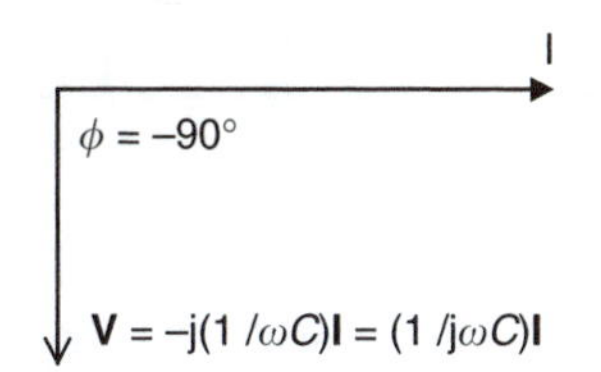

Figure 2.9 Phasor diagram for capacitance

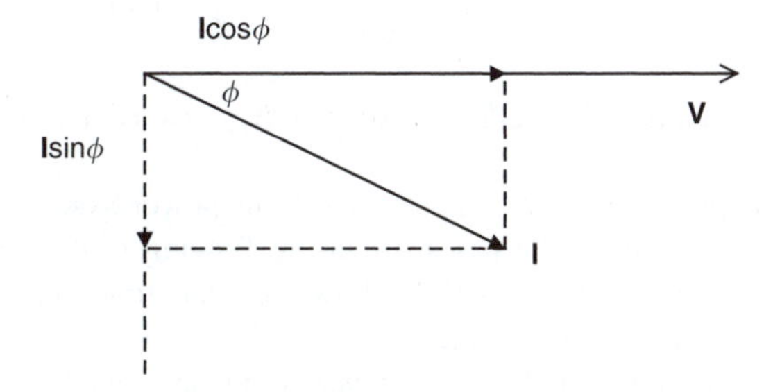

Figure 2.10 Real and reactive components of current

Note that a capacitive load absorbs negative reactive power, which is equivalent to generating positive reactive power.

2.3.2.2 Summary

The impedances for resistance, inductance and capacitance are therefore as follows:

Component	Symbol	Impedance
Resistance	R	R
Inductance	L	$\mathrm{j}\,\omega L$
Capacitance	C	$1/\mathrm{j}\,\omega C$

2.3.3 Power in AC systems

Consider the voltage and current phasors depicted in Figure 2.10. The current has been resolved into a real or active component in phase with voltage and a reactive component lagging the voltage by 90°.

Multiplying the currents in Figure 2.10 by the voltage magnitude results in active power in phase with the voltage phasor and reactive power lagging the voltage phasor by 90°, as shown in Figure 2.11. The diagram clarifies the relationship between active and reactive power, on the one hand, and the *apparent power VI* on the other. This relationship is often referred to as the *power triangle*, for obvious reasons.

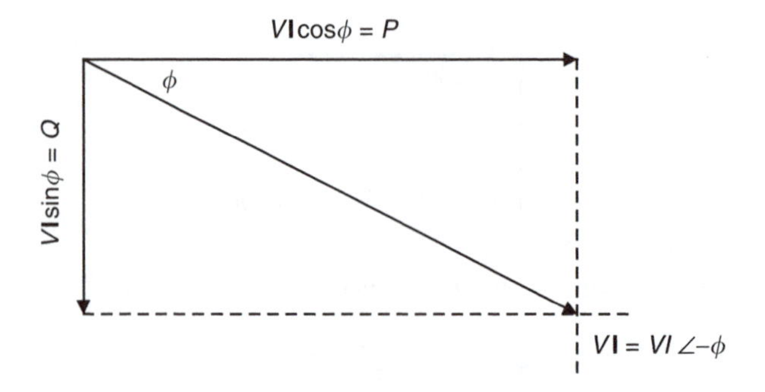

Figure 2.11 Triangular relationship between active, reactive and apparent power

The apparent power plays an important role in power system engineering. It defines, for a given voltage, the physical current flowing in the circuit. Apparent power is measured in volt-amperes (VA). Power engineering equipment is usually rated in terms of VA as well as voltage.

From Figure 2.11 it is seen that $VI =$ *apparent* power S and that

$$S = \sqrt{P^2 + Q^2}$$

S is measured in volt-amperes (VA, kVA or MVA).

2.4 Introduction to power systems

2.4.1 Three-phase systems

So far, it has been assumed that supply systems consist of a generator supplying a load through *go* and *return* conductors. This is a single-phase system. In fact, AC transmission invariably uses three phases. The reasons for this will be summarised below.

Consider the idealised synchronous generator shown in Figure 2.12. A magnetic field is produced by passing DC through the rotor or *field* winding. This may be obtained from a small generator, known as the *exciter*, on the same shaft. The exciter's windings are arranged in the opposite way to the main machine. Thus its field winding is on the stator and carries DC. The field will induce alternating voltages in the rotating coils on the rotor. This AC supply is rectified and fed to the main field winding. The net effect of this arrangement is that a change of current in the exciter field will cause the main field current to change in proportion. This in turn will create a change of synchronous machine-generated voltage – something which is essential for the smooth running of the power system as consumer demand and current vary.

The voltages induced in the stator coils A, B and C will be displaced by one third of an electrical cycle from each other, with A leading B and B leading C. Assuming that the machine has been designed to give sinusoidal voltages, these *three-phase* voltages may be depicted by the three phasors shown in Figure 2.13.

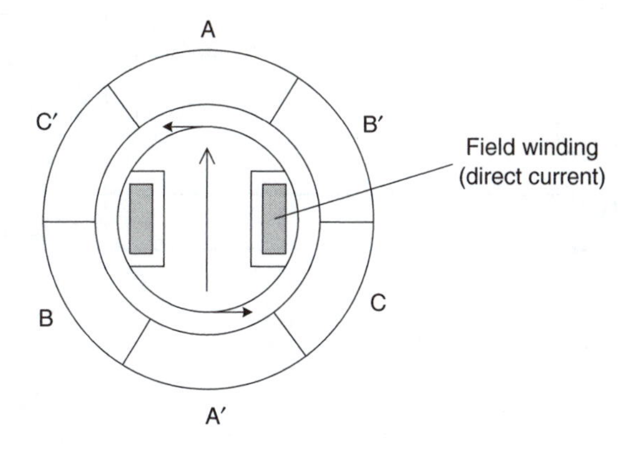

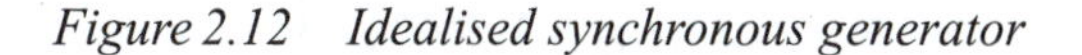
Figure 2.12 Idealised synchronous generator

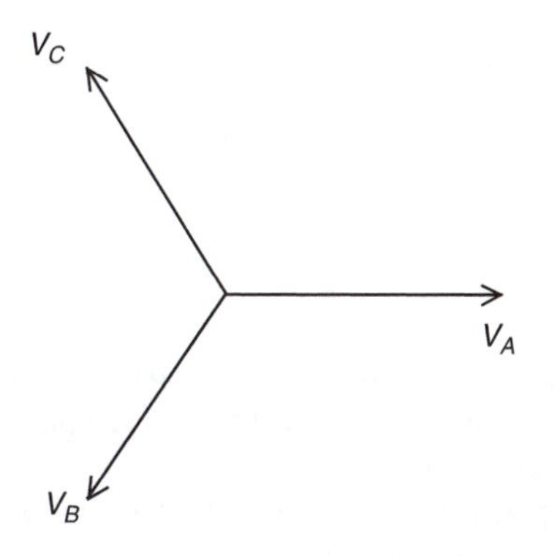

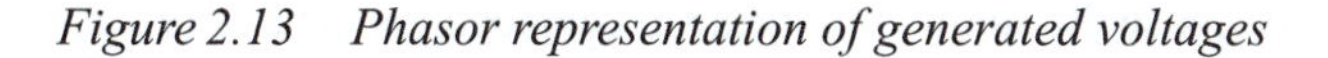
Figure 2.13 Phasor representation of generated voltages

This is a *balanced, three-phase supply*. We will assume balanced operation, except when considering unbalanced faults. The use of three phases has certain advantages:

- Enables better use to be made of the space available for machine windings.
- Provides the possibility of rotating magnetic fields in electric motors, and hence simpler starting.
- Reduces conductor material for a given power transmission capacity.

This will now be explained.

2.4.2 Comparison of single- and three-phase systems

Consider supply of a load of x kVA at a distance of l metres from a single-phase source, as shown in Figure 2.14.

The conductor material requirement is 2kxl, where k is the constant of proportionality relating conductor cross-sectional area to kVA for the system voltage.

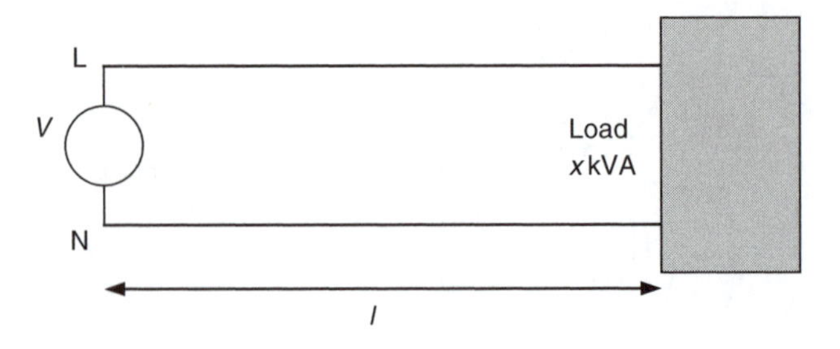

Figure 2.14 Single-phase supply

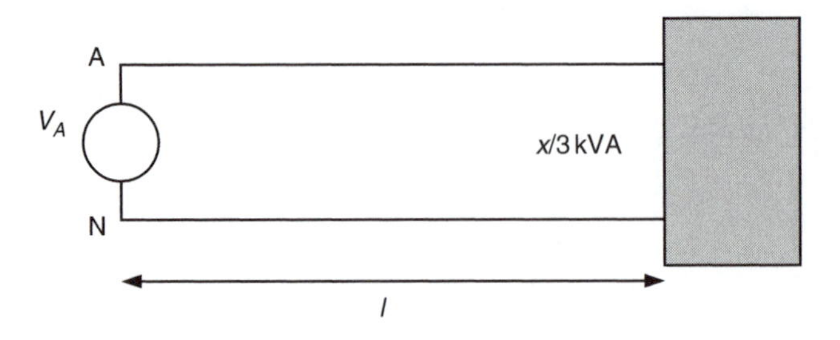

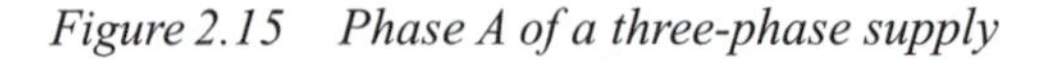

Figure 2.15 Phase A of a three-phase supply

Suppose the single-phase system to be replaced by a three-phase system delivering the same apparent power over the same distance. Each phase will supply one third of the load, as shown in Figure 2.15 for phase A.

The material requirement for phase A is $2kxl/3$. Phases B and C will be similar, giving a total material requirement of $2kxl$, as before. However, consider the *total* neutral (return) current. Because the supply voltages are balanced, the phase currents will be equal in magnitude but displaced by one third of a cycle from each other. The corresponding phasors will mimic the voltage phasors shown in Figure 2.13. Hence the total return current will be zero. We may take advantage of this by connecting the three phases with a common return, as shown in Figure 2.16.

The three neutrals have been merged. Because the resulting current is zero for a balanced system, the neutral connection is not required. Hence, conductor material is halved. In practice, a neutral conductor is provided, with 50–100 per cent of the rating of a line conductor. This carries current imbalance and higher harmonics.

2.4.3 Three-phase supply

A three-phase supply will always have three phase conductors, A, B and C, as shown in Figure 2.17. There will normally also be a neutral conductor N.

A balanced three-phase supply may be defined in terms of (a) line-to-line or *line* voltage V_L and (b) line-to-neutral or *phase* voltage V_{PH}.

It is important to establish the relationship between line and phase values of voltage and current for common three-phase connections.

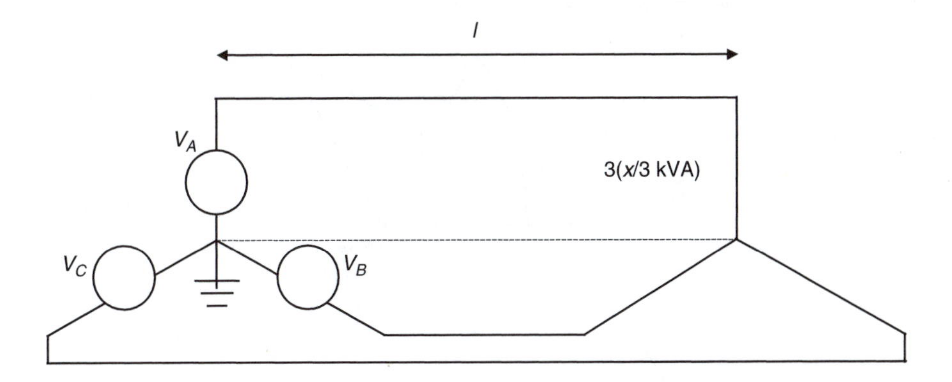

Figure 2.16 Three-phase supply

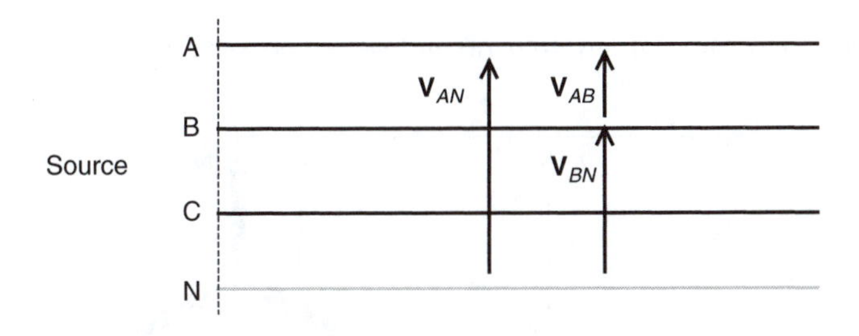

Figure 2.17 Three-phase supply

It may be seen from Figure 2.17 that the A to B line voltage is

$$\mathbf{V}_{AB} = \mathbf{V}_{AN} - \mathbf{V}_{BN}$$

Adding the corresponding phasors of Figure 2.18 results in the following relation for line voltage magnitude in terms of phase voltage magnitude:

$$V_{AB} = \sqrt{3}V_{AN}$$

In general, for a balanced three-phase supply:

$$V_L = \sqrt{3}V_{PH} \qquad (2.15)$$

2.4.4 Balanced star-connected load

The connections for a star-connected load are shown in Figure 2.19.

In general, for a balanced, star-connected load:

$$V_L = \sqrt{3}V_{PH}$$

$$I_L = I_{PH}$$

$$P = 3V_{PH}I_{PH}\cos\phi = \sqrt{3}V_L I_L \cos\phi$$

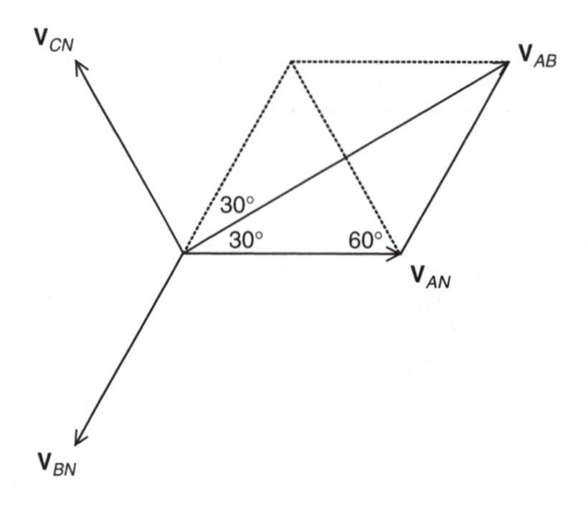

Figure 2.18 Phasor relationship between line and phase voltages

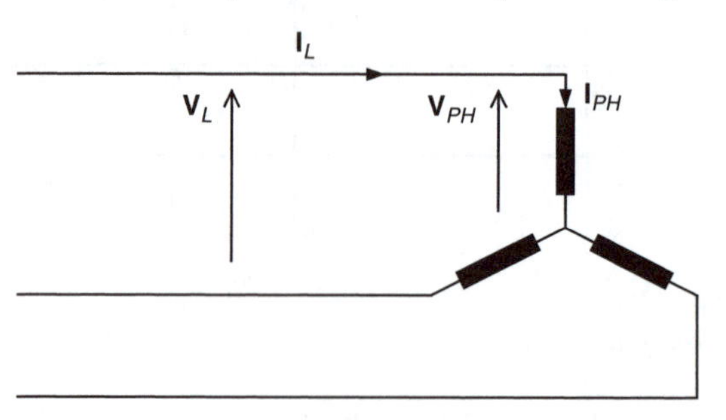

Figure 2.19 Star-connected load

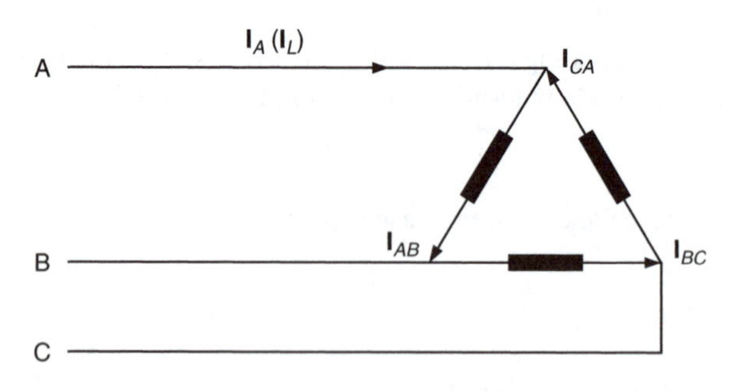

Figure 2.20 Delta-connected load

2.4.5 Balanced delta-connected load

The connections for a delta-connected load are shown in Figure 2.20.

The current in line A, for example, is related to the currents in adjacent phases by

$$\mathbf{I}_A = \mathbf{I}_{AB} - \mathbf{I}_{CA}$$

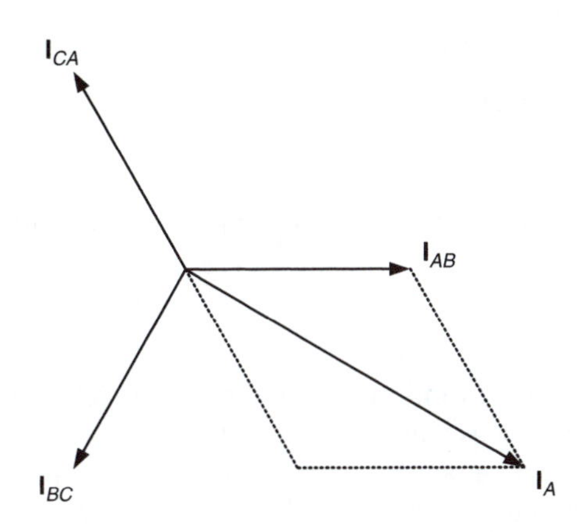

Figure 2.21 Phasor relationship between line and phase currents for a delta-connected load

This is shown in phasor form in Figure 2.21.

From Figure 2.21 it is clear that the magnitude of the current in line A is related to an adjacent phase current magnitude by

$$I_A = \sqrt{3}\,I_{AB}$$

In general, for a balanced, delta-connected load:

$$V_L = V_{PH}$$

$$I_L = \sqrt{3}\,I_{PH}$$

$$P = 3\,V_{PH}\,I_{PH}\cos\phi = \sqrt{3}\,V_L\,I_L\cos\phi \tag{2.16}$$

Note that the expression for power in Equation (2.16) is the same regardless of the load connection.

It follows from Equation (2.16) that reactive power in a balanced three-phase system is given by

$$Q = \sqrt{3}\,V_L\,I_L\sin\phi$$

The quoted voltage for a three-phase system is the line value.

2.4.6 Some useful conventions

For the most part we will be considering power transmission systems under *balanced* conditions. This embraces balanced three-phase faults as well as normal operation. Three key simplifications are:

1. The complex VA product.
2. The equivalent single-phase approach.
3. The per unit system.

2.4.7 The complex VA product

We now introduce complex power, defined as

$$\mathbf{S} = P + \mathrm{j}Q$$

where the real and reactive power, for a single-phase system, are given by

$$P = VI\cos\phi$$

$$Q = VI\sin\phi$$

It is important to note at this point that the *sign* of active power P is the same whether current leads or lags voltage by an angle ϕ, that is, $\cos(\phi) = \cos(-\phi)$. This reflects the nature of active power – it is a real quantity corresponding to a flow of energy in a particular direction during each electrical cycle.

The situation for reactive power Q is less clear. The sign of Q depends entirely on how we view the sign of ϕ, since $\sin(\phi) = -\sin(-\phi)$. This reflects the nature of reactive power, which describes the oscillating component of power. There is no net flow of Q during the electrical cycle. Hence the sign allocated to reactive power is optional.

When deriving the impedances of resistance, inductance and capacitance earlier, we took current as reference and measured the angle of the resulting voltage anti-clockwise from this datum. In the case of inductance, the voltage led the current by $90°$ ($\phi = 90°$) and the reactive power was positive. However, we could equally have taken voltage as reference; in that case the current would have lagged by $90°$, giving $\phi = -90°$ and negative reactive power.

Hence we need to decide on the sign of reactive power, based on whether the current is leading or lagging the voltage. As it happens, most loads take a lagging or inductive current. The associated reactive power manifests itself as extra current and losses in the utility's cables. Thus utilities tend to charge the consumer for reactive as well as for active power and energy. By assigning a positive sign to *lagging current* or inductive reactive power, utilities avoid the embarrassment of charging for a negative quantity. This supports the convention that

reactive power is deemed to be positive for a lagging/inductive current

The question now arises: how is complex power related to complex voltage and current? It is tempting to set

$$\mathbf{S} = \mathbf{V} \times \mathbf{I}$$

Assume that the voltage phasor leads the current phasor by the angle ϕ, and that the current phasor has an arbitrary angle θ. Taking $\mathbf{V} = Ve^{\mathrm{j}(\theta+\phi)}$ and $\mathbf{I} = Ie^{\mathrm{j}\theta}$, the complex power will then be

$$\mathbf{S} = VIe^{\mathrm{j}(2\theta+\phi)} = VI\cos(2\theta + \phi) + \mathrm{j}VI\sin(2\theta + \phi)$$

This bears no obvious relationship to power and reactive power. On the other hand, suppose we set

$$\mathbf{S} = \mathbf{V}\mathbf{I}^*$$

This equation uses the *complex conjugate* of the current, $\mathbf{I}^*$. The complex conjugate of a complex quantity is obtained by reversing its angle, or by changing the sign of its imaginary part. Thus

$$\mathbf{I}^* = Ie^{-\mathrm{j}\theta}$$

and

$$\mathbf{S} = VIe^{\mathrm{j}\phi} = VI\cos\phi + \mathrm{j}VI\sin\phi = P + \mathrm{j}Q$$

as required.

Thus complex power is given by

$$\mathbf{S} = \mathbf{VI}^* \tag{2.17}$$

2.4.8 Equivalent single-phase

Three-phase real and reactive power are given by

$$P = \sqrt{3}\,V_L I_L \cos\phi$$

$$Q = \sqrt{3}\,V_L I_L \sin\phi$$

For balanced conditions it is convenient to define an equivalent single-phase system with voltage and current as follows:

$$V = V_L$$

$$I = \sqrt{3}\,I_L$$

Equivalent single-phase power and reactive power are then given by

$$P = VI\cos\phi$$

$$Q = VI\sin\phi$$

It is always possible to retrieve the actual line current by dividing by $\sqrt{3}$. However, we are more likely to be interested in quantities such as (1) voltage profile, (2) real and reactive power flow and (3) losses/efficiency.

These quantities may be obtained directly from the equivalent single-phase model.

2.4.9 The per unit system

The per unit system has been devised to remove two difficulties:

1. The large numbers that would occur in power systems work if we restrict ourselves to volts, amperes and ohms.
2. The analysis of networks with several voltage levels.

For example, it is more meaningful to say that a 275 kV system node voltage is 1.05 per unit (pu) than 289 kV line-to-line. The per unit number tells us immediately that the voltage is 5 per cent above nominal – often that is all we need to know.

Per unit quantities may be expressed as 1.05 pu, or just 1.05 if it is obvious that it is a per unit quantity.

The basis of the per unit system is that we choose voltage, current and impedance bases such that

$$V_b = Z_b I_b$$

We can see immediately that only two of these bases can be chosen independently. The per unit quantities are then

$$\mathbf{V}_{\mathrm{pu}} = \mathbf{V}/V_b$$
$$\mathbf{I}_{\mathrm{pu}} = \mathbf{I}/I_b$$
$$\mathbf{Z}_{\mathrm{pu}} = \mathbf{Z}/Z_b$$

In power systems work we are more likely to work to MVA and kV bases:

$$\mathrm{MVA}_b$$
$$\mathrm{kV}_b$$

Assuming that we use the equivalent single-phase approach, the impedance base Z_b may be obtained as follows:

$$Z_b = V_b/I_b = \left(\mathrm{kV}_b \times 10^3\right)/\left(\left(\mathrm{MVA}_b \times 10^6\right)/\mathrm{kV}_b \times 10^3\right) = \mathrm{kV}_b^2/\mathrm{MVA}_b$$

Therefore

$$\mathbf{Z}_{\mathrm{pu}} = \frac{\mathbf{Z} \times \mathrm{MVA}_b}{\mathrm{kV}_b^2} \tag{2.18}$$

Also

$$\mathbf{S}_{\mathrm{pu}} = \frac{\mathbf{S}}{\mathrm{MVA}_b} = \frac{P}{\mathrm{MVA}_b} + \mathrm{j}\frac{Q}{\mathrm{MVA}_b} = P_{\mathrm{pu}} + \mathrm{j}Q_{\mathrm{pu}}$$

Note that the voltage, current and impedance bases are real, while the corresponding per unit quantities are complex.

2.4.9.1 Networks with multiple voltage levels

Practical power networks contain several voltage levels. For example, wind generators often have a nominal line voltage of 690 V. These feed a distribution network of 11 or 33 kV (in the United Kingdom) through a transformer. The distribution network is in turn connected, at a substation, to a transmission network of perhaps 275 kV. The per unit system provides a convenient means of analysing such multi-voltage networks, because the entire network may be considered to be a single voltage network with a nominal voltage of 1.0 pu. The actual voltages may be obtained easily from the calculated per unit values and the base values at the nodes of interest.

If the voltage base is changed in proportion to the nominal turns ratios of transformers, an equivalent *transformer-less* network with a nominal voltage level of 1.0 pu is created.

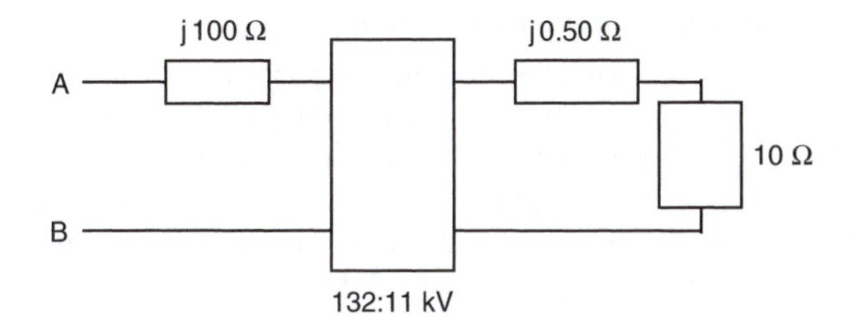

Figure 2.22 Simple multi-voltage network

Example
The simple network in Figure 2.22 could be used to model a length of 132 kV line of impedance j 100 Ω/phase supplying a 10 Ω load through a 132/11 kV step-down transformer and a line of impedance j 0.50 Ω.

Suppose we wish to determine the impedance between A and B, perhaps to assess the load on a generator connected between A and B. The conventional approach is to *refer* the impedances to the high or low voltage side. Suppose we refer the impedances to the 132 kV level. The impedances on the low voltage side must be modified to dissipate the same power. Taking the step-down ratio as n, the impedances referred to the high-voltage side will now experience a current $1/n$ th the low voltage value. Hence the impedance values must be increased by a factor of n^2 to ensure that the power dissipations are as before. The total effective impedance between A and B is thus

$$\mathrm{j}\,100 + n^2(10 + \mathrm{j}\,0.50) = (1440 + \mathrm{j}\,172) \quad \Omega/\text{phase}$$

This effective impedance may be expressed in per unit. Taking an MVA base of 100 and a kilovolt base of 132, this gives

$$\mathbf{Z}_{\mathrm{AB}} = (1440 + \mathrm{j}\,172)\frac{\mathrm{MVA}_b}{\mathrm{kV}_b^2} = (8.264 + \mathrm{j}\,0.987)\,\mathrm{pu}$$

If we now change the voltage base in proportion to the turns ratio of any transformer encountered, the voltage bases will be 132 and 11 kV. The MVA base will remain the same throughout the network, giving a per unit impedance of

$$\mathbf{Z}_{\mathrm{AB}} = \mathrm{j}\,100\frac{100}{132^2} + (10 + \mathrm{j}\,0.50)\frac{100}{11^2} = (8.264 + \mathrm{j}\,0.987)\,\mathrm{pu}$$

as required.

Use of the second method, based on modification of the base voltage with transformer turns ratios, is of little benefit in this simple case. However, it is much more convenient than the first method – referring all impedances to a single voltage level – for more realistic power network studies.

2.4.9.2 Conversion to a common MVA base

A common MVA base is required when using the per unit system. Plant impedances $\mathbf{Z}_p$ are usually quoted in per unit (or per cent) to the MVA rating of that particular item, say MVA_p. If the plant is to be included in a system study with a common system base of MVA_s, then the per unit impedance must be modified. It may be seen from Equation (2.18) that the per unit impedance is directly proportional to MVA. Hence the per unit impedance for the plant will become

$$\mathbf{Z}'_p = \mathbf{Z}_p \times \frac{\text{MVA}_s}{\text{MVA}_p}$$

for use in the system study.

2.5 Power transmission

The basic process of power transmission is that current flows round a loop consisting of source, line and load. The current flow is impeded by line resistance and loop inductance. There is also capacitance between the lines. Thus the line parameters are:

- Series resistance
- Series inductance
- Shunt capacitance.

In the power transmission relevant to wind farm connection, the significant parameters are overhead line resistance and inductance. Shunt capacitance cannot be ignored for underground cables. However, the capacitive effect of cabling within a wind farm is small, and will be ignored here.

2.5.1 Line parameters

The following discussion will be based on *single-phase* power transmission. This simplification is justified here – our objectives are merely to estimate resistance and inductance at a given voltage level and, in particular, to appreciate the factors that determine their relative magnitudes. The extension to three-phase transmission is fairly intuitive. A sound derivation of line parameters from the line's physical dimensions may be found in the well-established power systems textbook by Grainger and Stevenson (1994).

2.5.1.1 Line resistance

Consider the single-phase line shown in Figure 2.23, consisting of parallel conductors of length l and cross-sectional area a in air. The resistance of each conductor is $R = \rho l/a$ where ρ is the conductor resistivity. The resistance is often quoted per unit length:

$$R = \rho/a \tag{2.19}$$

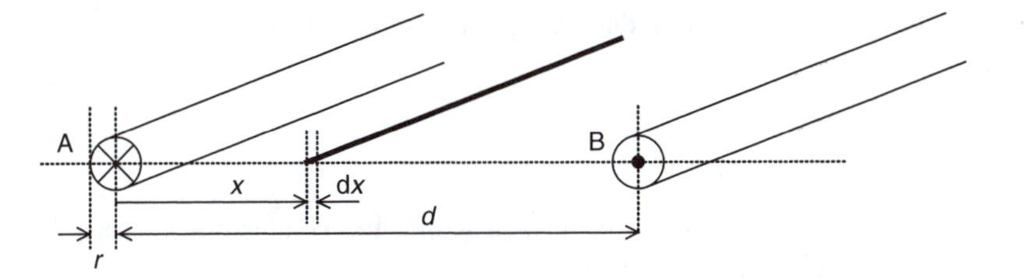

Figure 2.23 Single-phase line

2.5.1.2 Line inductance

The line loop inductance may be found by considering the flux through the elemental rectangle shown in Figure 2.23, co-planar with the conductors, x from conductor A and of width dx. The total flux linking the current path may then be found by integration. The inductance is then the flux linkage divided by the current, as seen earlier.

Assume that the loop current is i. The magnetic field strength normal to the elemental rectangle is obtained by applying Equation (2.1) to both conductors:

$$H = \frac{i}{2\pi x} + \frac{i}{2\pi(d-x)}$$

The corresponding flux density is, from Equation (2.2),

$$B = \mu_0 \left(\frac{i}{2\pi x} + \frac{i}{2\pi(d-x)} \right)$$

The flux enclosed by the rectangular element is the flux density multiplied by the area normal to the field:

$$\mathrm{d}\Phi = \frac{\mu_0 l i}{2\pi} \left(\frac{1}{x} + \frac{1}{d-x} \right) \mathrm{d}x$$

The total flux linked by the current is obtained by considering all rectangular elements between the conductors:

$$\begin{aligned}
\Phi &= \frac{\mu_0 l i}{2\pi} \int_r^{d-r} \left(\frac{1}{x} + \frac{1}{d-x} \right) \mathrm{d}x \\
&= \frac{\mu_0 l i}{2\pi} [\ln(x) - \ln(d-x)]_r^{d-r} \\
&= \frac{\mu_0 l i}{\pi} \ln\left(\frac{d-r}{r} \right)
\end{aligned}$$

The loop inductance is then

$$L_l = \Phi / i = \frac{\mu_0 l}{\pi} \ln\left(\frac{d-r}{r} \right)$$

It is more usual to work with conductor inductance, which will be half of the loop inductance. Also, r is small in comparison with d, giving the following approximate expression for the conductor inductance per unit length (Henry/m):

$$L = \frac{\mu_0}{2\pi} \ln \frac{d}{r} \tag{2.20}$$

This analysis ignores the flux linkage within the conductors, but any loss of accuracy is of little importance here.

In the case of three-phase transmission, the expression for inductance per phase is very similar. The only difference is that the distance between conductors d is replaced by the geometric average of the distances between each pair of phase conductors, D_m. This result assumes that each phase conductor is located in each of the three possible positions for one third of the line length – a condition known as equal transposition.

2.5.1.3 Typical line parameter values

The behaviour of transmission and distribution overhead lines is governed mainly by their series resistance and inductance parameters. Shunt capacitance becomes a significant issue for transmission lines longer than 100 km, and for cables of any length. We will confine our attention to the distribution lines and shorter transmission lines relevant to wind farm connection. Hence shunt capacitance will be ignored in the following discussion.

It is clear from Equation (2.19) that the resistance of a line is proportional to its length and inversely proportional to its cross-sectional area. The cross-sectional area a must be such as to enable the expected current to be carried without excessive heating or sag. The cross-sectional area a will clearly be a much higher value for transmission lines rated at several hundred MVA than for distribution lines designed to carry a few MVA. Thus distribution line resistance tends to be higher than transmission line resistance for a given length.

Line inductance per unit length was given in Equation (2.20). The value of inductance is governed by the ratio of the geometric average of the distances between conductors, D_m, and conductor radius r. D_m will be greater for the higher voltage levels found in transmission, to ensure adequate insulation between phases under extreme conditions. However, conductor radius will also be greater in transmission to cope with the higher power and current requirements. Hence the ratio of D_m/r does not vary greatly as we move from transmission to distribution. As the inductance per unit length is the natural log of this ratio, the variation in the parameter is even less marked than the D_m/r ratio.

It is convenient to express the reactance $X = \omega L$ of the line rather than its inductance. It is found that the X/R ratio of transmission lines, with line voltages greater than, say, 100 kV, are generally greater than 2.5. The same ratio for distribution lines (11, 20 and 33 kV) is often close to unity. The following analysis should be interpreted in the light of these very different ratios.

Typical line resistance and reactance values are shown in Table 2.1 (Weedy and Cory, 1998). It may be seen from these data that *reactance values vary little between voltage ranges* and *X/R ratios increase with voltage level.*

Table 2.1 Overhead line parameters at 50 Hz (per phase, per km)

Voltage (kV)	33	132	275
Number of conductors	1	1	2
Area of conductor (mm^2)	100	113	258
Thermal rating, 5–18 °C (MVA)	20	100	620
Resistance $R(\Omega)$	0.30	0.16	0.034
Reactance $X(\Omega)$	0.31	0.41	0.32

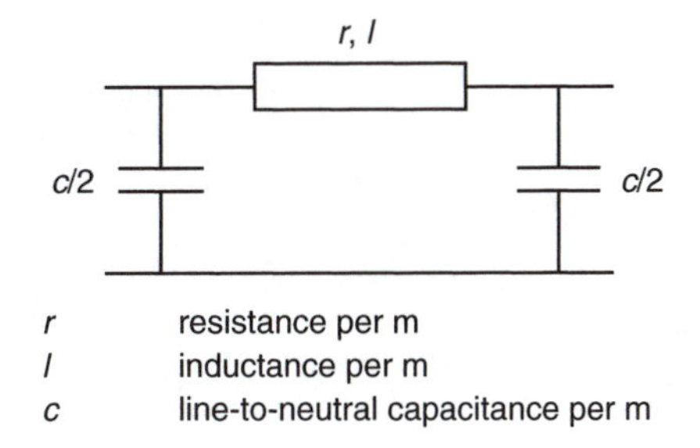

Figure 2.24 Transmission line parameters

2.5.2 Line models

A 1-m length of transmission line may be represented by a π section, as shown in Figure 2.24.

For a known length and frequency, the parameters become

R series resistance, $r \times$ length

X series reactance, $\omega l \times$ length

B shunt susceptance, $\omega c \times$ length

2.5.2.1 The short line model

The shunt capacitive effect is negligible up to around 100 km for overhead lines (but not for cables!), so a short line model can ignore B. The resulting line model is shown in Figure 2.25.

The behaviour of the line under various loading conditions may be described by expressing the sending-end voltage in terms of the receiving-end voltage and the voltage drop in the line series impedance:

$$\mathbf{V}_S = \mathbf{V}_R + (R + jX)\mathbf{I} \tag{2.21}$$

This equation leads to the phasor diagram shown in Figure 2.26. Assume the load current to lag the receiving-end voltage by ϕ. It may be seen that the angle, δ, by which the sending-end voltage leads the receiving-end voltage increases with load.

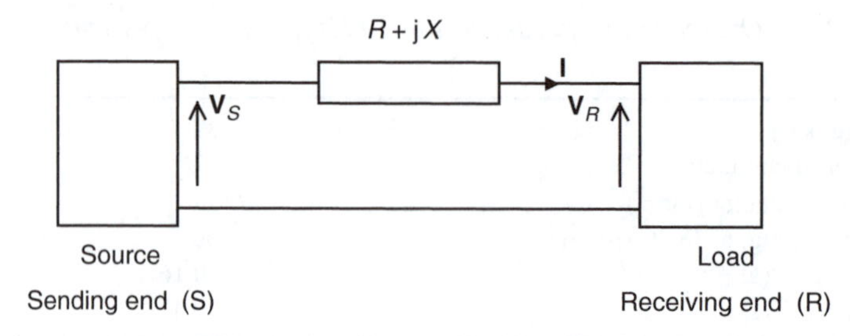

Figure 2.25 Short line model

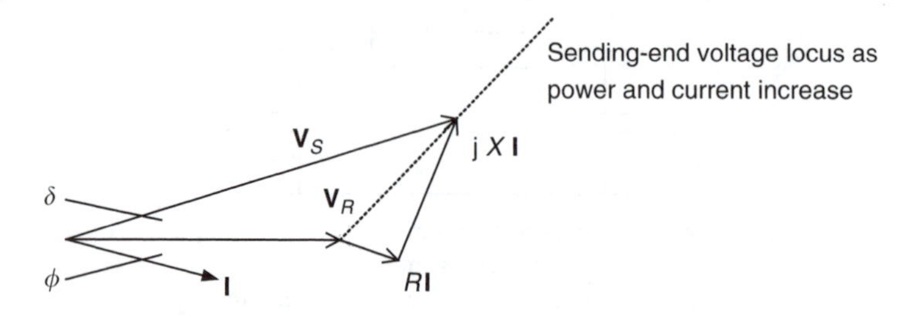

Figure 2.26 Voltage phasor diagram for a short transmission line

This angle is known as the load or power angle. Also, the sending-end voltage phasor terminates on a straight line locus whose orientation depends on ϕ. In practice, the sending-end voltage may be fixed, implying that the receiving-end voltage decreases with load for the assumed ϕ. However, a leading load current could result in receiving-end voltage increasing with load.

We also introduce a line *impedance* angle φ (not to be confused with the power angle) such that

$$\mathbf{Z} = R + \mathrm{j}X = Z\angle\varphi$$

The essential features of power transmission may be seen by obtaining an equation for the power received by the load. From the complex VA product,

$$P_R + \mathrm{j}Q_R = \mathbf{V}_R\mathbf{I}_R^*$$
$$P_R = \Re e\{\mathbf{V}_R\mathbf{I}_R^*\}$$

Let the receiving-end voltage be the reference:

$$\mathbf{V}_R = V_R\angle 0^\circ = V_R$$

Since $\mathbf{I} = \mathbf{I}_R$, we may use Equation (2.21) to express $\mathbf{I}_R^*$ in terms of the sending- and receiving-end voltages:

$$\mathbf{I}_R^* = \left(\frac{\mathbf{V}_S - \mathbf{V}_R}{\mathbf{Z}}\right)^*$$

After some manipulation we obtain

$$P_R = \frac{V_R V_S \cos(\varphi - \delta)}{Z} - \frac{V_R^2 \cos\varphi}{Z} \tag{2.22}$$

2.5.3 Power transmission

For high-voltage lines, the series reactance is several times the series resistance. Equation (2.22) can therefore be simplified by assuming $R = 0$.

$\therefore \quad \varphi = 90^\circ$

$\therefore \quad \cos\varphi = 0$

$\therefore \quad \cos(\varphi - \delta) = \cos\varphi\cos\delta + \sin\varphi\sin\delta = \sin\delta$

$$\therefore \quad P_R = \frac{V_S V_R \sin\delta}{X} \tag{2.23}$$

Voltages need to be kept within, typically, 6 per cent of their nominal values. Therefore the transmitted power is determined mainly by the power angle δ – hence its name.

Sending-end voltage does not need to be greater than receiving-end voltage for power to flow to the receiving-end. This can be seen from Figure 2.26 for a load with *leading* power factor.

2.5.3.1 Maximum transmissible power

Note that the power transmitted for a loss-less line is a maximum when the power angle is 90°. The maximum transmissible power will then be, from Equation (2.23),

$$P_{RM} = \frac{V_S V_R}{X}$$

The sending- and receiving-end voltages will be close to the rated values. Thus the key factor limiting transmissible power, apart from the conductor current rating, is reactance. This is determined by line length ($X = \omega l \times$ length).

2.5.4 Voltage regulation

Consider the case of a generator supplying a large system through an overhead line, as shown in Figure 2.27.

It is assumed that the system is large, and that its voltage is fixed. We wish to study the variation of the generator's voltage with generated active and reactive power. The generator voltage is given by

$$\mathbf{V}_g = \mathbf{V} + (R + \mathrm{j}X)\mathbf{I}$$

We will examine the effect of active and reactive power separately. Considering active power first, the current **I** will be in phase with the generator voltage. The relationship between generator voltage and system voltage may be seen from the phasor diagram shown in Figure 2.28.

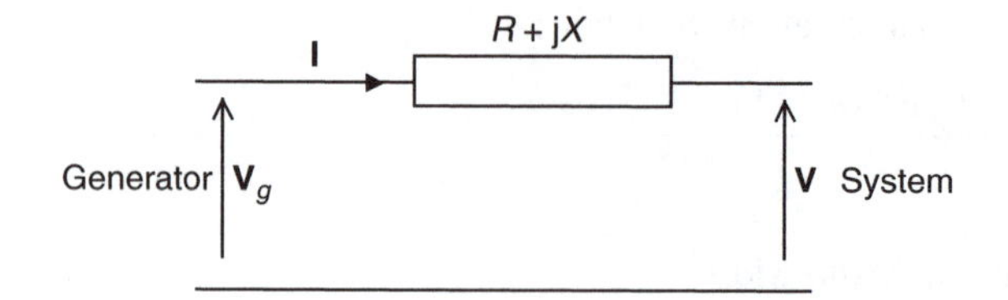

Figure 2.27 *Generator feeding large system*

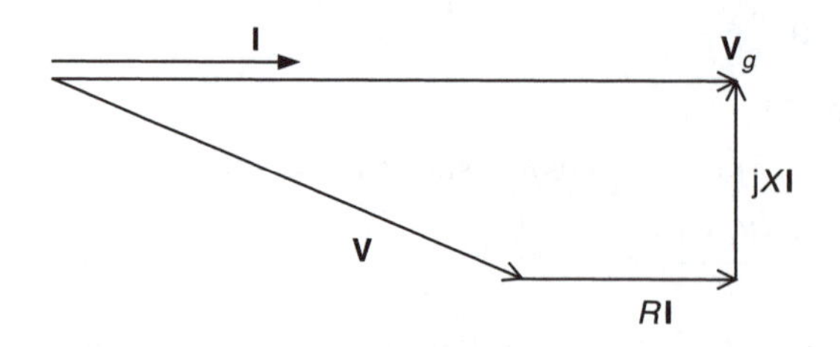

Figure 2.28 *Voltage regulation for active power*

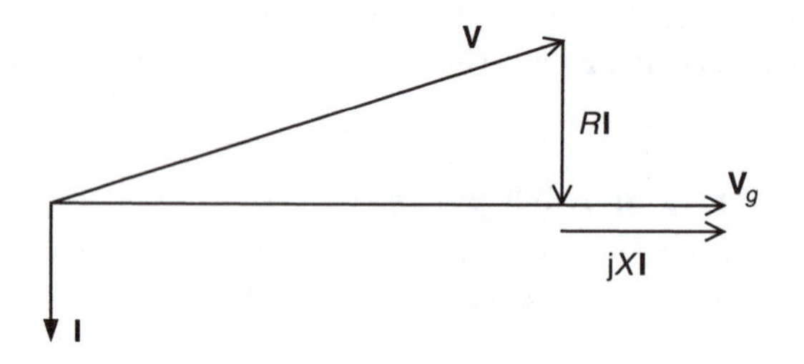

Figure 2.29 *Voltage regulation for reactive power*

It may be seen from Figure 2.28 that the voltage rise due to *active* power generation is given approximately by

$$\Delta V_a \approx RI = \frac{RV_g I}{V_g} = \frac{RP}{V_g}$$

Now consider *reactive* power generation. Positive reactive power consumption for a load implies current lagging voltage by 90°. Adoption of the generator convention reverses the current phasor. Positive reactive power generation reverses it again. This is depicted in the phasor diagram of Figure 2.29.

It may be seen from Figure 2.29 that the voltage rise due to *reactive* power generation is given approximately by

$$\Delta V_r \approx XI = \frac{XV_g I}{V_g} = \frac{XQ}{V_g}$$

The total voltage rise from generation is given by

$$\Delta V = \Delta V_a + \Delta V_r \approx \frac{RP + XQ}{V_g} \tag{2.24}$$

Equation (2.24) can be used iteratively to estimate the generator voltage. For example, we could:

1. set $V_g = 1.0$ pu
2. obtain ΔV
3. set $V_g = 1.0 + \Delta V$
4. if V_g is within a given tolerance of its previous value, exit; otherwise return to step 2

This is a primitive example of a *load flow* problem.

In transmission networks, R is small relative to X, and the voltage profile is determined mainly by reactive power flows. Wind farms are usually connected at distribution voltage level. As we saw above, R and X may be comparable at these voltages. Hence real power will often be more influential in determining the voltage at the wind farm terminals than reactive power. The resulting *voltage rise* problem, and possible solutions, will be considered in Chapter 4.

Chapter 3
Wind power technology

3.1 Introduction

The main purpose here is to explain the technology in common use for large-scale wind power generation. The evolution of turbine design is first of all placed in its historical context. The current design choices facing the industry are then considered. A simple analysis of the basic process of wind energy extraction is presented, leading to a discussion of the critical area of power regulation. It will be seen that the options here depend on whether fixed- or variable-speed technology is adopted.

The description of fixed-speed wind turbine technology includes a review of relevant induction machine theory. The induction (or asynchronous) machine is a key component of fixed-speed wind generators, as well as forming the basis for partial variable-speed designs. This section considers some of the shortcomings of the fixed-speed designs which dominated the industry's initial growth surge.

The final section of the chapter reviews the arguments that have led to the growing dominance of variable-speed technology. The section includes a detailed treatment of partial variable-speed (doubly fed induction generator [DFIG]) and full variable-speed wind turbine generator configurations.

3.2 Historical review of wind power technology

The wind has been used as a source of power for pumping water and grinding corn for more than a thousand years and, by the eighteenth century, the traditional European windmill had evolved into a sophisticated device capable of developing up to 25 kW in strong wind (Golding, 1955). It is estimated that before the industrial revolution there were some 10,000 windmills in England, but these fell into disuse with the introduction of reliable steam engines. Water-pumping windmills were used widely in the United States from around 1850 for a hundred years (Spera, 1994), only being superseded by electrical pumps following the extensive rural electrification programmes of the 1940s.

The success of the 'American' water-pumping windmill stimulated investigation of how the wind could be used to generate electrical energy. Dr C.F. Brush built a 17-m diameter, 12 kW direct current (DC), multi-bladed wind turbine in Cleveland, Ohio in 1888, while Prof. P. LaCour conducted important experiments that led to the construction of several hundred wind generators in the range 5–25 kW in Denmark in the early 1900s. From the 1920s two- and three-bladed, battery-charging wind turbines, rated at up to 3 kW, were used for domestic supplies in the United States. Again, these could not compete with the expanding rural electrification programme (Johnson, 1985).

Large wind turbine development began with the 30-m diameter, 100 kW turbine erected in the Crimea in 1930, but perhaps the most impressive early machine was the 53-m diameter, 1,250 kW, Smith Putnam wind turbine erected in 1941 (Putnam, 1948). This remarkable two-bladed, upwind, full span pitch-regulated machine exhibited many of the features of modern wind turbines and ran successfully for around 1,000 hours before shedding a blade. Although considered a technical success, the project was then abandoned, as the manufacturer could not see a market for large electricity-generating wind turbines in the face of cheap fossil fuel. Research continued in Europe and a number of important large wind turbines were constructed. In 1957, the Gedser three-bladed, 24-m diameter, 200 kW turbine was built in Denmark, while Dr U. Hutter in Germany constructed an advanced 100 kW lightweight machine. Around this time, individual wind turbines of around 100 kW were also constructed in France and the United Kingdom, but the low price of oil led to limited commercial interest in wind energy.

However, in the early 1970s the world price of oil tripled, stimulating wind energy research and development programmes in a number of countries including the United States, United Kingdom, Germany and Sweden. In general, these programmes supported the development of large, technically advanced wind turbines by major aerospace manufacturing companies. Typical examples included the Mod 5B (97.5-m diameter, 3.2 MW) machine in the United States, and the LS-1 (60-m diameter, 3 MW) machine in the United Kingdom, but similar prototypes were constructed in Germany and Sweden. The technical challenges of building reliable and cost-effective wind turbines on this scale were under-estimated and these prototypes did not lead directly to successful commercial products, although much useful information was gained.

In a further response to high oil prices, tax and other financial incentives were put in place to support the deployment of wind turbines, most notably in California. These measures provided a market for manufacturers to supply much smaller, simpler wind turbines. Initially a wide variety of designs were used, including vertical axis wind turbines, but over time the so-called 'Danish' concept of a three-bladed, upwind, stall regulated, fixed-speed wind turbine became dominant. Initially these turbines were small, sometimes rated at only 30 kW, but were developed over the next 15 years to around 40-m diameter, 800–1,000 kW. However, by the mid-1990s it was becoming clear that for larger wind turbines it would be necessary to move away from this simple architecture and to use a number of the advanced concepts (e.g. variable-speed operation, pitch regulation, advanced materials) that had been investigated in the earlier government funded research programmes. Thus, large wind turbines (up to 100 m in diameter, rated at 3–4 MW) are now being developed using the concepts of the large prototypes of the 1980s but building upon the experience gained from over twenty years of commercial operation of smaller machines.

3.3 Design choices for large wind turbine generators

There are a large number of choices available to the designer of a wind turbine and, over the years, most of these have been explored. However, commercial designs for electricity generation have now converged to horizontal axis, three-bladed, upwind turbines. The larger machines tend to operate at variable-speed while smaller, simpler turbines operate at fixed-speed using stall regulation.

The very earliest windmills used a vertical axis rotor. These were simple to construct and used aerodynamic drag to produce only low output power. Their principle of operation was similar to a cup anemometer, being based on the differential drag on each side of the rotor. However, with careful arrangement of the blades, lift forces can be made much larger than drag forces and so the axis of rotation of windmills was changed from vertical to horizontal and their principle of operation changed to using the lift force.

In modern times, the simplicity, lack of requirement to orientate the rotor into the wind and ability to operate at low wind speeds has made vertical axis, drag-based machines attractive for small wind turbines that are integrated into buildings. The Savonius rotor is perhaps the best-known example of this class of wind turbine (Le Gourieres, 1982). However, the high solidity (and hence high cost) and relatively low output of the rotor means that machines operating using drag forces are unlikely to be cost-effective for large-scale electricity generation.

Although for electricity generation it is clear that the rotor should operate using lift forces, the question of whether the axis of rotation should be vertical or horizontal was still actively debated until around 1990. A large number of Darrieus *egg-beater* vertical axis wind turbines were installed in California and a multi-megawatt prototype constructed in Canada. In addition, the United Kingdom constructed two prototypes based on a straight bladed, vertical axis design proposed by Dr P. Musgrove. Among the advantages claimed for the vertical axis rotor was that the gearbox and generator could be located at ground level and that there was no requirement to yaw the rotor into the wind. However, after detailed assessment, it became clear that vertical axis wind turbines have to be made heavier, and hence more expensive, than horizontal axis machines, and so this line of development came to an end.

With a horizontal axis wind turbine, there is an obvious choice as to whether the rotor should be upwind or downwind of the tower. Locating the rotor downwind allows the blades to flex without the danger of them hitting the tower and, in principle, downwind rotors can be arranged to yaw into the wind without requiring yaw motors or brakes. Early prototypes of the US Government programme in the 1970s used downwind rotors, and experimental machines continued with this concept until the

1990s. However, downwind operation leads to greater tower shadow effects, when the wind stream seen by the rotor is impeded by the tower, and hence causes reduced electrical power quality and increased aerodynamic noise.

Modern electricity-generating wind turbines now use three-bladed upwind rotors although two-bladed, and even one-bladed, rotors were used in earlier commercial turbines. Reducing the number of blades means that the rotor has to operate at a higher rotational speed in order to extract the wind energy from the rotor disc. Although a high rotor speed is attractive in that it reduces the gearbox ratio required, a high tip speed leads to increased aerodynamic noise and increased drag losses. Most importantly, three-bladed rotors are visually more pleasing than other designs and so these are now always used on large electricity-generating turbines.

The simplest arrangement for electricity-generating wind turbines is to operate at essentially constant rotational speed using an induction (sometimes known as asynchronous) generator. It is not practical to use a directly connected synchronous generator as the tower shadow effect causes large pulsations in the mechanical torque developed by the aerodynamic rotor. Practical synchronous generators cannot damp these oscillations and so an induction machine, which has much greater damping, is used. An alternative approach is to couple the generator to the network through a frequency converter using power electronics and so allow its speed to vary. The advantages of this arrangement are: it reduces mechanical loads; makes the control of transient torque easier; and enables the aerodynamic rotor to operate at its maximum efficiency over a wide range of wind speeds. The penalty is the cost and electrical losses of the power electronics. Most of the largest wind turbines now being installed operate at variable-speed, as the power electronic converters also allow much greater control of the output power and it is easier to comply with the requirements of the power system network operator.

3.4 Energy extraction and power regulation

3.4.1 Energy extraction across the rotor disc

A wind turbine operates by extracting energy from the swept area of the rotor disc (Burton *et al.*, 2001; Manwell *et al.*, 2002), as shown in Figure 3.1.

Power in the airflow is given by

$$P_{\text{air}} = \frac{1}{2}\rho A V^3 \tag{3.1}$$

where ρ = air density (approx. 1.225 kg/m^3); A = swept area of rotor; V = free wind speed.

This may be seen to be true by considering the kinetic energy of the air passing through the rotor disc in unit time.

$$P_{\text{air}} = \frac{1}{2}\rho A V \cdot V^2 \tag{3.2}$$

where $\rho A V$ = mass flow rate of air.

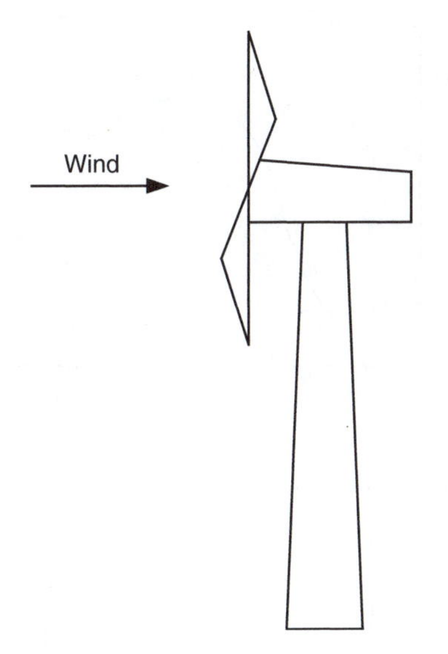

Figure 3.1 Horizontal axis wind turbine

However, not all the power can be extracted by the turbine and so a power coefficient (C_P) is defined. The power coefficient is simply the ratio of power extracted by the wind turbine rotor to the power available in the wind.

$$C_P = \frac{P_{\mathrm{wt}}}{P_{\mathrm{air}}} \tag{3.3}$$

$$P_{\mathrm{wt}} = C_P \cdot P_{\mathrm{air}} = C_P \cdot \frac{1}{2}\rho A V^3 \tag{3.4}$$

It can be shown that for any fluid turbine there is a maximum power that can be extracted from the fluid flow as given by the equation:

$$C_{P_{\mathrm{max}}} = 16/27 \text{ or } 0.593$$

This is known as the Betz limit, which states that a turbine can never extract more than 59 per cent of the power from an air stream.

It is also conventional to define a tip speed ratio (λ)

$$\lambda = \frac{\omega R}{V} \tag{3.5}$$

where ω = rotational speed of rotor; R = radius to tip of rotor and V = free wind speed.

λ and C_P are dimensionless and so can be used to describe the performance of any size of wind turbine. Figure 3.2 shows that the maximum power coefficient is only achieved at a single tip speed ratio, and for a fixed rotational speed of the wind turbine this only occurs at a single wind speed. Hence one argument for operating a

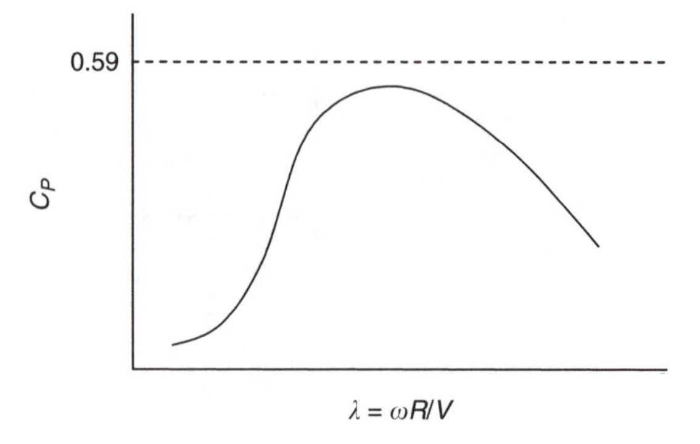

Figure 3.2 Illustration of power coefficient/tip speed ratio curve (C_P/λ)

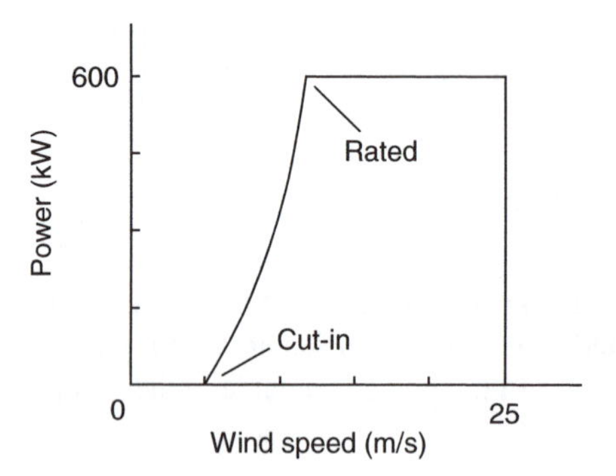

Figure 3.3 Power curve of a wind turbine

wind turbine at variable rotational speed is that it is possible to operate at maximum C_P over a range of wind speeds.

The overall performance of the wind turbine is described by its power curve (Fig. 3.3).

The power curve relates the steady-state output power developed by the wind turbine to the free wind speed and is generally measured using 10-minute average data. Below the cut-in wind speed, of about 5 m/s, the wind turbine remains shut down as the power in the wind is too low for useful energy production. Then, once operating, the power output increases following a broadly cubic relationship with wind speed (although modified by the variations in C_P) until rated wind speed is reached. Above rated wind speed the aerodynamic rotor is arranged to limit the mechanical power extracted from the wind and so reduce the mechanical loads on the drive train. Then, in very high wind speeds, the turbine is shut down.

The choice of cut-in, rated and shutdown wind speeds is made by the wind turbine designer who, for typical wind conditions, will try to balance maximum energy extraction with controlling the mechanical loads (and hence the capital cost) of the turbine. For a mean annual site wind speed V_m of 8 m/s typical values will

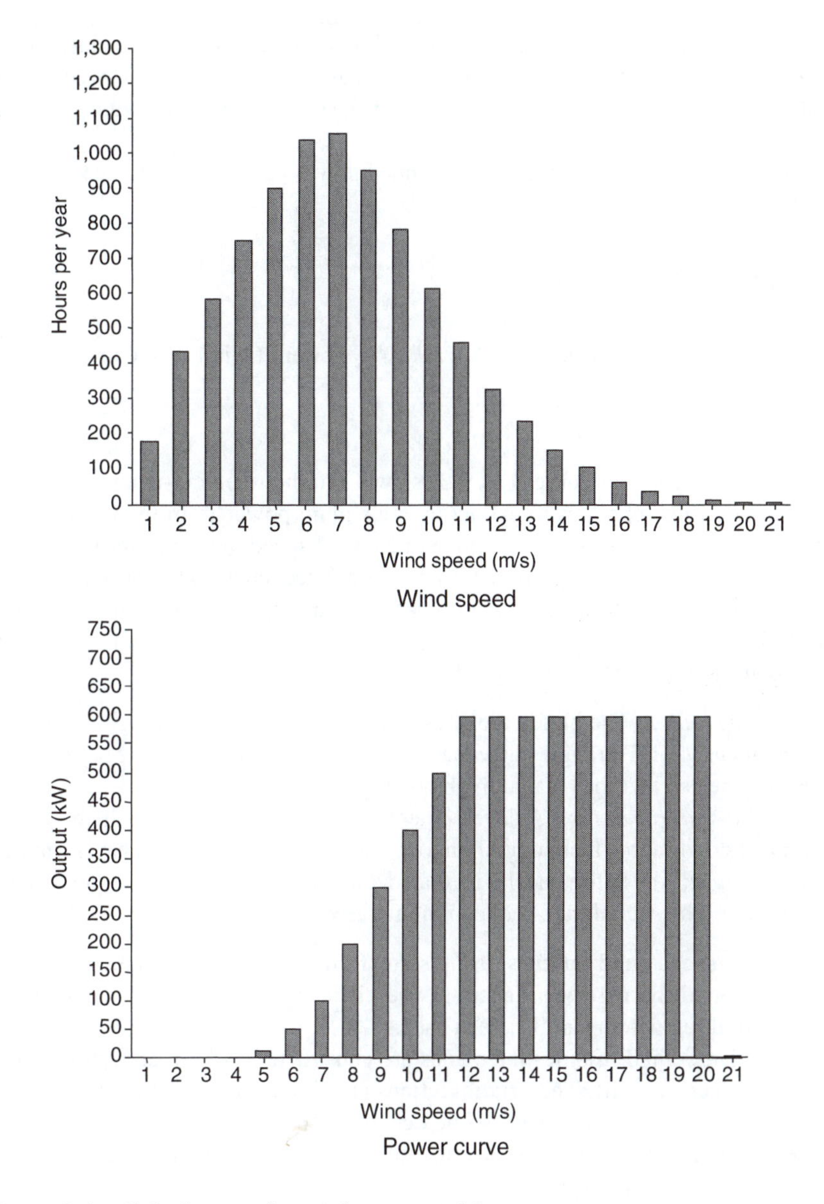

Figure 3.4 Calculation of annual energy yield

be approximately: cut-in wind speed 5 m/s, 0.6 V_m; rated wind speed 12–14 m/s, 1.5–1.75 V_m and shut-down wind speed 25 m/s, 3 V_m.

The energy extracted over a year by the wind turbine is, of course, determined by the power curve and the site wind resource. Figure 3.4 shows binned wind speed data and a power curve. The wind speed bins are of 1 m/s, for example, 5.5–6.5 m/s, 6.5–7.5 m/s.

By multiplying the hours per year for each wind speed bin by the binned power curve of the turbine, the annual energy yield may be calculated. It is necessary to deduct losses, including electrical losses within the wind farm and aerodynamic array losses.

The calculation of energy from the binned power curve and wind speed data is simply:

$$\text{Energy} = \sum_{i=1}^{i=n} H(i) \cdot P(i) \tag{3.6}$$

where $H(i)$ = hours in wind speed bin i; $P(i)$ = power at wind speed bin i.

3.4.2 Power regulation

The wind turbine power curve (Fig. 3.3) shows that, between cut-in and rated wind speeds, the turbine operates to extract the maximum power from the wind passing across the rotor disc. However, at above rated wind speed the mechanical power on the rotor shaft is deliberately limited in order to reduce loads on the turbine.

Fixed-speed wind turbines, that is, those using induction (or asynchronous) generators directly connected to the network, may use one of a number of techniques to limit power at above rated wind speed:

- *Pitch regulation*: The blades are physically rotated about their longitudinal axis.
- *Stall regulation*: The angle of the blades is fixed but the aerodynamic performance is designed so that they stall at high wind speeds.
- *Assisted-stall regulation*: A development of stall regulation where the blades are rotated slowly about their longitudinal axis but the main control mechanism is stall.
- *Yaw control*: The entire nacelle is rotated about the tower to yaw the rotor out of the wind. This technique is not commonly used.

Variable-speed wind turbines, using some form of power electronics to connect the generator to the network, generally use pitch regulation at high wind speeds, although stall regulation has also been used.

On all large wind turbines the blades are constructed to form an airfoil section designed to generate lift. The airfoil sections are similar to those found on aircraft wings. These sections have well-defined characteristics depending on the angle of incidence, sometimes known as the angle of attack. A typical blade element airfoil section is shown in Figure 3.5.

The usual definitions are as follows:

- A: leading edge
- B: trailing edge
- L: chord line (a straight line between the leading and trailing edges)
- α: the angle of incidence between the free wind velocity and the chord line.

As the wind passes over the blade its speed increases and pressure reduces on the top surface. Conversely, the speed reduces and the pressure increases on

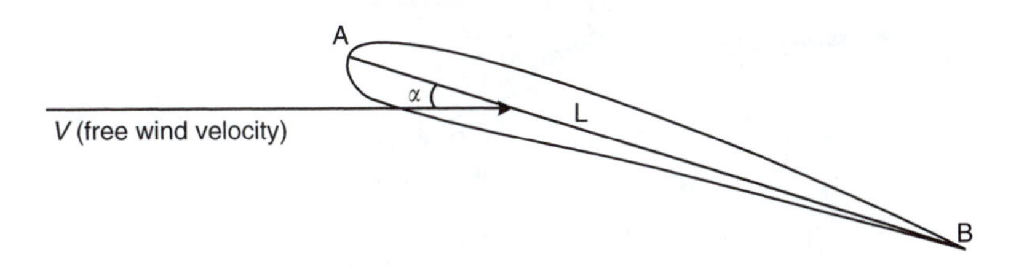

Figure 3.5 Airfoil section

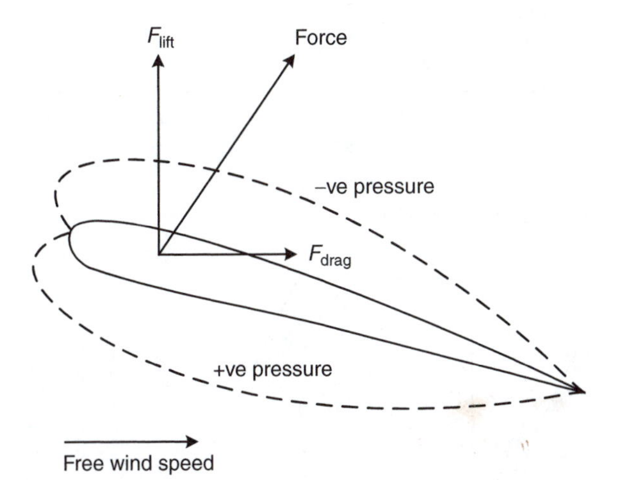

Figure 3.6 Lift and drag forces on an airfoil

the lower surface. This follows from Bernoulli's theorem where for a horizontal airflow:

$$\frac{1}{2}\rho V^2 + p = \text{constant} \tag{3.7}$$

where V is the free wind speed, ρ is the air density and p is the pressure. The pressure difference can be represented as shown in Figure 3.6.

The force resulting from this pressure difference can be split into two components at right angles to each other, F_{lift} and F_{drag}, the lift and drag forces. The lift force is perpendicular to the free wind velocity and the drag force is parallel to it.

These forces are then described by lift and drag coefficients C_l and C_d. The lift force is proportional to C_l and the drag force to C_d.

Both C_l and C_d change with the angle of incidence of the blade in the air stream. The lift coefficient is approximately linear up to its maximum and the airfoil then stalls and the lift coefficient reduces (Fig. 3.7). Stalling is caused by the boundary layer on the upper surface of the airfoil separating and a circular wake forming. The drag coefficient increases more slowly with α.

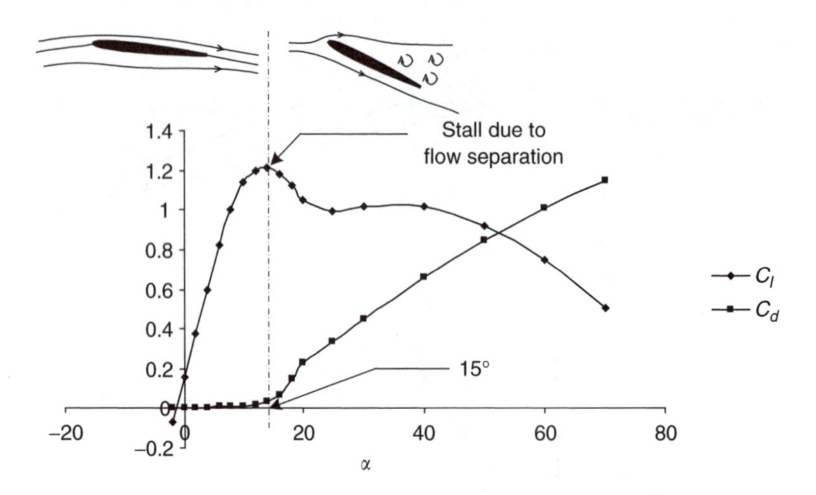

Figure 3.7 Airfoil characteristics

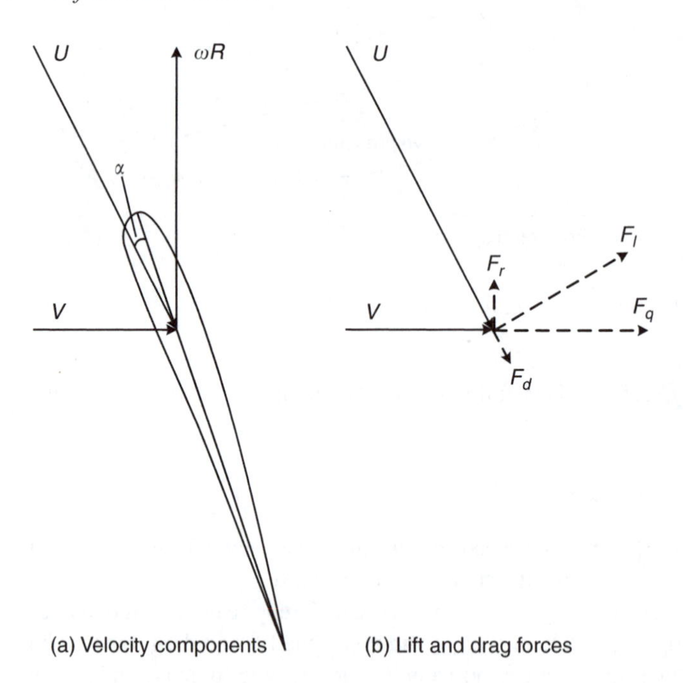

Figure 3.8 Operation of wind turbine

Figure 3.8 illustrates how a wind turbine operates and how the two usual mechanisms of controlling the power – stall and pitch regulation – work. Figure 3.8(a) shows a wind turbine blade viewed from its tip. It is useful to consider the blade vertical with the viewer looking down on to the tip.

U is the relative wind speed seen by the blade and is formed by the incident-
wind speed V and the rotational speed of the blade ωR. By definition F_d (the drag
force) is parallel and F_l (the lift force) is perpendicular to U (Figure 3.8(b)). The lift and drag forces can then be resolved into the rotational force that turns the blades, F_r, and the thrust force on the rotor that has to be resisted by the structure, F_q.

For a fixed-speed wind turbine the angular velocity of the rotor ω is fixed by the electrical generator, which is locked on to the network. Hence it may be seen that there are two ways of reducing the rotational force F_r at above rated wind speeds. The angle of incidence α may be reduced by mechanically turning the blade. As shown in Figure 3.7 this reduces the lift coefficient C_l, with only a small effect on drag coefficient, C_d. Reducing the lift coefficient C_l then reduces the lift force F_l, which in turn reduces the rotation force, F_r. This is *pitch regulation*, as it requires a change in the pitch of the blades by mechanical rotation.

Alternatively, the blades can be fixed to the hub at a constant angle. Recalling that ωR is constant, then as the free wind speed V increases it can be seen that the angle of incidence α also increases. As shown in Figure 3.7, once the blade has stalled the lift coefficient C_l and hence the torque decreases. This is *stall regulation*, which does not require any physical change in the pitch angle of the blades.

It is obvious that stall regulation is attractive, as no moving parts are required. Unfortunately, it has proved rather difficult to predict the wind speed at which blades stall, as it appears that the stall effect is a complex three-dimensional phenomenon involving a degree of hysteresis. Pitch regulation is easier to predict but requires a control mechanism, as shown in Figure 3.9, to alter the pitch angle of the blades. The output power of the wind turbine is measured with an electrical power transducer and used as the input to a control system that alters the pitch angle of the blades. Because stall occurs at various wind speeds at differing radial locations along the

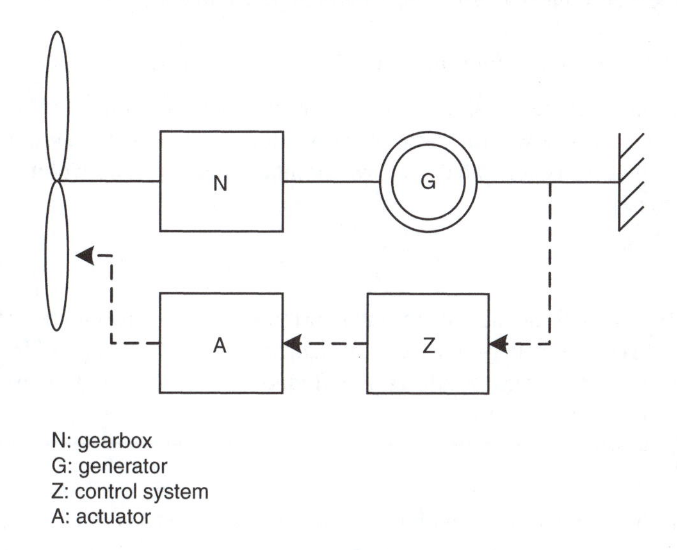

Figure 3.9 Pitch regulation control system

rve of a stall-regulated wind turbine is less steep than that of ade, with a consequent loss in energy capture. However, both ılated fixed-speed wind turbines up to 80-m diameter and 2 MW now commercially available.

has been significant interest in using *assisted stall* control (sometimes known ... tive stall). In this technique a slow blade pitch actuator is used to position the blades, but the main control mechanism remains aerodynamic stall. One reason for using assisted stall is that the braking requirements for very large wind turbines require blade actuators, and so these can be used for assisted stall without additional equipment.

Variable-speed wind turbines generally employ pitch regulation to limit the power into their rotors, but stall regulation has also been used (Burton *et al.*, 2001). However, for stall regulation to be effective, the rotational speed, ωR, must be kept constant. This can be done using the power electronic converters, but a key advantage of variable-speed operation of large wind turbines is that the rotor may accelerate when hit by a gust of wind and so reduce mechanical loads. Stall regulation, with the requirement to maintain constant speed, reduces this advantage.

3.5 Fixed-speed wind turbines

Fixed-speed wind turbines are electrically quite simple devices consisting of an aerodynamic rotor driving a low-speed shaft, a gearbox, a high-speed shaft, and an induction or asynchronous generator. From the electrical system viewpoint they are perhaps best considered as large fan drives with torque applied to the shaft from the wind flow. Before describing the main components and characteristics of these turbines, the basic characteristics of an induction machine are reviewed.

3.5.1 Review of the induction (asynchronous) machine

The induction machine consists of the stator and rotor windings. When balanced three-phase currents flow through the stator winding a magnetic field rotating at synchronous speed, n_s, is generated. The synchronous speed, n_s, in revolutions/minute is expressed as

$$n_s = \frac{120 f_s}{p_f} \tag{3.8}$$

where f_s (Hz) is the frequency of the stator currents, and p_f is the number of poles. If there is a relative motion between the stator field and the rotor, voltages of frequency f_r (Hz) are induced in the rotor windings. The frequency f_r is equal to the slip frequency sf_s, where the slip, s, is given by

$$s = \frac{n_s - n_r}{n_s} \tag{3.9}$$

where n_r is the rotor speed in revolutions/minute. The slip is positive if the rotor runs below the synchronous speed and negative if it runs above the synchronous speed (Krause, 2002; Kundur, 1994).

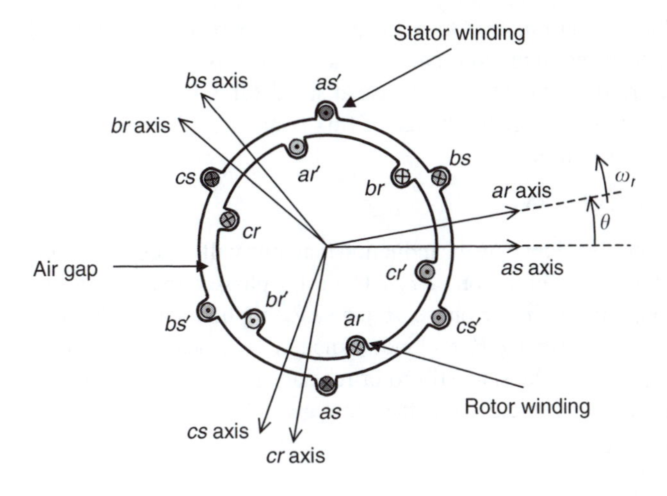

Figure 3.10 Schematic diagram of a three-phase induction machine (Kundur, 1994)

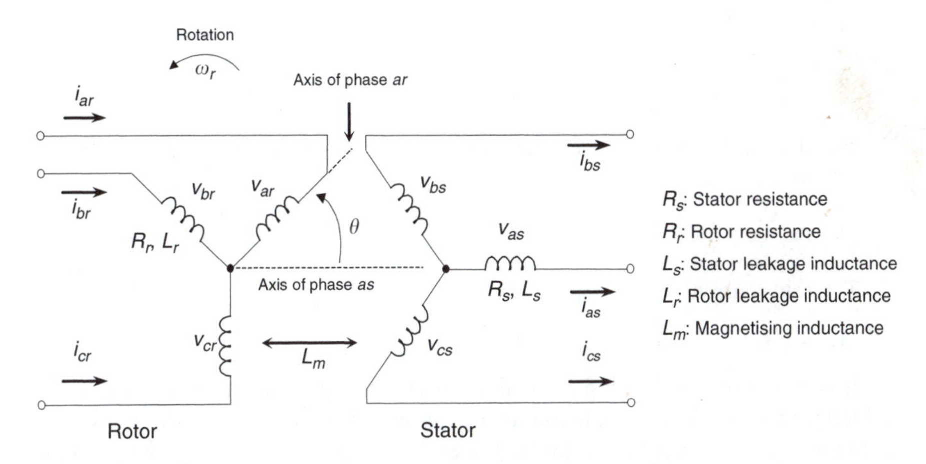

Figure 3.11 Stator and rotor electrical circuit of an induction machine (Kundur, 1994)

Figure 3.10 shows the schematic of the cross-section of a three-phase induction machine with one pair of field poles, and Figure 3.11 illustrates the stator and rotor electrical circuits. The stator consists of three-phase windings *as*, *bs* and *cs* distributed 120° apart in space. The rotor circuits have three distributed windings *ar*, *br* and *cr*. The angle θ is given as the angle by which the axis of the phase *ar* rotor winding leads the axis of phase *as* stator winding in the direction of rotation, and ω_r is the rotor angular velocity in electrical rad/s. The angular velocity of the stator field in electrical rad/s is represented by $\omega_s = 2\pi f_s$.

Voltages are induced in the rotor phases by virtue of their velocity relative to the stator field, in accordance with Faraday's Law (Equation (2.10)). The magnitude of the induced electromotive force (e.m.f.) is proportional to the slip. If the rotor is stationary, then the induction machine may be regarded as a transformer. Suppose the induced rotor voltage in each phase at standstill is $\boldsymbol{V_r}$. Since the induced voltage is proportional to the rate of change of flux (Equation (2.10)), the rotor voltage at a particular slip s will be

$$\text{rotor voltage} = s\mathbf{V_r}$$

The torque developed by the induction machine will depend on the current flowing in each rotor phase (Equation (2.7)). The rotor phase current for a given slip s will be determined by the rotor phase voltage $s\mathbf{V_r}$ applied across the rotor impedance. This consists of resistance R_r and inductance L_r. The reactance depends on the rotor current frequency. If the standstill rotor reactance is $X_r = \omega_s L_r$, then its value at slip s will be sX_r. The rotor current is therefore given by

$$\mathbf{I_r} = \frac{s\mathbf{V_r}}{R_r + \mathrm{j}sX_r}$$

Note that the slip-dependent rotor voltage may be replaced by the standstill rotor voltage as follows:

$$\mathbf{I_r} = \frac{\mathbf{V_r}}{\frac{R_r}{s} + \mathrm{j}X_r}$$

The standstill rotor voltage can be referred to the stator circuit, given the effective rotor/stator turns ratio N:

$$\mathbf{V_r'} = \frac{\mathbf{V_r}}{N}$$

The rotor current referred to the stator is

$$\mathbf{I_r'} = N\mathbf{I_r}$$

It is instructive to develop an equivalent circuit representing each phase of the induction machine. This is analogous to the equivalent circuit for a transformer shown in Figure 2.3, with machine rotor in place of transformer secondary and machine stator in place of transformer primary. It is convenient to refer all quantities to the stator. The current in a rotor phase will be $1/N$ times the current in a stator phase, to achieve the magnetomotive force (m.m.f.) balance required for transformer action (see Section 2.2.5). Rotor impedance referred to the stator must therefore be scaled by $1/N^2$ to give the same relative voltage drop and power loss in the stator. Thus

$$R_r' = R_r/N^2$$

$$X_r' = X_r/N^2$$

The rotor circuit referred to the stator is shown to the right of aa′ in Figure 3.12. The stator resistance R_s and stator leakage reactance X_s are also shown in Figure 3.12. The equivalent circuit is completed by the magnetising reactance X_m.

It is convenient to place the magnetising reactance across the supply voltage $\mathbf{V_s}$, as shown in Figure 3.13. This simplifies analysis with minimal loss of accuracy.

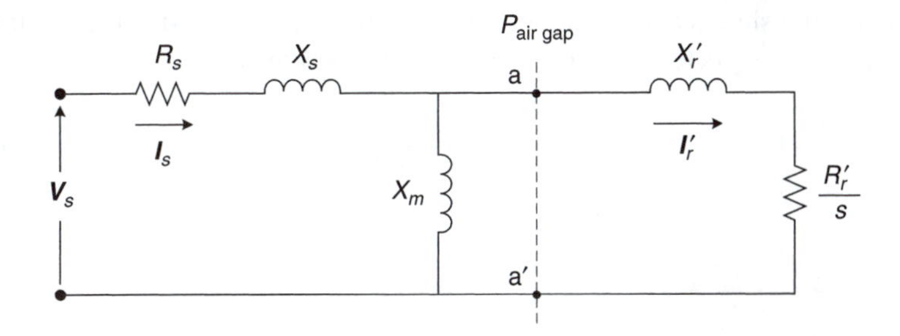

Figure 3.12 Single-phase equivalent circuit of an induction machine (Kundur, 1994)

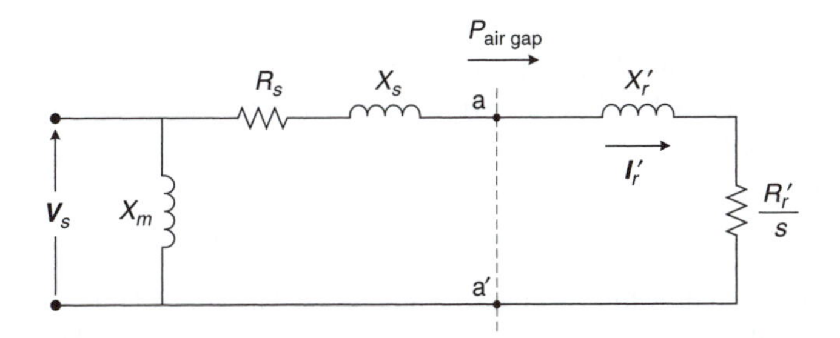

Figure 3.13 Equivalent circuit suitable for evaluating simple torque-slip relationships

The power transferred across the air gap to the rotor (of one phase) is

$$P_{\text{air gap}} = \frac{R_r'}{s}(I_r')^2 \tag{3.10}$$

The torque developed by the machine (three-phase) is obtained from Equation (3.8) and $P_{\text{air gap}} = 2\pi n_s T/60$:

$$T = 3\frac{p_f}{2}\frac{R_r'}{s\omega_s}(I_r')^2 \tag{3.11}$$

where $\omega_s = 2\pi f_s$. As seen in Equation (3.11), the torque is slip-dependent. For simple analysis of torque–slip relationships, the equivalent circuit of Figure 3.13 may be used.

From Figure 3.13, the rotor current is

$$\boldsymbol{I}_{\mathbf{r}}' = \frac{\boldsymbol{V}_{\mathbf{s}}}{(R_s + R_r'/s) + \mathrm{j}(X_s + X_r')} \tag{3.12}$$

Then from Equation (3.11), the torque is

$$T = 3\frac{p_f}{2}\left(\frac{R_r'}{s\omega_s}\right)\frac{V_s^2}{(R_s + R_r'/s)^2 + (X_s + X_r')^2} \tag{3.13}$$

The relationship between torque and slip is readily estimated if the small stator resistance R_s is neglected. The torque may then be written as

$$T = 3\frac{p_f}{2}\left(\frac{R_r'}{\omega_s}\right)\frac{sV_s^2}{R_r'^2 + s^2X^2} = k\frac{sR_r'}{R_r'^2 + s^2X^2} \tag{3.14}$$

where

$$k = 3\frac{p_f}{2}\cdot\left(\frac{V_s^2}{\omega_s}\right)$$

$$X = X_s + X_r'$$

It may be seen from Equation (3.14) that, when the slip s is small, the torque is directly proportional to it. Also, when the slip is large, the torque will be given approximately by

$$T \simeq k\frac{R_r'}{sX^2}$$

The torque will therefore be inversely proportional to slip for larger values of slip.

It follows that there must be an intermediate value of slip for which the torque is a maximum. This value of slip may be obtained by differentiating the torque with respect to slip and equating to zero:

$$\begin{aligned}\frac{\mathrm{d}T}{\mathrm{d}s} &= \frac{\mathrm{d}}{\mathrm{d}s}\{ksR_r'(R_r'^2 + s^2X^2)^{-1}\}\\ &= kR_r'(R_r'^2 + s^2X^2)^{-1} - ksR_r'(R_r'^2 + s^2X^2)^{-2} \times 2sX^2\\ &= k\frac{R_r'(R_r'^2 + s^2X^2) - 2s^2R_r'X^2}{(R_r'^2 + s^2X^2)^2}\\ &= k\frac{R_r'^3 - s^2R_r'X^2}{(R_r'^2 + s^2X^2)^2}\\ &= 0\end{aligned}$$

The torque will therefore be a maximum when

$$R_r'^3 - s^2R_r'X^2 = 0$$

$$\therefore \quad s = \frac{R_r'}{X} \tag{3.15}$$

The value of the maximum torque is, from Equations (3.14) and (3.15),

$$T_{\max} = \frac{k}{2X} = \frac{3V_s^2}{4\pi(n_s/60)(X_s + X_r')} \tag{3.16}$$

Note that the maximum torque is independent of the value of rotor resistance.

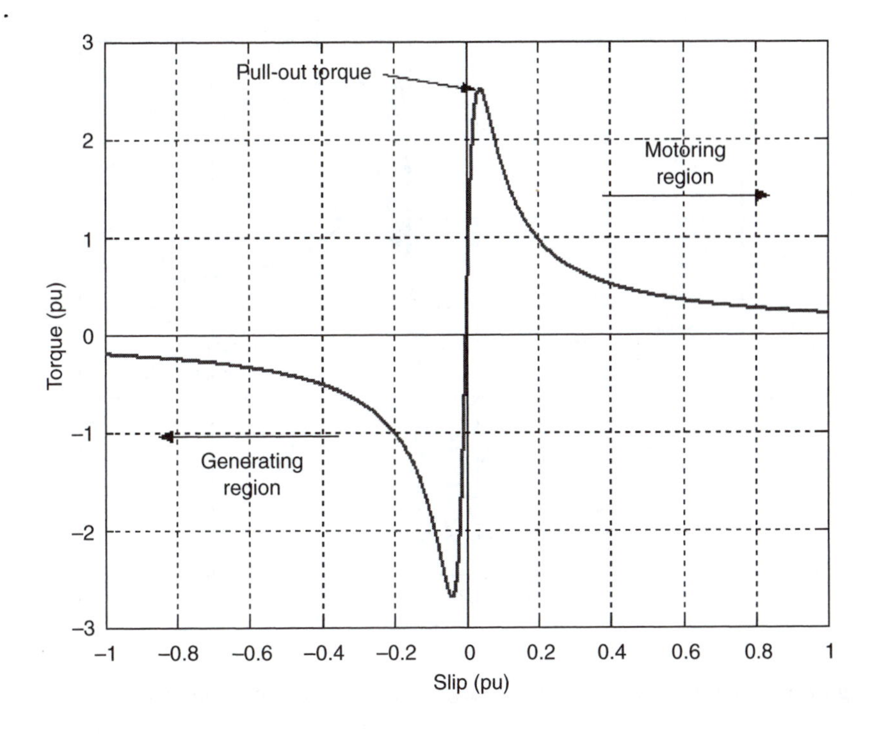

Figure 3.14 Typical torque-speed characteristic of an induction machine

A typical relationship between torque and slip is presented in Figure 3.14. At standstill the speed is zero and the slip, s, is equal to 1 per unit (pu). Between zero and synchronous speed, the machine performs as a motor. Beyond synchronous speed the machine performs as a generator.

Figure 3.15 illustrates the effect of varying the rotor resistance, R_r, on the torque of the induction machine. On the one hand, a low rotor resistance is required to achieve high efficiency at normal operating conditions. On the other hand, a high rotor resistance is required to produce high slip.

One way of controlling the rotor resistance (and therefore the slip and speed of the generator) is to use a wound rotor connected to external variable resistors. The rotor resistance is then adjusted by means of electronic equipment. The wound rotor may be connected to the external variable resistors through brushes and slip rings. However, a more innovative solution is to locate both the resistors and the electronic control equipment on the rotor itself. The required slip-controlling signal is then passed on to the rotor via optical fibre communications.

3.5.2 Fixed-speed induction generator-based wind turbine

In a fixed-speed wind turbine the induction generator, operating typically at 690 V, transmits power via vertical pendant cables to a switchboard and local transformer, usually located in the tower base (see Fig. 3.16). Switched power factor correction

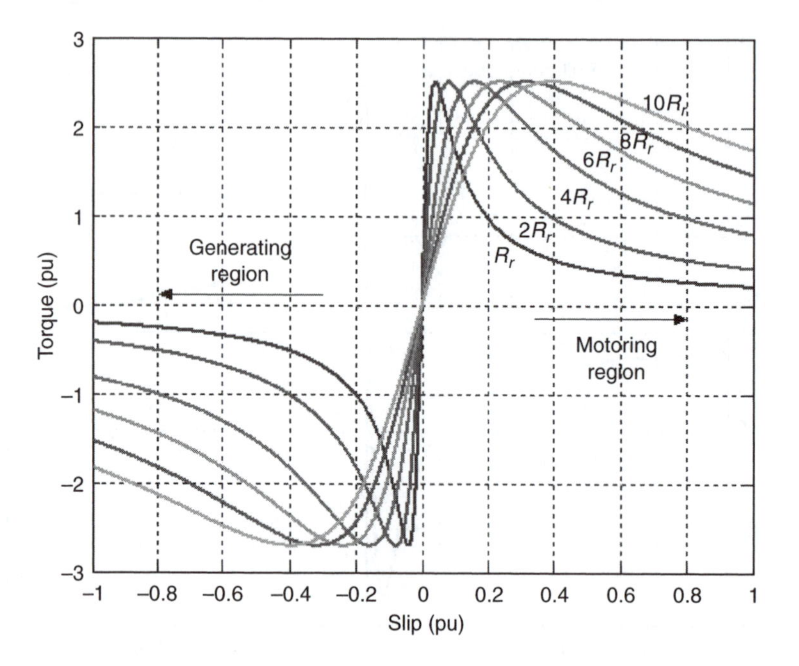

Figure 3.15 Torque-slip curves showing the effect of rotor circuit resistance

capacitors are used to improve the power factor of the induction generator while an anti-parallel thyristor soft-start unit is used to energise the generator once its operating speed is reached. The function of the soft-start unit is to build up the magnetic flux slowly and so minimise transient currents during energisation of the generator. Also, applying the network voltage slowly to the generator, once energised, brings the drive train slowly to its operating rotational speed.

A pitch-regulated rotor is able to control the generator speed during this starting period but a fixed-pitch, stall-regulated turbine is allowed to run up, driven by the wind, and the generator connected at slightly below synchronous speed.

Large steam-turbine generators supplying national electrical power systems all use synchronous machines. Their advantages include high efficiency and the ability to control independently the real output power (P) through adjusting the torque on the shaft by a mechanical governor and the reactive output power (Q) by varying the rotor field current. However, wind turbine aerodynamic rotors develop a significant torque pulsation at the blade passing frequency, caused by tower shadow and wind shear effects. By an unfortunate coincidence these aerodynamic torque variations are often close to the natural frequency of oscillation of the connection of a synchronous generator to the network. Thus, it is not possible to use synchronous generators directly connected to the network and a simple mechanical drive train in fixed-speed wind turbines. Some early wind turbines, using synchronous generators, included mechanical dampers in the drive train but modern fixed-speed wind turbines all use induction machines.

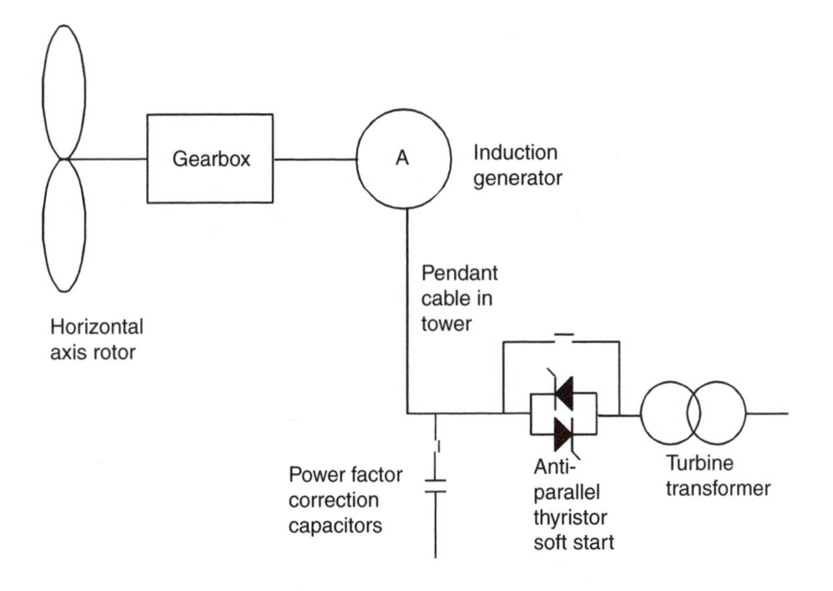

Figure 3.16 Schematic of fixed-speed wind turbine

Although supplying significant damping, induction generators suffer from a number of important disadvantages. The damping is proportional to the slip (the difference between rotor speed and that of the stator field) and is developed by power losses in the rotor. Thus a high slip (say 1 per cent) that is desirable to damp drive train oscillations results in 1 per cent of the generator output being generated as heat in the rotor.

An induction machine does not have a separate field circuit and so there is no direct control over reactive power. There is a fixed relationship between real and reactive power (Fig. 3.17). The operating locus in the generating region is shown as the line A–B. Even at zero real power output, reactive power (MVAr import) is required to energise the magnetic circuits of the machine. As the real power export is increased, then additional reactive power is drawn from the network. The effect of the power factor correction capacitors is to translate the operating curve vertically downwards. When the generator and capacitors remain connected to the power system then the network determines their voltage. However, if a network fault occurs and the generator and capacitors are isolated, then there is the possibility of a resonant condition, known as self-excitation, which can result in significant over-voltages. This can be avoided either by limiting the size of the capacitor bank or by arranging protection to trip the capacitors rapidly in such an event.

On an electrical power system, network short-circuits are usually detected by sensing fault current from large synchronous generators. Induction generators only provide fault current into three-phase short-circuits during the so-called sub-transient period and this is too short for reliable operation of over-current relays. Hence induction generators cannot be considered as a reliable source of fault current. Thus it is conventional practice, in the event of a network fault, to rely on the short-circuit

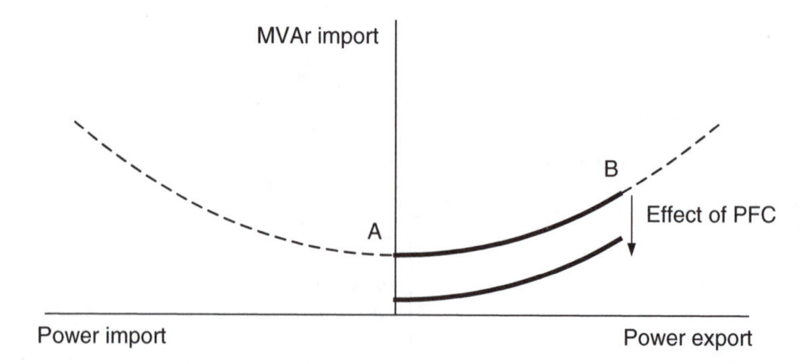

Figure 3.17 Circle diagram of induction machine (showing effect of power factor correction)

current from the network to operate protection to isolate the wind farm and then to use under/over-voltage or frequency relays to trip the wind turbines.

An important limitation of fixed-speed wind generators is that they can over-speed and lose stability if the network voltage is depressed. Voltage depressions can occur over a wide geographical area due to short-circuits on the main transmission network. In this case the low terminal voltage of the induction generator allows the generator to over-speed and draw high values of reactive power. This in turn lowers the network voltage further and leads to voltage collapse. It is clearly very undesirable that, just when the power system is potentially stressed due to a short-circuit, wind turbines, over a wide area, will trip. Hence the transmission system operators, who are responsible for the security of the power system, are imposing so-called 'fault ride-through' requirements. These require that, in the event of a fault on the high voltage transmission system (275 or 400 kV), which depresses the transmission network voltage to zero, the wind turbines continue to operate. These requirements are difficult to meet with simple fixed-speed wind turbines.

3.6 Variable-speed wind turbines

In recent years the size of wind turbines has become larger and the technology has switched from fixed-speed to variable-speed. The drivers behind these developments are mainly the ability to comply with connection requirements and the reduction in mechanical loads achieved with variable-speed operation. Variable-speed wind turbines provide the following key advantages (Müller *et al.*, 2002):

- 'They are cost effective and provide simple pitch control. At lower wind speed, the pitch angle is usually fixed. Pitch angle control is performed only to limit maximum output power at high wind speed.
- They reduce mechanical stresses; gusts of wind can be absorbed, i.e., energy is stored in the mechanical inertia of the turbine, creating an 'elasticity' that reduces torque pulsation.

- They dynamically compensate for torque and power pulsations caused by pressure of the tower. This back pressure causes noticeable torque pulsations rate equal to the turbine rotor speed times the number of rotor wings.
- They improve power quality; torque pulsations can be reduced due to the elasticity of the wind turbine system. This eliminates electrical power variations, i.e. less flicker.
- They improve system efficiency; turbine speed is adjusted as a function of wind speed to maximise output power. Operation at the maximum power point can be realised over a wide power range.
- They reduce acoustic noise, because low-speed operation is possible at low power conditions.'

Presently the most common variable-speed wind turbine configurations are:

- DFIG wind turbine
- Wide-range variable-speed wind turbine based on a synchronous generator

3.6.1 DFIG wind turbine

A typical configuration of a DFIG wind turbine is shown schematically in Figure 3.18. It uses a wound rotor induction generator with slip rings to take current into or out of the rotor winding and variable-speed operation is obtained by injecting a controllable voltage into the rotor at slip frequency (Holdsworth *et al.*, 2003). The rotor winding is fed through a variable frequency power converter, typically based on two alternating current (AC)/direct current (DC) insulated gate bipolar transistor (IGBT) based Voltage Source Converters (VSCs), linked by a DC bus. The power converter decouples the network electrical frequency from the rotor mechanical frequency, enabling the variable-speed operation of the wind turbine. The generator and converters are protected by voltage limits and an over-current *crowbar*. VSCs and related power electronic devices are described in the appendix.

A DFIG system can deliver power to the grid through the stator and rotor, while the rotor can also absorb power. This is dependent upon the rotational speed of the generator. If the generator operates in super-synchronous mode, power will be delivered from the rotor through the converters to the network, and if the generator operates in sub synchronous mode then the rotor will absorb power from the network through the converters.

These two modes of operation are illustrated in Figure 3.19, where ω_s is the synchronous speed and ω_r is the rotor speed.

Operating in steady state the relationship between mechanical power, rotor electrical power and stator electrical power in a DFIG is shown in Figure 3.20. In this figure P_m is the mechanical power delivered to the generator, P_r is the power delivered by the rotor, $P_{\text{air gap}}$ is the power at the generator air gap and P_s is the power delivered by the stator. P_g is the total power generated and delivered to the grid.

From Figure 3.20, it is seen that if the stator losses are neglected then:

$$P_{\text{air gap}} = P_s \tag{3.17}$$

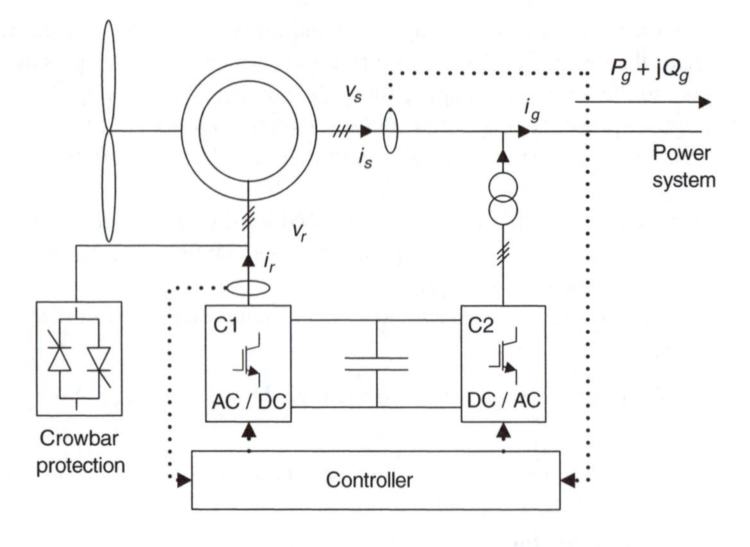

Figure 3.18 Typical configuration of a DFIG wind turbine

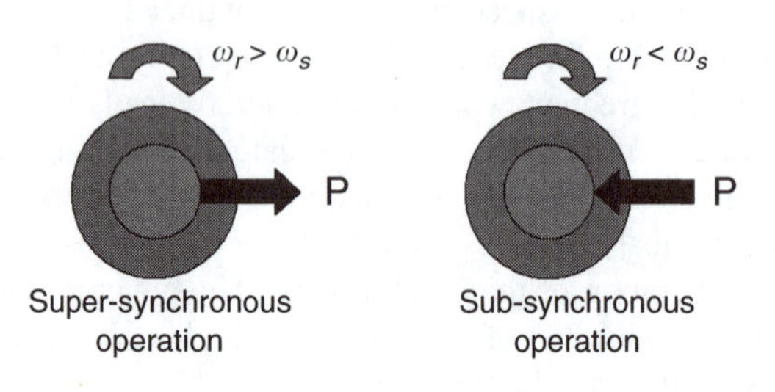

Figure 3.19 Super-synchronous and sub-synchronous operation of the DFIG wind turbine

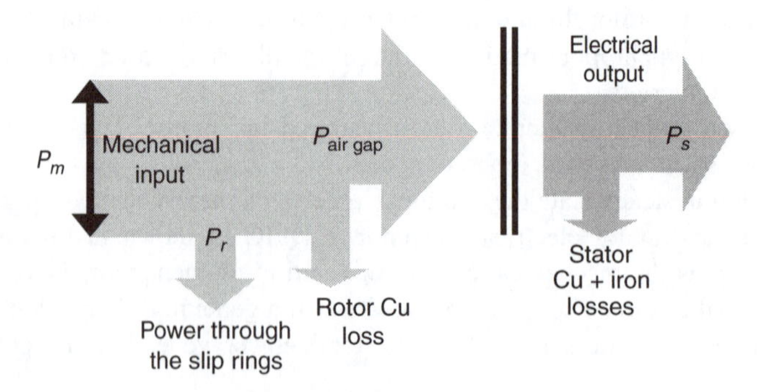

Figure 3.20 DFIG power relationships

and if we neglect the rotor losses then:

$$P_{\text{air gap}} = P_m - P_r \tag{3.18}$$

From Equations (3.17) and (3.18), the stator power P_s is expressed by

$$P_s = P_m - P_r \tag{3.19}$$

Equation (3.19) can be written in terms of the generator torque, T, as:

$$T\omega_s = T\omega_r - P_r \tag{3.20}$$

where $P_s = T\omega_s$ and $P_m = T\omega_r$. Rearranging terms in Equation (3.20),

$$P_r = -T(\omega_s - \omega_r) \tag{3.21}$$

The stator and rotor powers can then be related through the slip s as

$$P_r = -sT\omega_s = -sP_s \tag{3.22}$$

where s is given in terms of ω_s and ω_r as

$$s = \frac{(\omega_s - \omega_r)}{\omega_s} \tag{3.23}$$

Combining Equations (3.19) and (3.22) the mechanical power, P_m, can be expressed as,

$$\begin{aligned} P_m &= P_s + P_r \\ &= P_s - sP_s \\ &= (1 - s)P_s \end{aligned} \tag{3.24}$$

The total power delivered to the grid, P_g, is then given by

$$P_g = P_s + P_r \tag{3.25}$$

The controllable range of s determines the size of the converters for the DFIG. Mechanical and other restrictions limit the maximum slip and a practical speed range may be between 0.7 and 1.1 pu.

3.6.1.1 DFIG wind turbine control

The control of a DFIG wind turbine is achieved through converters C1 and C2 (Fig. 3.18). A control scheme implemented by a number of manufacturers uses converter C1 to provide torque/speed control, together with terminal voltage or power factor (PF) control for the overall system. Converter C2 is used to maintain the DC link voltage and provide a path for rotor power flow to and from the AC system at unity power factor. Although reactive power injection can also be obtained from the stator-side converter C2, for DFIG voltage-control schemes the rotor-side converter C1 is likely to be preferred to converter C2. This is largely due to the fact that the reactive power injection through the rotor circuit is effectively amplified by a factor of $1/s$.

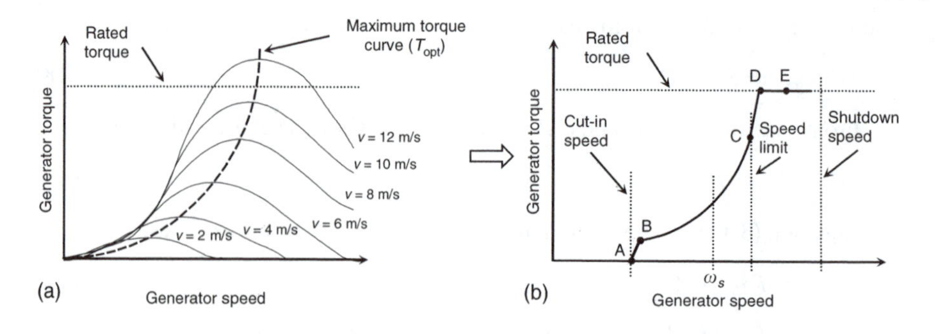

Figure 3.21 Wind turbine characteristic for maximum power extraction

For purposes of analysis, simulation and control, the favoured way of representing a DFIG is in terms of direct and quadrature (dq) axes, which form a reference frame that rotates at synchronous speed ($\omega_s = 2\pi f_s$). Adjustment of the dq-axis components of the rotor voltage provides the capability of independent control over two generator variables. This can be achieved in a variety of control schemes. A control methodology known as current-mode control is commonly employed where the d-axis component of the rotor current is used to control terminal voltage (reactive power), and the q-axis component is used to control the torque of the generator (active power).

3.6.1.2 Torque control

The aim of the torque controller is to optimise the efficiency of wind energy capture in a wide range of wind velocities, keeping the power generated by the machine equal to the optimal defined value. A typical wind turbine characteristic with the optimal torque-speed curve plotted to intersect the $Cp_{\max}$ points for each wind speed is illustrated in Figure 3.21(a). The curve T_{opt} defines the optimal torque of the device (i.e. maximum energy capture), and the control objective is to keep the turbine on this curve as the wind speed varies. The curve T_{opt} is defined by:

$$T_{opt} = K_{opt}\omega_r^2 \tag{3.26}$$

where K_{opt} is a constant obtained from the design of the wind turbine.

The complete generator torque-speed characteristic used for control purposes is shown in Figure 3.21(b). The optimal torque-speed curve is characterised by Equation (3.26), which corresponds to section B–C. Within this operating range, during low-medium wind speeds, the maximum possible energy is obtained from the turbine. Due to power converter ratings, it is not practicable to maintain optimum power extraction from cut-in up to the rated wind speed. Therefore, for very low wind speeds the model operates at almost constant rotational speed (A–B). Rotational speed is also limited, for example, by aerodynamic noise constraints, and at point C the controller allows the torque to increase, at essentially constant speed, to rated torque (C–D). If the wind speed increases further the control objective follows D–E, where pitch regulation limits aerodynamic input power. For very high wind speeds, the pitch control will regulate input power until the wind shutdown speed limit is reached.

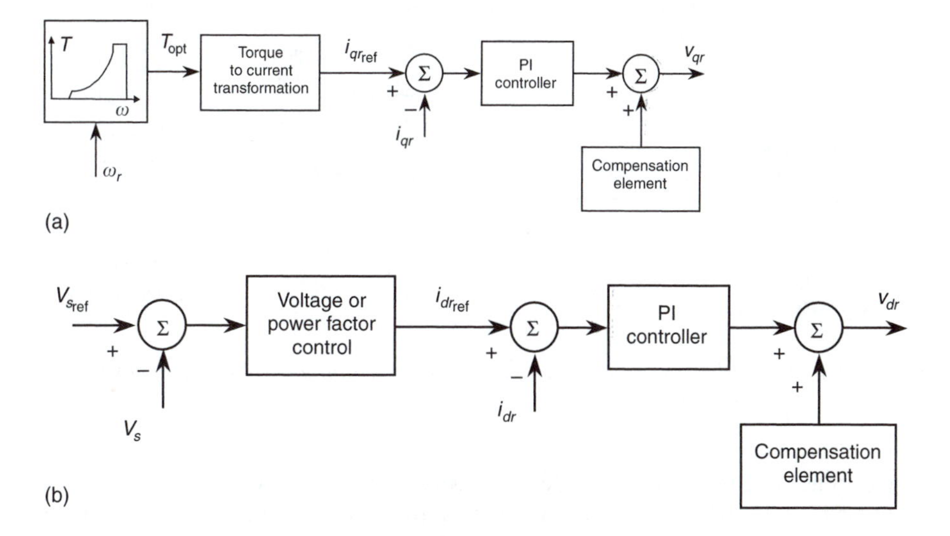

Figure 3.22 Current-mode control scheme for a DFIG wind turbine (a) torque-control loop and (b) voltage-control loop

The implementation of the DFIG torque control is illustrated in Figure 3.22(a). Given a rotor speed measurement, ω_r, the torque-speed characteristic (Fig. 3.21(b)) is used to obtain the optimal torque, T_{opt}, which after some manipulation generates a reference current $i_{qr_{\text{ref}}}$. Comparing the reference current, $i_{qr_{\text{ref}}}$, to the actual value of the rotor current in the q-axis, i_{qr}, an error signal is obtained. Although i_{qr} imposes the effect of torque control, the converter C1 (Fig. 3.18) is a controlled voltage source. Hence, the required rotor voltage v_{qr} is obtained by processing the error signal with a standard PI controller and adding to the output a compensation term to minimise cross coupling between torque- and voltage-control loops.

3.6.1.3 Voltage control

A basic implementation of the DFIG voltage controller is shown in Figure 3.22(b). In this scheme the difference in magnitude between the terminal voltage reference, $V_{s_{\text{ref}}}$, and the actual terminal voltage, V_s, is manipulated to generate the reference of the rotor current in the d-axis, $i_{dr_{\text{ref}}}$. The reference current, $i_{dr_{\text{ref}}}$, is compared with the actual value of the rotor current in the d-axis, i_{dr}, to generate an error signal, which is then processed by a standard PI controller. The required rotor voltage v_{dr} is obtained as the addition of the PI controller output and a compensation term used to eliminate cross coupling between torque- and voltage-control loops.

3.6.2 Wide-range variable-speed synchronous generator wind turbine

The wide-range variable-speed wind turbine based on a synchronous generator is shown schematically in Figure 3.23. The aerodynamic rotor and generator shafts may be coupled directly (i.e. without a gearbox), in which case the generator is a multi-pole

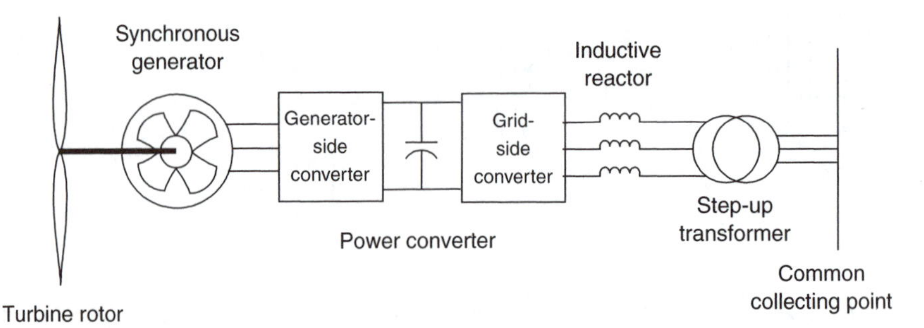

Figure 3.23 Wide-range variable-speed wind turbine based on a synchronous generator

synchronous generator designed for low speeds; or they can be coupled through a gearbox, which allows for a generator with a lower number of poles. The generator can be either an electrically excited synchronous generator or a permanent magnet machine (Heier, 1998). To permit variable-speed operation, the synchronous generator is connected to the grid through a variable frequency power converter system, which completely decouples the generator speed from the grid frequency. Therefore the electrical frequency of the generator may vary as the wind speed changes, while the grid frequency remains unchanged. The rating of the power converter system in this wind turbine system corresponds to the rated power of the generator plus losses.

The power converter system consists of the grid-side and the generator-side converters connected back-to-back through a DC link. The grid-side converter is a Pulse Width Modulated-VSC (PWM-VSC), and the generator-side converter can be a diode-based rectifier or a PWM-VSC.

3.6.2.1 Wind turbine arrangement with diode-based rectifier

Figure 3.24 illustrates the schematic of the wind turbine with a diode-based rectifier as the generator-side converter. The diode rectifier converts the output of the generator to DC power and a PWM-VSC converts the DC power available at the rectifier output to AC power.

One way to control the operation of the wind turbine with this arrangement (and assuming a permanent magnet generator) is illustrated in Figure 3.24. A DC–DC converter is employed to control the DC link voltage (controller-1), and the grid-side converter controls the operation of the generator and the power flow to the grid (controller-2). With appropriate control, the generator and turbine speed can be adjusted as wind speed varies so that maximum energy is collected.

3.6.2.2 Wind turbine arrangement with PWM-VSCs

In this arrangement both the generator- and the grid-side converters are PWM-VSCs as shown in Figure 3.25. The generator can be directly controlled by the generator-side converter (controller-1) while the grid-side converter (controller-2) maintains the

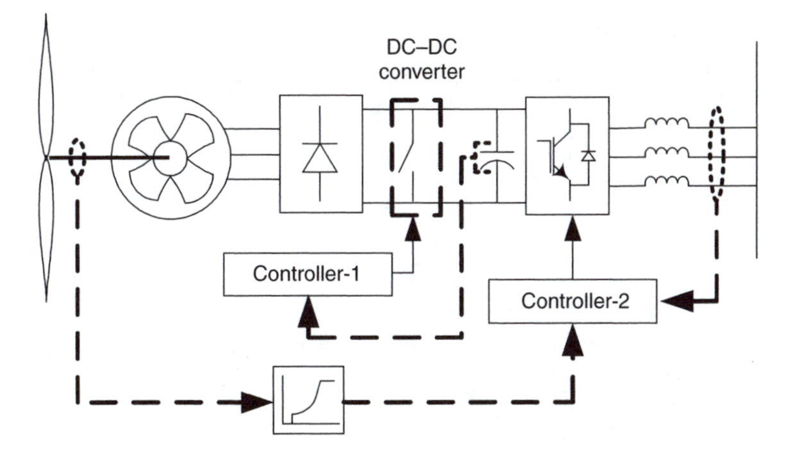

Figure 3.24 *Wide-range synchronous generator wind turbine with a diode-based rectifier as the generator-side converter*

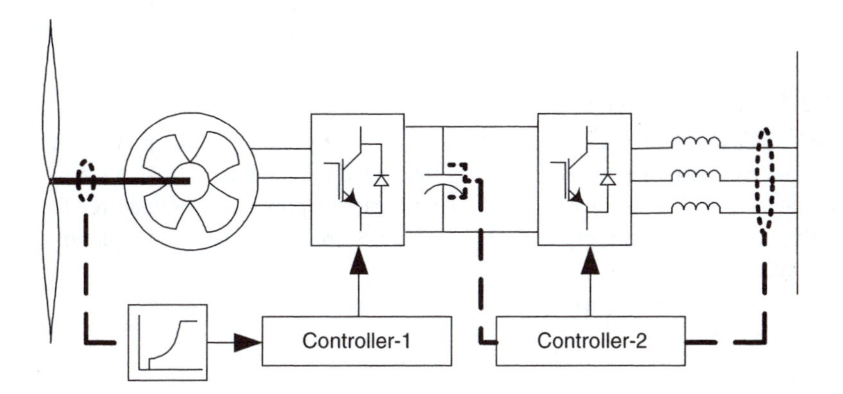

Figure 3.25 *Wide-range synchronous generator wind turbine with a PWM-VSC generator-side converter*

DC link voltage at the desired value by exporting active power to the grid. Controller-2 also controls the reactive power exchange with the grid.

3.6.2.3 Wind turbine control

Control over the power converter system can be exercised with different schemes. The generator-side converter can be controlled using load angle control techniques (Akhmatov *et al.*, 2003) or it can be controlled by means of more accurate but also more sophisticated techniques such as vector control (Morimoto *et al.*, 1994; Tan and Islam, 2004). The grid-side converter is normally controlled using load angle control techniques. In this section we explain the implementation of the load angle control scheme assuming a permanent magnet generator.

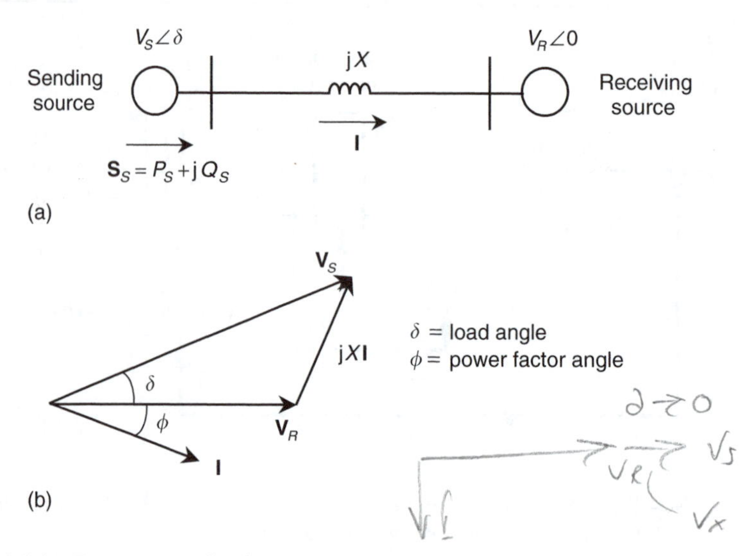

Figure 3.26 Power transfer between two sources (Kundur, 1994) (a) equivalent circuit diagram and (b) phasor diagram

3.6.2.4 Load angle control

The load angle control technique can be explained by analysing the transfer of active and reactive power between two sources connected by an inductive reactance as shown in Figure 3.26(a). The active power, P_S, and reactive power, Q_S, transferred from the sending-end to the receiving-end may be obtained from application of Equation (2.17) as detailed below:

$$\mathbf{S}_S = \mathbf{V}_S \mathbf{I}_S^* = \mathbf{V}_S \left(\frac{\mathbf{V}_S - \mathbf{V}_R}{\mathrm{j}X} \right)^*$$

$$= \mathbf{V}_S \frac{(\mathbf{V}_S^* - \mathbf{V}_R^*)}{-\mathrm{j}X} = \mathrm{j}\frac{V_S^2}{X} - \mathrm{j}\frac{\mathbf{V}_S \mathbf{V}_R^*}{X}$$

Noting from Figure 3.26(b) that $\mathbf{V}_S = V_S e^{\mathrm{j}\delta}$ and $\mathbf{V}_R^* = V_R$ we have

$$\mathbf{S}_S = P_S + \mathrm{j}Q_S = \mathrm{j}\frac{V_S^2}{X} - \mathrm{j}\left(\frac{V_S V_R \cos\delta + \mathrm{j} V_S V_R \sin\delta}{X} \right)$$

Hence

$$P_S = \frac{V_S V_R}{X} \sin\delta \tag{3.27}$$

$$Q_S = \frac{V_S^2}{X} - \frac{V_S V_R}{X} \cos\delta \tag{3.28}$$

where V_S is the voltage magnitude of the sending source, V_R is the voltage magnitude of the receiving source, δ is the phase angle between these two sources, and X is the inductive reactance between them.

The dependence of active power and reactive power flows on the source voltages can be determined by considering the effects of differences in voltage magnitudes and angles.

- If $\delta = 0$ then the active power flow is zero (Equation (3.27)) and the reactive power flow is determined by the magnitudes of V_S and V_R (Equation (3.28)).
- If $V_S = V_R$ but $\delta \neq 0$, then the active power flow is determined by the angle δ (Equation (3.27)), and the reactive power flow is small (Equation (3.28)).

Additionally, if we assume that the load angle δ is small, then $\sin\delta \approx \delta$ and $\cos\delta \approx 1$. Hence, Equations (3.28) and (3.29) can be simplified to:

$$P_S = \frac{V_S V_R}{X}\delta \tag{3.29}$$

$$Q_S = \frac{V_S^2}{X} - \frac{V_S V_R}{X} \tag{3.30}$$

From Equations (3.29) and (3.30) it is seen that the active power transfer depends mainly on the phase angle δ by which the sending source voltage leads the receiving source voltage. The reactive power transfer depends mainly on voltage magnitudes, and it is transmitted from the side with higher voltage magnitude to the side with lower magnitude.

Equations (3.29) and (3.30) represent the basic steady-state power flow equations used by the load angle control technique.

3.6.2.5 Control of the generator-side converter

The operation of the generator and the power transferred from the generator to the DC link are controlled by adjusting the magnitude and angle of the voltage at the AC terminals of the generator-side converter. This can be achieved using the load angle control technique where the internal voltage of the generator is the sending source ($V_S\angle 0$), and the generator-side converter is the receiving source ($V_R\angle\delta$). The inductive reactance between these two sources is the synchronous reactance of the generator, X_{gen}, as shown in Figure 3.27.

The voltage magnitude, V_R, and angle magnitude, δ, required at the terminals of the generator-side converter are calculated using Equations (3.29) and (3.30) as:

$$\delta = \frac{P_{S_{\text{gen}}}^{\text{ref}} X_{\text{gen}}}{V_S V_R} \tag{3.31}$$

$$V_R = V_S - \frac{Q_{S_{\text{gen}}}^{\text{ref}} X_{\text{gen}}}{V_S} \tag{3.32}$$

where $P_{S_{\text{gen}}}^{\text{ref}}$ is the reference value of the active power that needs to be transferred from the generator to the DC link, and $Q_{S_{\text{gen}}}^{\text{ref}}$ is the reference value for the reactive power. The reference value $P_{S_{\text{gen}}}^{\text{ref}}$ is obtained from the characteristic curve of the machine for maximum power extraction for a given generator speed, ω_r. As the generator

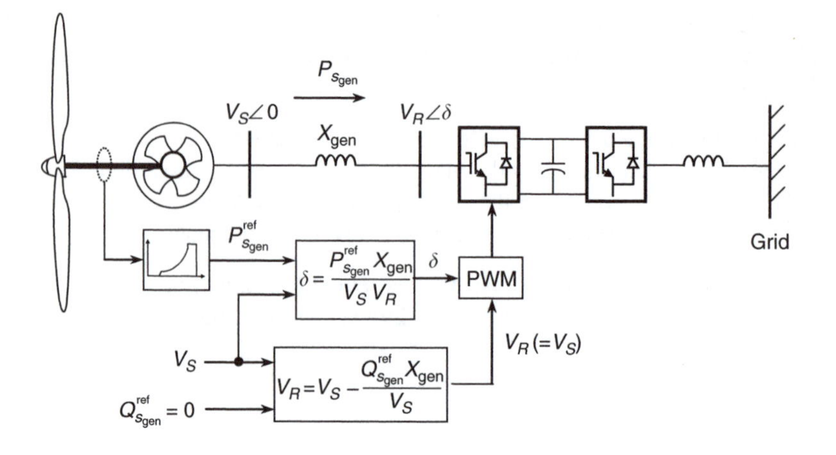

Figure 3.27 Load angle control of the generator-side converter

has permanent magnets, it does not require magnetising current through the stator, thus the reactive power reference value can be set to zero, $Q_{S_{gen}}^{ref} = 0$ (i.e. V_S and V_R are equal in magnitude). The implementation of this load angle control scheme is illustrated in Figure 3.27.

The major advantage of the load angle control is its simplicity. However, as the dynamics of the generator are not considered it may not be very effective in controlling the generator during transient operation. Hence an alternative way to control the generator is to employ vector control techniques (Morimoto *et al.*, 1994; Tan and Islam, 2004).

3.6.2.6 Vector control

Vector control techniques are implemented based on the dynamic model of the synchronous generator expressed in the dq frame. The torque of the generator is defined as the cross product of the stator flux linkage vector, $\boldsymbol{\psi}_s$, and the stator current vector, $\mathbf{I}_s$, giving

$$T = \mathrm{k} \cdot (\boldsymbol{\psi}_s \times \mathbf{I}_s) \tag{3.33}$$

where k is a constant related to the generator parameters. The stator flux linkage vector $\boldsymbol{\psi}_s$ is given by

$$\boldsymbol{\psi}_s = L_s\mathbf{I}_s + \boldsymbol{\psi}_{fd} \tag{3.34}$$

where L_s is the stator self inductance and $\boldsymbol{\psi}_{fd}$ is the field flux vector. Thus Equation (3.33) can be written as

$$T = \mathrm{k} \cdot (L_s\mathbf{I}_s + \boldsymbol{\psi}_{fd}) \times \mathbf{I}_s = k \cdot (\boldsymbol{\psi}_{fd} \times \mathbf{I}_s) \tag{3.35}$$

Assuming a permanent magnet synchronous generator (with constant field flux $\boldsymbol{\psi}_{fd}$), the torque of the generator can be controlled by adjusting the stator current vector $\mathbf{I}_s$ (Equation (3.35)). When modelling the synchronous generator it is usual

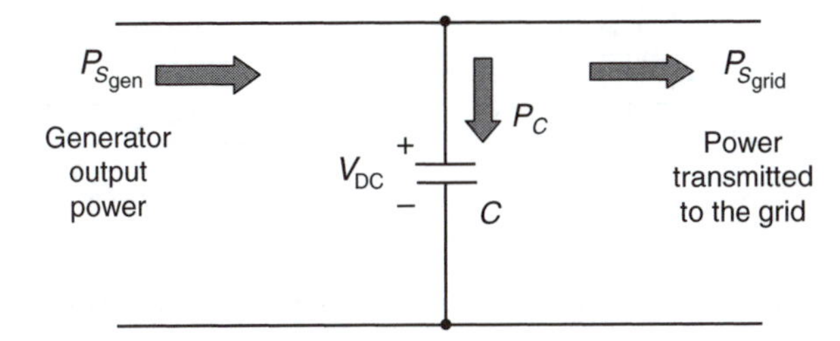

Figure 3.28 Power flow in the DC link

to select a dq frame where the d-axis is aligned with the magnetic axis of the rotor (field). Hence, as the d component of the stator current, i_{ds}, is aligned with the field flux vector, ψ_{fd}, the cross product of these two quantities is zero, and the torque is controlled only by means of the q component of the stator current, i_{qs}, as,

$$T = \mathrm{k} \cdot \psi_{fd} i_{qs} \tag{3.36}$$

The reference value of the stator current in the q-axis, $i_{qs_{\mathrm{ref}}}$, is calculated from Equation (3.36) and compared with the actual value, i_{qs}. The error between these two signals is processed by a PI controller whose output is the voltage in the q-axis, v_{qs}, required to control the generator-side converter. The required voltage in the d-axis, v_{ds}, is obtained by comparing the reference value of the stator current in the d-axis, $i_{ds_{\mathrm{ref}}}$, with the actual current in the d-axis, i_{ds}, and by processing the error between these two signals in a PI controller. The reference $i_{ds_{\mathrm{ref}}}$ may be assumed zero for the permanent magnet synchronous generator.

3.6.2.7 Control of the grid-side converter

The objective of the grid-side converter controller is to maintain the DC link voltage at the reference value by exporting active power to the grid. In addition, the controller is designed to enable the exchange of reactive power between the converter and the grid as required by the application specifications.

A widely used converter control method is also the load angle control technique, where the grid-side converter is the sending source ($V_S\angle\delta$), and the grid is the receiving source ($V_R\angle 0$). As the grid voltage is known it is selected as the reference, hence the phase angle δ is positive. The inductor coupling these two sources is the reactance X_{grid}.

For simulation purposes the reference value for the active power, $P_{S_{\mathrm{grid}}}^{\mathrm{ref}}$, that needs to be transmitted to the grid can be determined by examining the DC link dynamics with the aid of Figure 3.28. This figure illustrates the power balance at the DC link:

$$P_C = P_{S_{\mathrm{gen}}} - P_{S_{\mathrm{grid}}} \tag{3.37}$$

P_C is the power across the DC link capacitor, C, $P_{S_{\mathrm{gen}}}$ is the active power output of the generator (and transmitted to the DC link), and $P_{S_{\mathrm{grid}}}$ is the active power transmitted from the DC link to the grid.

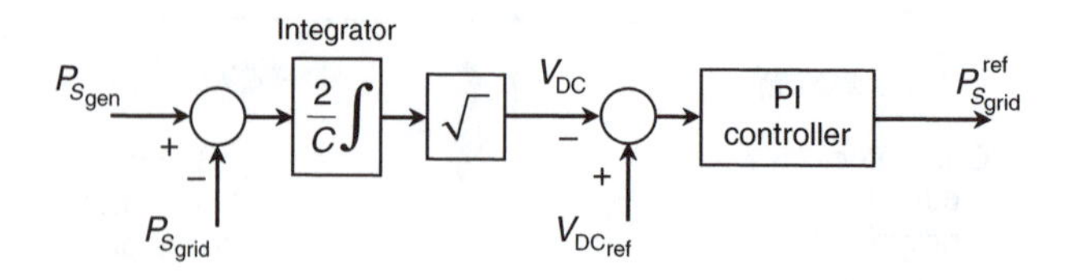

Figure 3.29 Calculation of active power reference, $P_{S_{grid}}^{ref}$, (suitable for simulation purposes)

The power flow across the capacitor is given as

$$\begin{aligned} P_C &= V_{DC} \cdot I_{DC} \\ &= V_{DC} \cdot C \frac{dV_{DC}}{dt} \end{aligned} \tag{3.38}$$

From Equation (3.38) the DC link voltage V_{DC} is determined as follows:

$$\begin{aligned} P_C &= V_{DC} \cdot C \frac{dV_{DC}}{dt} = \frac{C}{2} \cdot 2 \cdot V_{DC} \frac{dV_{DC}}{dt} \\ &= \frac{C}{2} \cdot \frac{dV_{DC}^2}{dt} \end{aligned} \tag{3.39}$$

Rearranging Equation (3.39) and integrating both sides of the equation,

$$V_{DC}^2 = \frac{2}{C} \int P_C dt \tag{3.40}$$

giving

$$V_{DC} = \sqrt{\frac{2}{C} \int P_C dt} \tag{3.41}$$

By substituting P_C in Equation (3.37) in Equation (3.41), the DC link voltage V_{DC} can be expressed in terms of the generator output power, $P_{S_{gen}}$, and the power transmitted to the grid, $P_{S_{grid}}$, as

$$V_{DC} = \sqrt{\frac{2}{C} \int (P_{S_{gen}} - P_{S_{grid}}) dt} \tag{3.42}$$

Equation (3.42) calculates the actual value of V_{DC}. The reference value of the active power, $P_{S_{grid}}^{ref}$, to be transmitted to the grid is calculated by comparing the actual DC link voltage, V_{DC}, with the desired DC link voltage reference, $V_{DC_{ref}}$. The error between these two signals is processed by a PI controller, whose output provides the reference active power $P_{S_{grid}}^{ref}$, as shown in Figure 3.29. It should be noted that in a physical implementation the actual value of the DC link voltage, V_{DC}, is obtained from measurements via a transducer.

To implement the load angle controller the reference value of the reactive power, $Q_{S_{grid}}^{ref}$, may be set to zero for unity power factor operation. Hence, the magnitude, V_S,

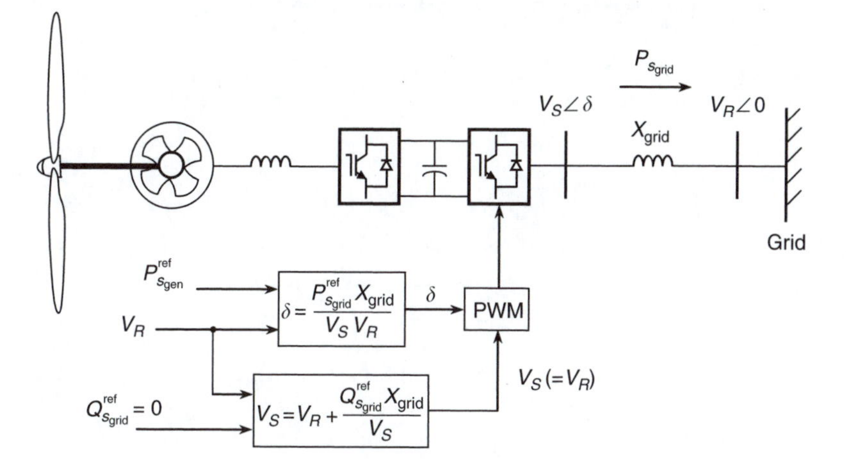

Figure 3.30 Load angle control of the grid-side converter

and angle, δ, required at the terminals of the grid-side converter are calculated using Equations (3.29) and (3.30) as:

$$\delta = \frac{P^{\mathrm{ref}}_{S_{\mathrm{grid}}} X_{\mathrm{grid}}}{V_S V_R} \tag{3.43}$$

$$V_S = V_R + \frac{Q^{\mathrm{ref}}_{S_{\mathrm{grid}}} X_{\mathrm{grid}}}{V_S}, \quad Q^{\mathrm{ref}}_{S_{\mathrm{grid}}} = 0 \tag{3.44}$$

Figure 3.30 illustrates the implementation of the load angle control scheme for the grid-side converter with unity power factor. If the reactive power exchange with the network is not zero it should be noted that Equation (3.44) is transcendental.

Chapter 4
Network integration of wind power

4.1 Introduction

This chapter examines the effect of wind generation on the transmission and distribution network to which it is connected. Topics include control of voltage and power flows, the quality of supply and the protection of plant and equipment. The principles underlying network development in the face of increasing wind generation are reviewed. The perspective is that of a network operator sympathetic to wind power development. It will be seen that wind power capacity may exceed strict technical limits, provided wind power operators are prepared to accept occasional energy curtailment. Achievement of this ideal balance will require enlightened negotiation between the two parties.

There are eight basic issues to be considered by network and wind farm planners. These aspects of wind farm connection are listed below, with the corresponding sections.

1. Wind farm starting (Section 4.2)
2. Network voltage management (Section 4.3)
3. Thermal/active power management (Section 4.4)
4. Network power quality management (Section 4.5)
5. Transient system performance (Section 4.6)
6. Fault level issues (Section 4.7)
7. Information (Section 4.8)
8. Protection (Section 4.9)

4.2 Wind farm starting

Fixed-speed wind turbine generators (WTGs) must be rotated within 1–2 per cent of their electrical synchronous speed. Acceleration from rest is generally achieved by using the kinetic energy in the wind to drive the turbine and electrically isolated generator close to synchronous speed, at which point the circuit breaker with the grid is closed. It is common to use an anti-paralleling thyristor soft start arrangement to reduce the fluxing surge experienced by the network when the generator circuit breaker is closed (see Fig. 3.16). When the generator is fully fluxed the anti-paralleling arrangements are by-passed. If not controlled, the fluxing period can lead to a large drain of reactive power from the grid, which creates an unacceptable voltage depression and a step change in voltage. The degree of voltage depression and step will be a function of the size of the induction machine and the strength of the network at the point of connection. Since wind turbines tend to be on sites remote from dense population, the network is often weak at the point of connection, that is, electrically remote from generation sources and hence of low fault level.

It is uncommon to have the power factor capacitance switched in during starting, which exacerbates the issue, but with accurate control the starting current can be reduced to between 1.6 and 1.0 times full-load current. It is nonetheless at very low power factor.

In the case of retrofit projects, it is possible, but expensive, to install dynamic reactive compensation. Devices such as static VAr compensaters (SVCs), static compensators (STATCOMs) and dynamic VAr compensators (DVArs) can be used. These are described in the appendix. Such devices may also be justified for new installations to improve fault ride-through (see Section 4.6). Combinations of static and dynamic voltage support are possible.

Doubly fed induction generator (DFIG) machines are also accelerated by wind energy and no major starting voltage dips are evident due to their ability to control reactive as well as active power generation (see Section 3.6).

Full speed-range machines are usually synchronous machines behind a fully rated converter (Section 3.6) and may be accelerated by the wind energy. The network effects are entirely dependent on the performance of the converter, which will be a voltage source device (see appendix), to cope with the absence of fault level on the rotor side. The converter will draw only active power from the grid system and will supply the required power factor to the generator during fluxing.

Wind farms contain a number of turbines, and can be subject to overall control by a wind farm management system. An objective of the system can be to control the wind farm so that only one machine may start at any time, or alternatively the voltage may not be allowed to fall below a threshold value during starting. This reduces the network impact considerably.

One unresolved issue relates to energising the wind farm feeder. Each wind turbine commonly has an associated network transformer (Fig. 3.16). Energising a large wind farm implies simultaneously fluxing a large number of transformers. This is similar to energising a distribution feeder, which also has a large number of transformation points to distribution load. It is common for standards to refer to a voltage limit for infrequent operation, which may be twice the maximum amount allowed for normal operation. It is fair to consider that level of voltage disturbance as a guide.

Note that a wind farm is often the sole user of a distribution feeder. Since, while the circuit is being energised, all wind farm circuit breakers would be open, the only connected apparatuses are the wind farm transformers, auxiliaries and site supplies. The voltage drop should not be such as to cause mal-operation of these units.

4.3 Network voltage management

Network voltage variation is a key design factor in assessing optimum wind farm connection arrangements. Clearly a wind farm can operate from no-load to full-load and may be located at any point in the network. The connection voltage level is likely to depend on the maximum output of the wind farm. In assessing the output, due regard should be taken of the short-term overload capability of the wind farm due to gusting. This could be typically 125 per cent of nominal rating.

A further variable is whether the farm has a dedicated connection or is embedded in a load serving circuit. If it is in a load serving circuit it might be located close to the source or close to the remote end. Load served might be point load or distributed. The circuit source may have no voltage tap changers, manual tap changers or automatic tap changers. Circuit real-time information may or may not be available. There may or may not be other embedded generation on that part of the network and the network may or may not have voltage support. It is the several permutations of these variables that provide the challenge to integrate wind generation with its high unpredictability and variability. Taken together these variables mean that network voltage control is likely to be successful only if fully automatic.

4.3.1 The voltage level issue

The major network components in the impedance path between the wind farm and grid are transformers, underground cables and overhead lines.

The main impact of transformers on voltage profile, apart from the obvious one of voltage transformation, is due to leakage reactance (see Section 2.2.5). Winding resistance is much smaller and less important in this context. The cabling within a wind farm has only a second-order effect on system voltage profile and will be ignored here. It should be noted, however, that long lengths of cable within a wind farm installation can give rise to voltage and resonance management issues for wind farm owners.

The main factors influencing network voltage profile are the parameters of the distribution or transmission system overhead lines adjacent to the wind farm. It was seen in Section 2.5 that transmission lines have an inductive reactance many times greater than the resistance – the X/R ratio is significantly greater than unity. On the other hand, distribution lines may have an X/R ratio close to unity. Typical values for transmission and distribution lines are given in Table 2.1.

When real power flows through a resistive element the current is in phase with the voltage and therefore an in-phase voltage drop occurs across the network. There is no angular shift between the voltages at the sending-end and the receiving-end of the circuit. When real power flows through a purely reactive component, no in-phase voltage drop occurs but there is an angular shift between the voltages at the two nodes. The opposite is true for reactive power. These relations are encapsulated in Equation (2.24).

The conclusion from the above is that the transfer of active power has a major effect on the voltage profile on low-voltage systems, whereas reactive power transfer is the dominant factor on high-voltage systems. Apart from thermal loading considerations, this is one reason why it is important to connect large wind farms to higher voltage networks. In all cases reactive power transfer is the dominant factor in transformer voltage drop (sometimes called regulation).

4.3.1.1 Large wind farms

Consider a 400 MW wind farm connected to a 275 kV network by a direct connection. The network impedance is mainly reactive, viewed from the wind farm. This allows the exchange of large amounts of active power with relatively little voltage drop and low losses. Any exchange of reactive power between the grid and wind farm will affect the voltage. Transmission elements generate and use reactive power. Thus in lightly loaded conditions, there is a surplus of reactive power on the network while in heavily loaded situations a deficit occurs. Generation absorbs the surplus and at a different time or location may have to generate the deficit of reactive power, or else the network voltage will be outside an acceptable range. System planners try to ensure that large wind farms and other embedded generation can contribute to this regulation of network voltage.

Generators are specified to have a leading or lagging reactive capability at full load. Ranges of 0.85 power factor lagging to 0.90 power factor leading are common. Target voltages could be as high as 1.08 pu or as low as 0.94 pu to achieve an even network voltage measured at various load nodes. It is common with traditional generation to augment the variation available from alternator field current adjustment with a wide tap range on the generator transformer to 'force' reactive power in one direction or the other. Put another way, the generator terminal voltage can be kept within limits while allowing the set to contribute or absorb a large amount of reactive power by adjusting the tap changer on the transformer.

In the case of wind farms, the generation/absorption of reactive power could be by switched devices, for example, reactors and capacitors or power electronic devices, or by control of converter firing as in the case of DFIGs. In the case of mechanically switched devices, these will be incapable of dealing with a rapid change in the network requirement for reactive power resulting from a change in topology. Large wind farms may also require a transformer with a wide tap range.

The voltage-control range and facilities available from generation connected at these high voltages is well regulated by Grid Codes. The practice has been to apply this requirement to each conventional generating unit. However, in the case of a wind farm, it may be effective to treat the generation node as providing the network voltage regulating capability. Where the farm is very large, it may be desirable to be able to control sections of the wind farm so that the failure of one control element does not affect the overall capability.

Network Grid Codes were prepared as part of the arrangements for liberalisation or open access. They relied largely upon earlier standards. Some have since undergone considerable modification while others are struggling to accommodate newer forms of generation within rules based on existing technologies. In this context, an appraisal is required as to whether the power factor range demanded of generators is appropriate.

The origins of the range are unclear, due to the passage of time, but it is widely believed that the reactive capability was based on what could be achieved for a reasonable cost with steam powered plant driving generators which are naturally or hydrogen cooled. Reconsideration of the problem requires a determination of how much static (slow acting) and dynamic (fast response) reactive power is needed on a particular power system.

The problem analysis steps are:

- Determination from load flow calculations of the reactive power requirements of an intact maximum loaded system in normal configuration.
- Determination of the additional requirements for worst-case secured generation and circuit outages.
- Determination from load flows of the reactive power requirements of an intact minimum loaded system in normal configuration.
- Determination from load flows of the reactive power requirements of a minimum loaded system with maximum foreseen maintenance.
- Determination from load flows of the reactive power requirements of a minimum loaded system with maximum foreseen maintenance and worst-case generation and circuit outages.

Examination of these cases should indicate the requirement for static and dynamic reactive compensation throughout the year. More refinement can be achieved by repeating the steps for a number of intermediate loads. Dynamic compensation is required to take account of unexpected occurrences. The limits applied in assessing the dynamic requirements throughout the studies are the maximum and minimum network voltage for normal and secured outage situations and the associated allowable step changes. Studies on small systems tend to indicate that the static and dynamic requirements are approximately equal, but network topology and the ratio of loading to capacity will have a bearing on the result.

Replacing traditional generation with wind farms will change this requirement to the extent that some wind generators may demand reactive power under reduced voltage conditions, hence increasing the dynamic burden. Also some plant will trip off if the voltage falls, causing lengthier energy flows to supply load and a consequent increase in dynamic burden. Therefore different penetration levels and different generation types should be treated as distinct development scenarios. These studies could lead to a table which shows the amount of reactive power required for different levels of penetration and types of wind generator for, say, a 15-year forecast. For worst-case investigation, it is necessary to assume that traditional plant will be switched-off to accommodate wind farms. It will therefore make no contribution to serving reactive requirements. This leads to a design requirement for static and dynamic reactive power for a given wind penetration and wind generator type mix, leaving the system safe against voltage instability.

How these increased reactive demands are to be met is another matter, and the subject of an optimisation study taking account of the costs of increased wind farm converter size, provision of centralised versus distributed reactive compensation,

Table 4.1 Cost implications of WTG power factor capability of 0.9

Method	Cost increase (%)
SVC at MV	100
Converter resizing	41–53
Passive elements at LV	47

mechanically switched technologies and even rotating synchronous compensators (which also add to system inertia and support fault level).

A further Grid Code legacy issue is how the range is specified. Traditional generation has an availability of about 98 per cent and is thus expected to be delivering full active power output when a system outage requires maximum reactive contribution. By contrast, wind turbines may not be rotating for 20 per cent of the time.

It has been reported (Gardner *et al.*, 2003) that some wind farms generate below 5 per cent of capacity for 50 per cent of the time. The system must therefore be designed and operated to do without some or all of their output. Variability suggests that even when the wind is optimal, allowance must be made for output variation. It is therefore less critical that wind farms can deliver rated reactive output contemporaneously with rated real power output. It is a matter of control to prioritise: (1) reactive contribution when voltage falls; and (2) reactive absorption when voltage rises over active power output.

There is reported to be little surplus cost in wind farm construction for a power factor range of 0.95 lagging to 0.95 leading at full output, as this has become the standard for most manufacturers. Irrespective of what is available, the generator is following a circle diagram (Fig. 3.17), albeit with limits represented by chords, and therefore the machine is being oversized by the manufacturer to supply reactive power. To extend the range beyond the above, it is necessary to either produce further oversized plant and converters or to forego some active output at the time. Vestas turbines have estimated that WTGs with a reactive power capability corresponding to a power factor of 0.9 at rated active power output incur the costs shown in Table 4.1 over turbines constructed for unity power factor.

Table 4.2 shows the October 2003 grid code positions for a number of countries.

4.3.1.2 Wind farms connected to sub-transmission systems

Consider a large wind farm connected to a 132 or 33 kV node. The network voltage control requirement needs to be more stringent if other network users are also connected electrically close to the wind farm.

If the connection is directly to the 132 kV network, the situation is relatively similar to the 275 kV connection above except that it is likely that the connection nodal voltage range will be more restricted. In practice about 0.95 to 1.05 pu voltage is likely to be the working range.

Table 4.2 Grid Code reactive power requirements, October 2003

Country	Absorbing reactive power		Generating reactive power	
	Rated power	Lower limit	Rated power	Lower limit
Australia	1.0 pf	Proportional	0.95 pf	Proportional
Denmark	1.0 pf	Proportional	1.0 pf	Proportional
Germany	0.95 pf	Proportional	0.925 pf	Proportional
Ireland	0.93 pf	0.7 pf, 0.4 fl	0.85 pf	0.35 pf, 0.4 fl
Scotland	0.85 pf	Proportional	0.95	Proportional

In the case of a connection to the 33 kV system, the worry will be that the voltage will fluctuate wildly. In the United Kingdom, a standard, P28 (Electricity Association, 1989), is used to transparently state the allowable voltage regulation. Other countries have similar approaches. The relationship between voltage variation and frequency of occurrence set out in P28 may be summarised as follows:

Voltage variation (%)	Time between each change (s)
0.4	1
0.8	10
1.6	100
3.0	1,000

The standard makes reference to different patterns of voltage change (e.g. step, ramp and motor starting). For most equipment other than the fastest load cycling systems a 3 per cent voltage step is seen as acceptable. As stated earlier, very infrequent occurrences giving rise to a step change up to 6 per cent may be acceptable.

If the wind farm can go from high output to low under gusting conditions then a reasonable test is whether the voltage at the 33 kV connection bus varies by more than 3 per cent. (It is always possible to control the rate of increase of power but not the rate of decrease, although if bad weather prediction is accurate enough, and evasive action taken, it may be possible to allow a 6 per cent step change for a full gust-induced shutdown.) To explore this, a study would take a minimum and maximum load on the 33 kV bus and the variation in voltage created by minimum and maximum wind.

4.3.1.3 Steady-state operation of a wind farm: example

We will examine the operation of a wind farm consisting of twenty 660 kW WTGs connected to a large 50 Hz system by a 33 kV line, as shown in Figure 4.1. The large system consists of a transmission network to which the line is connected by a transformer in a substation. As far as our example is concerned, we may assume that

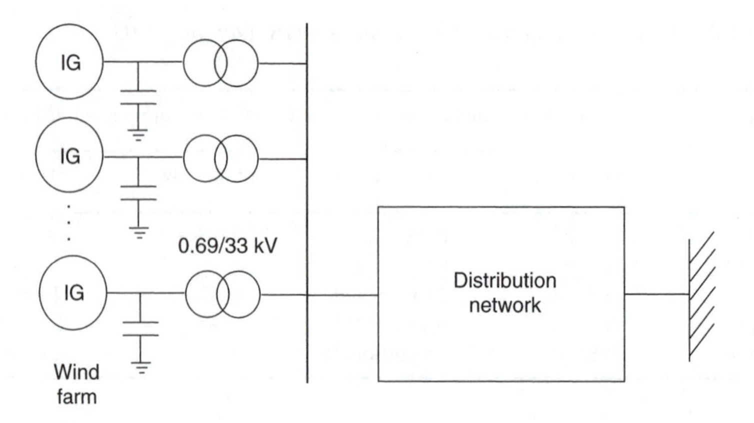

Figure 4.1 Wind farm connected to a large system through a distribution line

the line is connected to a node at 33 kV, backed by a source or sink of zero internal impedance and fixed frequency. This system model is known as an 'infinite busbar'.

The main points of interest will be the behaviour of the wind farm terminal voltage and transmission efficiency. We will consider also the use of power factor correction applied at the terminals of each WTG.

The relevant data for the WTG are as follows. All equivalent circuit parameters are referred to the stator, and are given in ohms/phase.

Rating	747 kVA
Power rating	660 kW
Voltage (line) rating	690 V
Stator resistance	0.0048 Ω
Stator reactance	0.0680 Ω
Rotor resistance	0.0127 Ω
Rotor reactance	0.0897 Ω
Magnetising reactance	2.81 Ω

We will assume that the WTGs are connected to the 33 kV line by a 16 MVA, 690 V/33 kV transformer with a leakage reactance of 0.058 pu and negligible losses. The 33 kV line is 20 km long and has the parameters given in Table 2.1.

The slip s corresponding to a power generation of 660 kW, or 220 kW per phase, is found from the induction machine model (Section 3.5.1) to be −0.018465 (a speed of 1.018465 pu). This may be checked using the equivalent circuit of Figure 3.13:

$$R_s + \frac{R_r'}{s} = -0.683\ \Omega$$

$$X_s + X_r' = 0.158\ \Omega$$

With a stator terminal phase voltage of $690/\sqrt{3} = 398$ V, the current is

$$\mathbf{I} = (-553 - \mathrm{j}\,270)\ \mathrm{A}$$

giving a complex power consumption per phase (see Section 2.4.7) of

$$\mathbf{S} = \mathbf{VI}^* = -220 \times 10^3 + \mathrm{j}107 \times 10^3$$

Considering all three phases, the WTG generates 660 kW and consumes 321 kVAr at full-load.

At no-load, with a slip of zero, the current is accounted for by the magnetising reactance alone and will be $\mathbf{I} = -\mathrm{j}142$ A, giving a complex power consumption per phase of

$$\mathbf{S} = \mathbf{VI}^* = 0 + \mathrm{j}\,56.5 \times 10^3$$

We may now use Equation (2.24) to assess the wind farm terminal voltage variation over the WTG operating range. The line resistance is

$$R = 20 \times 0.30 = 6.0\ \Omega$$

The transformer reactance in ohms per phase is, from Equation (2.18),

$$X_t = \frac{0.058 \times \mathrm{kV}_b^2}{\mathrm{MVA}_b} = 3.95\ \Omega$$

The total reactance of line and transformer is then

$$X = 20 \times 0.31 + 3.95 = 10.15\ \Omega$$

The wind farm terminal voltage will be considered at no-load and full-load. At no-load, the wind farm power generation P is zero and the reactive power generation per phase is due to the magnetising current alone, that is

$$Q = -20 \times 56.5 \times 10^3\ \mathrm{VAr}$$

From Equation (2.24), assuming an initial wind farm terminal phase voltage V_g of $33{,}000/\sqrt{3} = 19{,}052$ V, the phase voltage rise is given by

$$\Delta V \approx \frac{RP + XQ}{V_g} = \frac{0 - 10.15 \times 20 \times 56{,}500}{19{,}052} = -602\ \mathrm{V}$$

A second iteration, with an updated V_g, gives

$$\Delta V \approx -622\ \mathrm{V}$$

This gives a line voltage at the wind farm end of the line of 31.9 kV, or 3.3 per cent below nominal. As expected, the induction machines' reactive power requirements have depressed the voltage at this node of the system.

At full-load, the generated powers for each phase of each WTG are

$$P = 220\ \mathrm{kW}$$

$$Q = -107\ \mathrm{kVAr}$$

The application of Equation (2.24) for two iterations gives

$$\Delta V \approx 242 \text{ V}$$

This corresponds to a wind farm line voltage of 33.4 kV or 1.2 per cent above nominal.

It should be noted that the above procedure takes no account of the off-nominal terminal voltage on the WTG reactive power consumption. An accurate assessment of the voltage profile would require use of a suitable computer simulation package.

The foregoing analysis suggests a no-load to full-load variation in wind farm terminal voltage of 0.967–1.012 pu. This is perfectly acceptable, given that variations of 6 per cent above or below nominal would be allowed normally.

We have ignored power factor correction so far. The effect of power factor correction on the wind farm terminal voltage variation and online losses will now be considered.

4.3.1.4 Effect of power factor correction

It is usual to install power factor correction capacitance (PFCC) across the WTG terminals to reduce the reactive power consumption, as shown in Figure 3.16. A common criterion is to size the capacitance C to reduce the resulting no-load current to zero. The uncompensated no-load current is

$$\frac{V}{\mathrm{j}X_m}$$

The capacitance must be such that

$$\frac{V}{\mathrm{j}X_m} + \frac{V}{1/\mathrm{j}\omega C} = 0$$

$$\therefore \quad C = \frac{1}{\omega X_m}$$

In our case, with $X_m = 2.81\ \Omega$/phase, the required capacitance C is $1,133\ \mu$F/phase. With this amount of PFCC, the no-load current taken from the supply is zero. The line voltage at the wind farm end of the line will therefore be 33 kV.

The full-load generated powers per phase with PFCC will be

$$P = 220 \text{ kW}$$

$$Q = -(107 - 56.5) = -50.5 \text{ kVAr}$$

The voltage rise obtained from two applications of Equation (2.24) is

$$\Delta V \approx 811 \text{ V}$$

This corresponds to a 4.5 per cent voltage rise at the wind farm end of the line. This is acceptable, but much closer to the usual 6 per cent limit than was the case without PFCC. The results may be summarised as follows.

	No-load voltage rise	Full-load voltage rise
Without PFCC (%)	−3.3	1.2
With no-load PFCC (%)	0	4.5

It may be seen from these results that the effect of the PFCC is to elevate the wind farm terminal voltage by a fixed amount of 3.3 per cent. Loading imposes an extra voltage rise of 4.5 per cent.

The main purpose of PFCC is to reduce the current. This may reduce the required thermal capacity of the line – but distribution line ratings are often dictated by voltage regulation rather than current carrying capacity. PFCC should reduce line losses. The relevant WTG phase currents, with and without PFCC, have already been obtained. Their magnitudes are summarised below.

	Without PFCC	With PFCC
No-load current (A)	142	0
Full-load current (A)	615	568

The corresponding line phase currents are obtained by noting that there are 20 WTGs, and that the line phase currents will be transformed in inverse proportion to the voltages, (i.e. 690/33,000). The resulting line phase currents and line losses at full-load are summarised below.

	Without PFCC	With PFCC
Full-load line current (A)	257	237
Full-load line losses (MW)	1.19	1.01

The use of PFCC has reduced the losses, as a percentage of production, from 6.0 to 5.1 per cent.

It may be noted, in conclusion, that any attempt to achieve unity power factor at full-load in this case is misguided. The wind farm terminal voltage would then be an unacceptable 1.068 pu. Also, the minimum possible WTG current of 553 A still gives losses of 0.96 MW or 4.8 per cent of production. This is little better than was achieved with no-load compensation.

4.3.1.5 Wind farms connected directly to the sub-transmission system

When a wind farm is connected adjacent to the sub-transmission system, the impedance between farm and system will be dominated by transformer leakage reactance. The series resistance is negligible. Hence any voltage variation at the wind farm terminals will be due to reactive rather than active power flow. Thus unity power factor generation from the wind farm creates minimum voltage fluctuation at the 33 kV

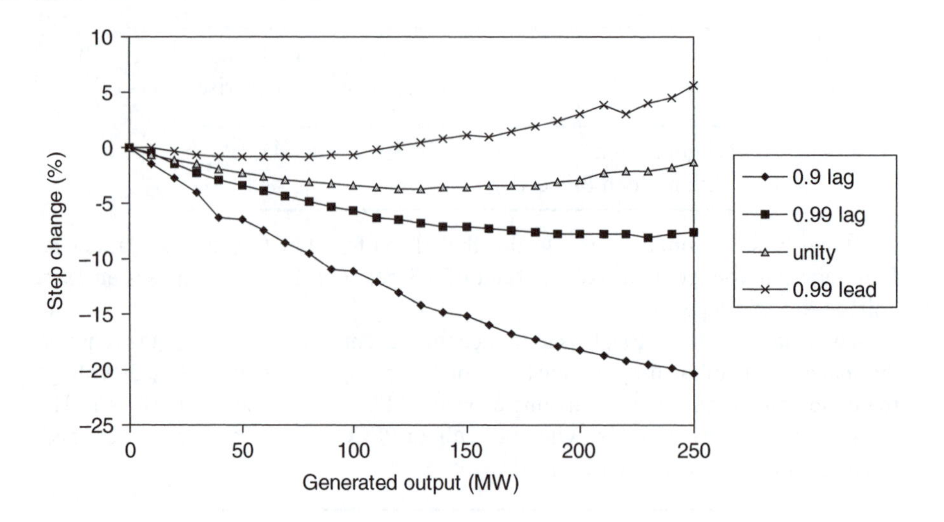

Figure 4.2 *Voltage step change on loss of generation plotted for different generating outputs and power factors*

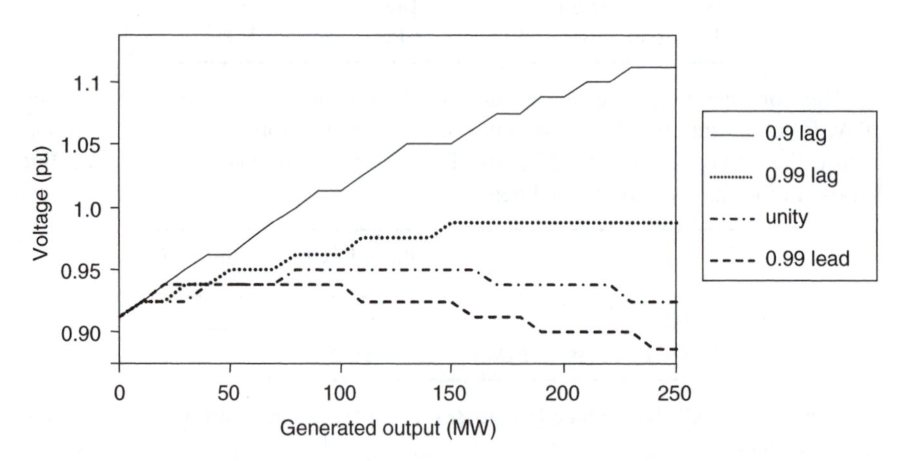

Figure 4.3 *System transformer tap position for different levels of generation and power factors*

busbar and similarly causes a much reduced number of tap change operations on the 132/33 kV transformers.

The graphs in Figures 4.2 and 4.3 were produced for a UK utility and show the voltage and tap change situation for a 200 MW wind farm connected directly to the 33 kV system (along with demand customers). The wind farm was to have multiple 33 kV cable connections and the 33 kV node was a direct connection point to the transmission system. In the first graph, the worst-case voltage step on variation of output is explored by losing all the wind farm output. It can be observed that overall a unity or leading power factor is preferred. The second curve shows that the tap

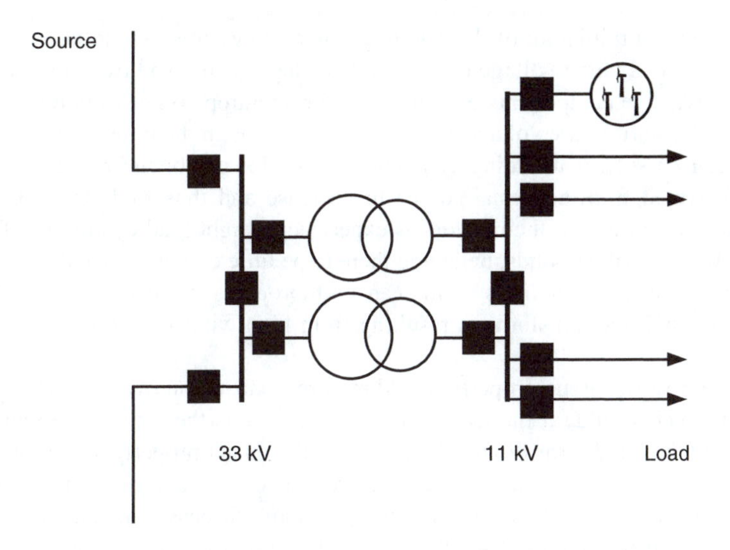

Figure 4.4 Wind farm connected directly to a substation low-voltage busbar

changer varies over a narrow range and thus operates infrequently at unity power factor generation. Separate studies have confirmed this, showing that tap change operations are reduced by orders of magnitude.

It is therefore likely that the objective of such a wind farm's automatic voltage regulation software will be to achieve unity power factor within an acceptable voltage band. Outside this band, the wind farm should endeavour to control voltage as the wind farm connected at 400 kV would. In this case, the normal control is power factor, reverting to out-of-range voltage control and finally emergency actions (e.g. load reduction and ultimately a wind farm trip). It should be noted that for some types of wind farm (e.g. DFIG) it may be inappropriate to trip the wind farm, since, by operating even at low output, it contributes positively to voltage control.

4.3.1.6 Rural network connected wind farms

Here, planners are breaking new ground. The network design does not lend itself easily to incorporation of significant generation at lower voltage levels. Voltage control is a significant issue.

Consider first generation connected directly to a substation lower voltage busbar, as shown in Figure 4.4.

Transformer automatic voltage regulators at this system level are likely to have an objective of adjusting the lower voltage busbar voltage to take account of:

- the stochastic absolute level of the higher voltage;
- the voltage drop through the transformer and to some extent the voltage regulation on connected circuits, both of which are likely to be stochastic in that they depend on time-varying demand.

There are a minimum of three settings on such systems, one to take account of an average load target voltage on the lower voltage busbar and two to represent the real and reactive components of impedance for line drop compensation (LDC). The measured quantities are voltage and current flow through the transformers.

It can be seen that adding generation to the lower voltage busbar causes the load delivered from a normal source to decrease and thus fools the control system into believing that the network is experiencing light load conditions. Thus the LDC element mal-responds, believing there to be little demand-related voltage drop on the load circuits. Some systems can be thoroughly confused by a reversal of power through the transformers resulting from high wind farm output at low load periods.

To correct this issue properly would require a substation overall voltage control system which would, at the least, establish the load as the sum of the transformer flows and the wind farm flows. The LDC could then act properly and, provided the wind farm objective was unity power factor, the tap change control would operate in a normal manner to control network voltage. The above discussion on direct connection to the sub-transmission system is then relevant in assessing the voltage variation seen by connected customers.

Finally, it is important to assess the ability of tap changers to deal with reverse power and/or reactive power flow conditions.

4.3.1.7 Generation embedded in distribution circuits

The key distinguishing feature of distribution circuits is that demand is connected directly, or connected through fixed-tap transformers. These circuits are designed on the basis of voltage standards, so that customers at the sending-end can receive up to the maximum allowable voltage while customers at the receiving-end can be supplied at the lowest allowable voltage. The voltage drop between the ends varies stochastically as a result of demand-time variation, and is also a function of distance and conductor impedance. These issues are illustrated in Figures 4.5 and 4.6.

The higher voltage network is assumed stiff (in the limit an infinite busbar), thus the voltage profile in the lower voltage network must change direction around this fixed voltage as a result of a change in power flow direction.

In summary, there is a sending-end voltage that rises and falls depending on the LDC perception of load, and a receiving-end voltage that increases or decreases depending on load level and on how strongly the wind is blowing. How can the receiving-end customers be protected from widely fluctuating voltages? In the absence of IT-based solutions, utilities install lower impedance circuits to reduce the range of voltage effect. They do not pretend that this is efficient, just expedient in fulfilling their licence conditions.

Clearly a controller incorporating a load model for the line, and which has knowledge of the receiving-end voltage and wind farm infeed, can estimate the sending-end voltage. However, since the load varies widely and randomly, this is dangerous. To be reliable, the controller should also know the sending-end voltage or, as a minimum, receive a signal of overall system load as a percentage of full-load. Even the latter is

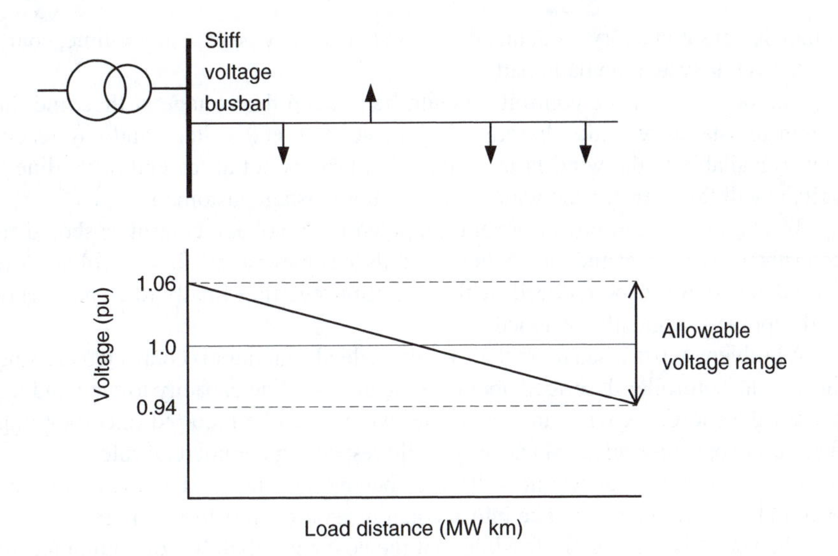

Figure 4.5 *Network design voltage profile without embedded generation (maximum load minimum generation scenario)*

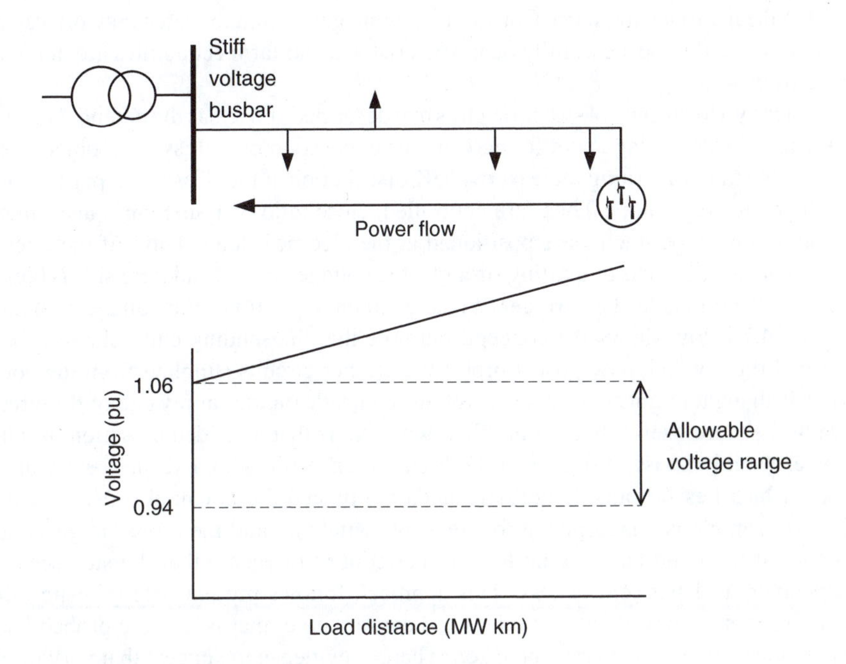

Figure 4.6 *Network voltage profile with tail-end embedded generation (minimum load maximum generation scenario)*

somewhat unsatisfactory, as demand in tourist areas may peak at lunch-time, counter to the overall system demand pattern.

The objective of the controller would be to keep the voltage at the wind farm within the statutory limits, by adjusting the sending-end voltage and any reactive power available at the wind farm. If the wind farm is not at the end of the line the voltage will fall between the wind farm and more distant customers.

To maximise wind power generation potential, a voltage controller should first control reactive power and, only when this fails to achieve a satisfactory voltage level, should active power be reduced, with the circuit reverting finally to a single-ended grid supply as originally designed.

A problem exists when more than one embedded generator is connected to a single circuit: the controller described above would allocate line capacity to the wind farm nearest the source. A wide-area controller would then be required, incorporating a function to optimise physical capacity while respecting commercial rules.

If several circuits at one node have embedded generation then the control may need to be wider again and take into account the source transformer limits.

The above is technically feasible, but the cost and reliability of communication and the establishment of agreed standards/protocols should not be underestimated. Embedding local network models in voltage controllers implies up-dating those models as connected load/generation changes. This may be resource intensive work unless it can be automated. Software and controllers may need to be changed several times within the life of a wind farm. Constraining wind generation to match network capacity would need to be a carefully quantified risk if wind farm economics are not to be undermined.

Clearly a local control system such as that described above has the potential to avail of demand-side management for network, as opposed to overall system, objectives.

At the time of writing, at least two UK distribution authorities are experimenting with voltage regulators. These are controlled, close ratio, variable tap transformers, which in the experiments are positioned at the electrical load centre of the circuit. When the wind farm is outputting strongly, the voltage on the wind farm side is higher than on the grid side. This triggers a change in the tap ratio of the voltage regulator. Figure 4.7 below shows the concept, but note that the sending-end voltage is here set to 1.0 pu, which may pose a problem for other circuits supplied from the node. A little thought might suggest that a voltage controller at the supply end of the circuit would be more useful, but the problem with this is that a sudden cessation of wind power would expose customers at both circuit extremities to a 12 per cent voltage step, which lies far outside normal standards. Indeed the potential 6 per cent step in this example is unacceptable for frequent switching, and therefore the generated output of the wind farm would have to be reduced to ensure that the step seen by consumers is 3 per cent or less. This approach follows present deterministic rules and experience may show that this is too conservative, that is, a more probabilistic approach to step voltage may be taken. There is evidence to suggest that very rarely does a wind farm actually drop its entire generated output over a very short period – apart from a network problem – and that in many cases the wind gusting which causes this could be predicted and the output pre-curtailed, as noted above.

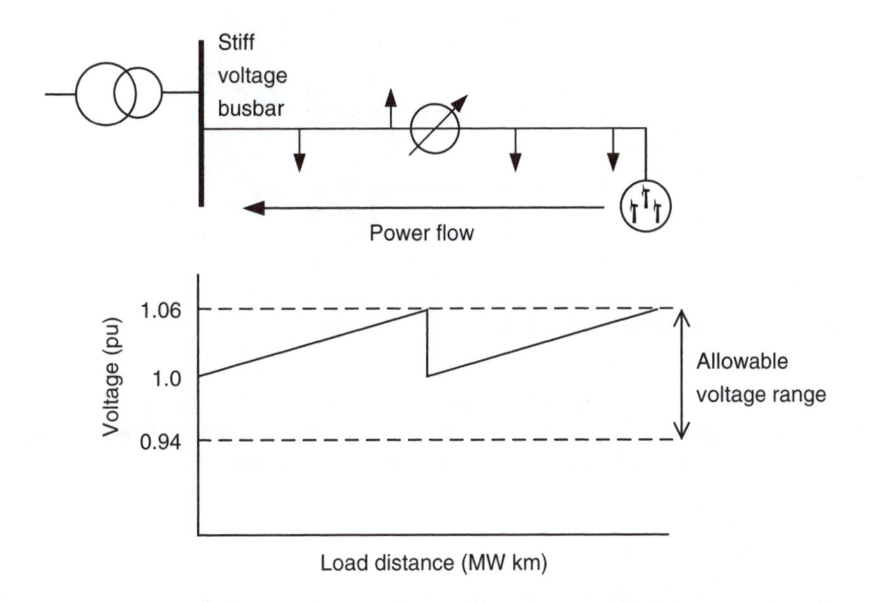

Figure 4.7 Network voltage profile with voltage regulator

4.4 Thermal/active power management

4.4.1 Planning approaches/standards

Utility plans are drawn up and developments authorised to maintain a security standard in the network. This is effectively a benchmark to ensure that customers in one part of the network experience the same conditions as customers in another similar part of the network. Common practice is that parts of the network that supply large load blocks are more stringently secured. This gives rise to the much quoted n-1, n-2, n-g-1, etc. security standards, where n refers to system normal running conditions and n-1 is a single circuit outage. n-2 is an ambiguous term, with the following possible meanings:

- n-d/c: system normal with the outage of a double-circuit tower line.
- n-1-1 or n-m-T: a circuit trip when another circuit is being maintained.
- true n-2: two unplanned and independent outages occurring simultaneously.
- n-g-1: a network outage arising during an unplanned generation outage. This could result in serious consequences. Since rotating machines are inherently less reliable than lines and cables, an unplanned outage of a machine could well occur during a line outage. This is tested by increasing output on all other generators to match load. The extra output might violate constraints on particular generation nodes. To prevent this happening, utilities have developed rules for generation security, for example, above 1,000 MW of generating capacity there should be four circuits on at least two routes.

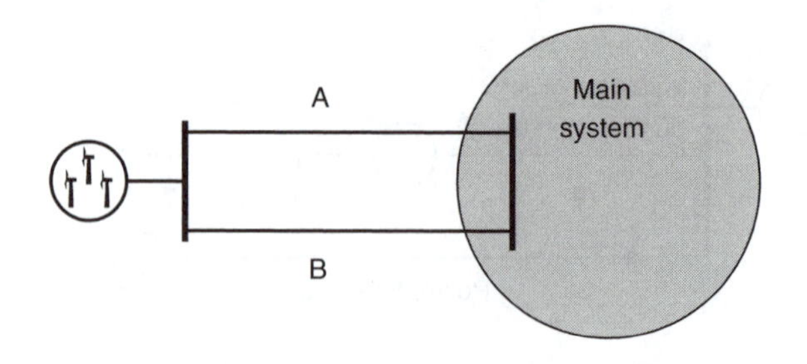

Figure 4.8 Wind farm with double-circuit connection

The above generation security standard, while seemingly expensive, does offer the comfort to project financiers that a new station has a very small chance of losing its route to market and thus helps to secure lower risk-premium funds.

4.4.2 Wind farm connection issues

In the case of wind farms, with an expected load factor of 30–40 per cent, it can be argued that the above thinking is an uneconomic luxury. It is argued that, if the wind gusts so that the farm is tripped and the system can survive this loss (as it must), then the system can stand the sudden loss of the wind farm for any other reason. Therefore the level of security of connection is a matter of wind farm operator choice where this does not affect the backbone system.

The issue then becomes one of complexity management. Consider the system shown in Figure 4.8.

There might be one or two connections to the wind farm. Clearly if there is one connection only (say A), the loss of the connection causes the complete outage of the wind farm. This could have serious economic consequences in the case of an undersea cable, where repair involves reserving a suitable vessel, ordering materials and awaiting a weather window.

Where there are two connections, the loss of, say, the B circuit reduces the wind farm output to the rating of A. The problem then becomes non-trivial in that the rating of A is an economic balance. If no benefit were obtained from covering the loss of another circuit, each cable would be rated at 50 per cent of the total requirement. However, in reality, the economic loss resulting from the outage of one circuit is calculated as:

(the risk of outage) $\times$ (the economic consequences)

Suppose the wind farm in Figure 4.8 has a capacity of 200 MW and an estimated life of 20 years. Cable A and B are rated at 100 MW and are each expected to suffer a 3-month outage in 30 years, giving an unavailability of $0.25/30 = 0.00833$.

In the event of cable A outage, the loss of wind production will depend on the probabilities of generation at various levels above 100 MW. The load duration curve above 100 MW is assumed to be a straight line joining the following points:

$$100 \text{ MW } 0.25 \times 8{,}760 = 2{,}190 \text{ h}$$

$$200 \text{ MW } 0 \text{ h}$$

The annual production above 100 MW is therefore

$$(100 \times 2{,}190)/2 = 109{,}500 \text{ MWh}$$

The potential loss of production is then $0.00833 \times 109{,}500 = 912$ MWh per annum. Taking the energy and incentive value of the lost production as £60/MWh, the economic loss over 20 years will be £1.1 million.

Consideration of cable B unavailability would lead to a similar result. It is likely that the economic rating for each cable would be between 50 and 100 per cent of wind farm capacity. Irrespective of the normal pricing mechanism, the wind farm developer should meet this extra cost, since this is where the averted risk decision must be made.

Research by Garrad Hassan (Gardner *et al.*, 2003) has been carried out mainly to determine the variability of wind power over short periods (as low as 10 minutes). As a by-product of this, and work on the incidence of calms, Garrad Hassan conclude that single wind farms in Northern Europe can be expected to produce less than 5 per cent of rated output for about 30 per cent of the year (high wind speed sites) to 50 per cent of the year (lower wind speed sites) (van Zuylen *et al.*, 1996). The figures drop to around 25–35 per cent of the time for aggregated wind farms. Perhaps more importantly, the results indicate that wind farms spend very little time (a few percent) above 95 per cent loading. Experience also indicates that the summated output of all wind generation in an area never reaches 100 per cent of total wind generation capacity. For example, a combination of data from 18 onshore wind farms in Ireland (recorded output data) and one offshore wind farm (simulated from wind speed data) shows that total output never exceeds 90 per cent of nominal capacity (ESB National Grid, 2004b). This is due partly to spatial averaging, but is also thought to be due to turbine availability and other loss factors. This indicates that there may be limited benefit in sizing connections and backbone systems to extract the last increments of capacity from a wind farm. It may be better to limit the wind farm output occasionally to relieve network constraints.

Note that the figures quoted above are representative of Northern European conditions, that is, influenced by the movement of large-scale weather systems such as depressions. Locations in, for example, the trade wind belts at lower latitudes exhibit much more constant winds, giving a stronger argument for sizing electricity systems for some fraction only of nominal wind generating capacity.

4.4.3 Backbone system issues

Backbone system problems may be complex. Supposing that due to the wind farm, flow is reversed or otherwise increased in some parts of the network and thus the outage of a certain circuit results in overloads or voltage problems, what is to be done? With a traditional generator it would be expected that the generator could continue generating during n-1 or n-m-T conditions. The above debate suggests the arguments to be significantly less strong for wind generators. Tidal stream generators, for example, might be treated more like traditional generation because they will be predictable and hence relied upon in dispatch. Thus if probabilistic analysis based on real wind farm performance shows that wind farms can be relied upon to a degree, the case for reinforcement is strengthened. If the critical outage could be communicated to a wind farm controller then the wind farm output could be curtailed to manage the network problem, hence averting the time delay of several years and the cost penalty of main system development.

Network planners would probably insist that the curtailment or remedial action scheme (RAS) meets the reliability standards of protection, since the consequences of failure might be system overload and widespread outage.

Two complications then arise:

- The embedded generated system input may come from a group of wind farms, in different ownership.
- There may be several different combinations of system contingencies, which give rise to the need for constraint. The amount of constraint needed may vary with system loading, other generation patterns and which contingency outage causes the problem.

Management of the problem is therefore complex, involving wide area security constrained re-despatch with a secondary objective of cost or equity management. In a market system, wind farms might pre-bid for constraints, but the system provider should not compensate them, since they agreed to a disconnectable network connection. The economic settlement would be internal within the affected group of wind farms. The cost of such a RAS, the associated protection standard communications and financial settlement system may be excessive.

In assessing maximum reaction time, voltage problems require rapid response, whereas thermal time constraints are likely to give a few minutes response time. Voltage problems can, however, be addressed by dynamic reactive power devices positioned strategically, whereas thermal issues must be addressed either by special protection schemes or RAS; or main network development, which may be prevented or delayed by permissions issues. RAS can be used as a temporary measure while awaiting new-build permissions. In the economic sense, the cost of the RAS is a hedge against main project delay.

The above discussion relates solely to alternatives to network development to *accommodate* wind generation. The presence of traditional reliable generation at a load centre defrays load serving network expenditure. This is definitely not the case for wind generation unless coupled with some form of firmness, e.g. diesel generator backup or direct acting demand-side management. Nonetheless, arguments about whether to carry out load related developments or allow load shedding can be influenced by the presence of wind generation. Suppose a combination of traditional generation outage and network outage is estimated at once in 40 years. Then the

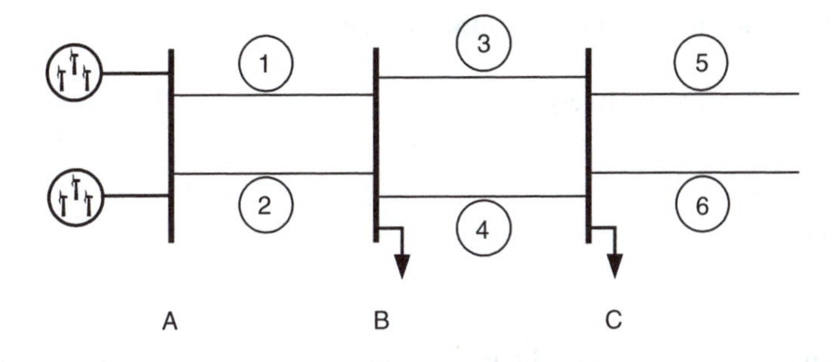

Figure 4.9 Assessment of line thermal ratings

presence of wind might mean that only when wind generation falls below, say, 60 MW is load shedding required. This is estimated to be 1/3 of the period of high load, so the risk moves to 1 in 120 years, making the development project less viable. There is as yet little published work on how network planners are dealing with these issues.

4.4.4 Equipment issues

Load flow analysis is undertaken to show whether the circuits (overhead lines) can cope thermally with the transfer. In the northern hemisphere, summer conditions may be more arduous than winter because of ambient temperature de-rating. In temperate climates this is combined with lower loads at B and C in Figure 4.9. If, for simplicity, we assume that lines 1–6 are rated equally, the outage of 1 or 2 represents a worst-case constraint for the wind farm – beyond B the load at B has reduced the flow in 3 and 4. The load at C further reduces the flow in 5 and 6. There is some merit in considering whether the rating of 1 and 2 can be viewed dynamically. It is known that the rating of lines increases dramatically over the first few km/h of wind speed.

There are three major consequences of thermal overloading:

- weak spots, e.g. overhead line jumper or joint connections, may fail
- sag will increase
- conductor grease may become fluid and be lost

The problem is that there are a large number of variables in dynamic line rating, viz.

- wind direction – axial or tangential to the line
- wind variability throughout the route
- line topography throughout the route
- tree or other shielding of conductors

Nonetheless, an argument is advanced which says that high line rating is only required to exit wind farm capacity in the presence of strong wind, therefore advantage should be taken of dynamic line rating. It is unlikely that generic research would be useful in this area, as the problem will be dominated by local factors. However, a study methodology and guidance note would prove useful to developers and planners.

4.5 Network power quality management

The issues of concern are likely to be:

1. Dips
2. Harmonics
3. Flicker

4.5.1 Dips

Other than starting conditions, the major source of dips in wind-powered networks is the variability of the wind. The discussion above has indicated how the voltage is altered by changes in the direction of load serving flows. Some research is required to consider the performance of mixed demand and embedded generation on real circuits against the national standard, e.g. P28 in the United Kingdom (Electricity Association, 1989). By and large the standards were constructed to facilitate design where the load switching is fixed or continuously variable. Wind variability is neither, and some is insulated gate bipolar transistor (IGBT) switched, therefore its performance should be assessed to determine whether the variation exceeds the standard at any point, and an assessment made as to whether this creates any customer difficulties. If no difficulties emerge, then standards can be relaxed.

4.5.2 Harmonics

Devices containing power converters are apt to create harmonic voltage distortion. The alternating current (AC) side harmonics are related to the number of pulse units in the converter according to $m.n \pm 1$, where n is the number of poles and m is any integral multiplier. Thus a 12-pulse converter will emit AC-side harmonics at 11, 13, 23, 25, 35, 37, …times fundamental frequency. The harmonic performance of voltage source converters depends entirely on whether the power order and Selective Harmonic Elimination (SHE) arrangements are met by PWM technology or capacitor voltage control and ancillary arrangements (see Appendix). Apart from the stress that these high frequencies place on utility and users' plant (capacitors are particularly susceptible to overload when higher harmonics are present), the harmonics can be induced into telecommunication circuits where they cause audible buzzing.

In general, power system planners will seek to ensure that installations do not worsen harmonics on the power system. Standards are quoted to comply with grid codes. There may be requirements not to exceed total harmonic distortion of the voltage waveform and limits on the distortion due to specific harmonics.

This however is not the whole story. Power systems experience resonance. Resonant frequency depends on network topology, connected generation and connected reactive power devices. It is difficult to predict harmonic resonance conditions, but simulation programmes exist. The challenge is to predict the higher frequency characteristics of plant based largely on 50 Hz information. Without a spectral analysis, the parameters will be approximate for higher frequencies.

As noted above, wind turbines with converters will emit harmonics, the severity depending upon the technology employed and the control algorithms. The practice is to negate them at source or construct filters to ensure that harmonic waveform distortion remains within limits. Nonetheless a very small voltage distortion can cause large harmonic currents at a resonant condition. The characteristic is similar to a notch pass filter. It will serve both utilities and generators well if there is a record of power system harmonic performance at a range of nodes and a model validated against known performance. Harmonic correction retrofit equipment can be very expensive, whereas capacitors rated as part of detuning networks can provide reactive compensation and harmonic detuning as a single unit. In theory, full speed range devices will emit more harmonics than DFIG devices, since the converter must transform the full output as against about 20 per cent needed for rotor power in a DFIG.

Where several converters are located side by side, there is also an opportunity for fractional interval harmonic beats (inter-harmonics) and internal resonances to occur. The detailed design of a wind farm should have regard to these phenomena.

4.5.3 Flicker

Flicker is hard to define, but is generally taken to be a discernible regular increase and decrease in the luminescence of incandescent luminaires connected to the system. Some wind farms exhibit a phenomenon known as 3P, that is, a power oscillation at three times the blade turning speed. The 3P frequency is typically about 1 Hz. The oscillation is thought to be due to the wind shielding effect of each blade of a three-blade turbine as it passes the tower. If the shielding effect were to reduce the blade torque output to zero, then the wind turbine output would decrease by one third at these times. The effect seems less than the theoretical maximum, resulting in a reduction of about 20 per cent. The physical mechanism is an impressed torque oscillation that is transmitted through the gearbox without frequency change. Reactive power demands will vary during this duty cycle. Where many turbines exist on a site, or many sites are connected to the system before the system supplies customers, diversity will reduce the impact of these oscillations on customers. If the oscillations cause significant distortion of customer voltage waveform, utilities are likely to receive complaints. It is incumbent upon utilities to manage power quality. For this reason, and because oscillations of this periodicity may excite small-signal instabilities in the power system, utilities will seek to ensure that wind farms minimise 3P. Measurement shows that DFIGs smooth 3P oscillations whereas ancillary energy storage equipment may be necessary to smooth the power oscillations of fixed-speed WTGs.

4.6 Transient system performance

4.6.1 Frequency performance and dynamic response

This area is perhaps the most difficult in which to obtain agreement between network engineers and wind farm providers. Network providers and operators seek to be reassured that the network will remain stable under all conditions.

Utilities carry out studies and tests to ensure that the loss of any infeed does not cause instability of other generation and that remaining generation provides a response in proportion to the decreasing system frequency. Plant must therefore be capable of

Table 4.3 Grid Code requirements for WTG operation at off-nominal frequencies

Endurance time		Australia (Hz)	Denmark (Hz)	Germany (Hz)	Ireland (Hz)	Scotland (Hz)
Normal limits (>1 h)	maximum	50.1	51	50.5	50.5	50.4
	minimum	49.8	49	49	49.5	47.5
1 h limits	over	50.1	51	50.5	52	52
	under	49.8	49	49	47.5	47.5
0.5 h limits	over	50.1	51	50.5	52	52
	under	49.8	48	49	47.5	47.5
Minutes	over	50.1	51	51.5	52	52
	under	49.8	47.5 (300 s)	48.5 (180 s) 48 (120 s)	47.5	47.5
Seconds	over	50.5	53 (60 s)	51.5	52	52
	under	49.5	47	47.5 (60 s)	47 (20 s)	47.5
<seconds	over	52 (400 ms)	53 (60 s)	51.5	52	52
	under	47 (400 ms)	47	47.5	47	47 (1 s)

operating within a frequency range and should be equipped with a control mechanism capable of adjusting the output in response to the frequency. Power systems operate ordinarily within a 1 per cent frequency bandwidth around the nominal frequency, and this poses no problems for wind generators. Loss of significant generation, interconnection or load can cause a system to operate, for a time, at up to 104 per cent or down to 94 per cent of nominal frequency. Load shedding or plant tripping may follow to restore the frequency to within the normal range. As a result, most Grid Codes require that plant is capable of remaining stably connected for defined periods throughout this frequency spectrum. Table 4.3, drawn from research carried out by Vestas, summarises the position.

Many wind turbine manufacturers claim that these frequency limits represent no significant issues. If capacitors form part of the installation it is important to ascertain whether these are rated for the higher frequencies. For fixed-speed devices, there may be a significant increase in mechanical loading at higher frequencies. DFIG machines isolate, to some degree, the speed of the turbine from the supply frequency and therefore do not experience extra mechanical forces.

Increasing the turbine output during times of system energy deficiency causes the generation–load energy balance to be restored, preventing the system frequency from falling to the point where there is automatic load disconnection. Grid Codes specify response, droop and dead-band characteristics. Typically load pick up over the pre-emergency condition is specified for 3- and 10-seconds post-event. Droop is usually specified as the percentage change in frequency that will cause a 100 per cent change in output of a unit. The droop refers only to the control system that issues the signal, and is no guarantee that the plant will respond eventually. The actual response is the load-lift measured in defined periods post-event. The dead-band is the normal operating range of frequency movement for which no emergency response is expected.

Utilities carry out compliance testing to ensure that new or re-commissioned plant meet the Grid Code requirements on the principle that plant must be capable of sharing the pain of a system disturbance. In the case of traditional plant, precise conditions are established and the governor set-point is adjusted up by say 0.5 Hz to simulate a frequency fall. The plant response is timed and values for, say, 3-seconds, 10-seconds and long-term load-lift are recorded. The latter gives the instantaneous droop at the plant pre-loading test point.

The problem with carrying out these tests for a wind turbine is that the source power is varying. It may therefore be necessary to record the plant performance over a range of real system events and assess average performance against accurately clocked wind speed while in spilled-wind operating mode. This makes Grid Code compliance a lengthy process during which a wind farm can only be entitled to temporary connection permission.

Table 4.4 shows the position in October 2003 as required by a range of Grid Codes.

Chapter 5 explores the need to spill wind in the context of system frequency management. Wind is a free resource once the wind farm costs have been met. There is thus a reluctance to spill wind rather than save fossil fuel to achieve spinning reserve. Nonetheless, with high penetration of wind, especially on smaller systems, there may be a need to ensure that the system still has adequate active power reserves. There is no major technical problem with this requirement other than the constantly varying response available depending upon wind strength.

4.6.2 Transient response

Three-phase power systems are prone to various kinds of fault, for example

- single line to ground
- line to line
- double line to ground
- three lines to ground

The detailed analysis of these fault types may be found in various standard texts (Weedy and Cory, 1998). The important issue here is to understand the behaviour of a WTG under the most severe *system* fault that may arise – namely a three-phase-to-ground fault in the transmission network. Such an event is rare, but it is the worst credible case and happens to be quite easy to analyse. If this fault is adjacent to the wind farm and distribution line of interest, the voltage seen by the wind farm will be virtually zero. Generators close to the fault cannot supply load energy due to the collapsed voltage. Since the load represents a braking torque on the generator, which has suddenly been removed, the generator will accelerate.

Generators far removed from the fault will see an increase in load and will tend to slow down. The objective of protection is to remove the fault quickly so as to prevent

Table 4.4 Grid Code frequency response requirements for WTGs

Frequency range		Australia	Denmark	Germany	Ireland	Scotland
Normal – no response	over	100%	102%	101%	101%	101%
	under	99.8% and 99% for 240 s	98%	99%	99%	95%
Action band	over	to 104% ↓ 2% per 0.1 Hz ($P_{initial} > 85\%$)	none	to 103% ↓ 4% per 0.1 Hz	none	to 104% ↓ 4% per 0.1 Hz
	under	to 94% ↑ 2% per 0.1 Hz ($P_{initial} > 85\%$)	none	time limited ↓ 1% per 0.1 Hz	none	none
Trip	over	>104% after 0.4 s	106% after 200 ms	>103%	none	>104% within 1 s
	under	<104% after 0.4 s	94% after 200 ms	<95%	none	<94% within 1 s

pole slipping between sets. Once the fault is cleared all generation should be present to continue to serve load.

A fault on a super-grid system can depress the voltage over a wide area of the network, hence creating widespread outage of WTGs. Studies in the Republic of Ireland have shown that a severe transmission fault creates a deep voltage disturbance over about one quarter of the island. Traditional synchronous generators deliver large quantities of reactive power as fault current, essentially because the voltage regulator attempts to restore the collapsed terminal voltage through an increase in excitation current (see Chapter 2). Prior to about 2003, wind generators were unable to mimic this behaviour, but the situation is changing.

The behaviour of fixed-speed and variable-speed WTGs during grid faults is considered below.

4.6.2.1 Induction generator performance

This type of plant absorbs more reactive power as the voltage falls, therefore instead of contributing to the reactive power needed into the fault, which helps the protection to clear the fault quickly, the plant does the opposite, drawing reactive power from the system and further depressing the voltage. Being unable to supply active power, the induction generator accelerates and will probably have increased in speed to greater than its breakout torque position (see Figure 3.14) when the fault is cleared. It may then be unable to supply load but will continue drawing reactive power. It will eventually trip from over-speed or over-current protection.

We need to review some simple mechanics in order to assess the change in velocity and hence slip during the fault. The net torque acting on the rotor is given by

$$T = I\alpha$$

where I is the rotor inertia and α is the rotor acceleration. Normally the accelerating torque is zero, with the applied turbine torque being balanced by the generator electromagnetic torque. However, the three-phase-to-ground fault mentioned above will reduce the WTG terminal voltage and hence, from Equation (3.14), the electromagnetic torque. The rotor acceleration may then be obtained, and thus the rotor speed and slip. The slip may be used in turn to determine the torque.

The WTG inertia is often quoted in terms of the so-called 'inertia constant' H. The H constant is in fact the ratio of kinetic energy at rated speed to the generator volt-ampere rating. It therefore has the dimensions of time, and is given by

$$H = \frac{\frac{1}{2} I\omega_0^2}{\text{rating}}$$

where I is the turbine generator inertia and ω_0 is the rated speed of rotation in rad/s. The inertia may then be expressed in terms of H:

$$I = \frac{2H \times \text{rating}}{\omega_0^2} \tag{4.1}$$

The behaviour of a WTG during and after a fault, and hence its transient stability, may be understood by considering the previous example.

Transient stability: example

The wind farm described in Section 4.3.1 will now be studied under fault conditions – a three-phase fault on the adjacent transmission network for 0.3 seconds. It is assumed that this fault depresses the wind farm terminal voltage to zero and that the wind farm was operating at full output – the most severe conditions possible. The following additional information is needed for each of the 20 WTGs:

Synchronous speed	1,500 revolutions/minute (157.1 rad/s)
Rated speed	1,575 revolutions/minute (165.0 rad/s)
H constant	2 s
Poles	4

The inertia I is then, from Equation (4.1), 110 kg m^2.

The slip corresponding to the full generator output of 660 kW is −0.018465, as seen previously. The pre-fault torque T applied by the wind turbine may be obtained from Equation (3.14). Note that the effective resistance (R'_r) and reactance (X) values should include the line and transformer resistances and reactances. This gives $T = -4,200$ Nm, the negative sign indicating that a torque is applied to the induction machine, as required for generation.

When the fault occurs, the electromagnetic torque collapses to zero, and the rotor accelerates at

$$\alpha = \frac{T}{I} = 39.32 \text{ rad/s}^2$$

The change in speed during the fault of 0.3 seconds is then $39.32 \times 0.3 = 11.8$ rad/s, giving a speed on fault clearance of

$$1.018465 \times 157.1 + 11.8 = 171.8 \text{ rad/s}$$

The corresponding slip is

$$1.0 - \frac{171.8}{157.1} = -0.09357$$

The equivalent circuit of Figure 3.14 may be combined with the distribution line impedance to determine the reduced induction machine terminal voltage. The calculated value is 315 V (phase), or 79 per cent of nominal, reflecting the significant reactive power now drawn by the induction generators. The torque is then, from Equation (3.14), −5,900 Nm. The magnitude of this electromagnetic torque just exceeds the applied turbine torque. The rotor will therefore decelerate and the WTG will remain stable. This may be a mixed blessing – the gearbox in particular could experience severe stresses during the transient.

A number of factors will modify this simple analysis. The machine has to be re-magnetised upon fault clearance, decreasing the initial decelerating torque and making instability more likely. On the other hand, the applied torque will decrease as the increasing speed moves the turbine away from the optimum aerodynamic condition. These factors can only be assessed properly with comprehensive simulations, combined with careful analysis of the results.

4.6.2.2 Mitigation measures

A possible solution to potential instability is to install dynamic reactive support. A commercial device manufactured for this purpose (DVAr) will produce three times its rated regulating output for 1 s. SVCs or STATCOMs are also possible solutions. By maintaining nominal voltage at the wind farm for the duration of the fault, these so-called FACTS (flexible AC transmission systems) devices prevent the turbines drawing excessive reactive power and tripping. This leaves them online and available to participate in the post-fault recovery. Clearly a fault close to the wind farm will absorb reative power from a SVC and the wind farm will trip, but this is to be expected and the system must cope. It is the network impedance between the wind farm and the fault that allows a SVC to support the local voltage. The operating characteristics of key FACTS devices are outlined in the appendix.

4.6.2.3 DFIG performance

Until about 2003, DFIGs, like fixed-speed WTGs, were unable to ride through faults, causing the wind farm terminal voltage to fall below about 70 per cent of the nominal voltage. This is because of the power requirements of the rotor. Effectively the power electronics self-protected through operation of the 'crowbar' (Fig. 3.18) and the turbine tripped as soon as 20 ms after the fault. The machine was then incapable of serving load following fault clearance.

The challenge of achieving fault ride-through with DFIGs is to supply reactive power to the system while protecting the WTG gearbox from mechanical shocks and the power electronics of the direct current (DC) link from over-currents. The following strategy goes some way to achieving these objectives:

- the WTG stator is disconnected from the grid on fault detection;
- the WTG pitch angle is adjusted to reduce turbine power to zero;
- the DC link capacitance is used in conjunction with the system-side converter to provide reactive power – essentially behaving as a STATCOM;
- the turbine speed is adjusted to the pre-fault value;
- when the fault is cleared, the WTG stator is re-connected to the grid.

In this way significant reactive power and close to zero active power can be supplied during the fault, helping to support system voltage and thus allowing ride-through down to 30 per cent of nominal terminal voltage. The most recent improvements claim that fault ride-through to 15 per cent of nominal voltage can be achieved for a short period.

4.6.2.4 Grid Codes

Some Grid Code operators claim that, to allow a significant penetration of wind turbines, a fault ride-through down to 5 or even 0 per cent of nominal voltage is

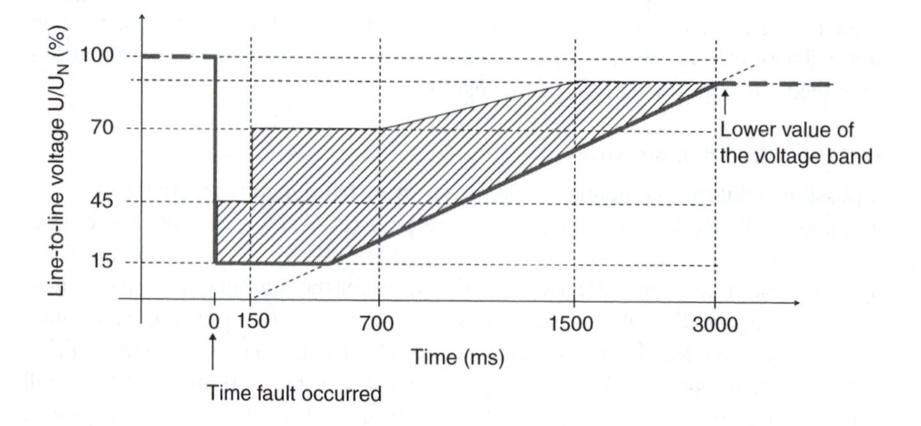

Figure 4.10 Typical fault ride-through requirement

required for a short period. The closer to the fault, the lower will be the voltage until the fault is cleared. The loss of a single wind farm, while exacerbating the position, is not seen as a problem. That loss will be met from the response of other plant. The worry is the widespread loss of generation that threatens the system's ability to supply all its customers.

For a transmission fault, as stated above, the voltage will be severely depressed over a wide area but the fault clearance time will be short. It is common on the highest level transmission systems to fit double main protection schemes so that there is practically no additional time lag if one scheme fails. Total fault clearance times of about 70–150 ms are to be expected. On the next voltage level, below the super-grid, worst-case clearance times of 300–450 ms are common. On the other hand, faults in distribution systems may take seconds to clear but they do not depress the voltage severely over a wide area. Figure 4.10 shows how Grid Code operators have expressed the fault ride-through requirement. Many intend to keep the requirements under review as wind penetration grows.

4.6.2.5 Dynamic modelling

To understand transient behaviour, validated models are required. In the case of any FACTS device, it is likely that high-order models will be needed to design the device. However, power system transient models consist of algebraic representations of the circuits and transformers and, typically, third-order representation of the synchronous generators. High-order models require short integration time steps, and therefore generally accommodate only a few network nodes, whereas 10,000 nodes may be present in a network model. The challenge is to reduce the high-order model complexity without materially affecting its validity. SVC models have been developed for standard network analysis programmes. The issue is to ensure that the control is properly modelled. In the final analysis, the performance of the model should be verified against the as-built device.

The issue of dynamic models is similar to that stated above for power electronic devices. Manufacturers have produced models in reduced-order industry standard formats for many devices, but many of these are unverified against actual wind farm performance. Utilities, installers and manufacturers need to work together to create an archive of performance data in real network environments.

4.7 Fault level issues

All embedded generation presents potential fault level issues. Fault level is a measure of the energy dissipated at a point by the worst possible fault. In practice the fault level – the product of per unit fault current at the point of interest and nominal per unit voltage (1.0 pu) and specified in MVA – is used to:

- assess the capability of equipment to make, break and pass the severe currents experienced under short circuit conditions
- establish suitable settings for protective relays and devices
- form the basis for many 'rule of thumb' calculations related to voltage depression during load or network switching, capacitor block sizing or harmonic distortion levels

If the fault level is excessive, switchgear may not be able to deal safely with the currents involved. If it is too low, network protection may not operate as designed, switching operations may result in large step voltages and harmonic distortion may be severe. DC drives and FACTS devices may experience commutation failures.

4.7.1 Equipment capability

Three quantities are needed to determine equipment capability:

- the maximum current which might flow
- the maximum time for which it will be allowed to flow
- the power factor of the current relative to the voltage at that point

Switchgear is usually specified as having symmetrical and asymmetrical load making and load breaking duties as well as a continuous load current rating. The make duty is to cope with a situation where a circuit breaker may be closed when a fault is already present. It must not suffer damage and explode, endangering the person closing it. The break duty is to ensure that the breaker can interrupt the fault safely.

Equipment experiences two physical effects when fault currents flow. The dominant factor in most equipment tends to be thermal. This is proportional to energy loss in the equipment due to current of magnitude I flowing for time t or I^2t. It is therefore important to remove the disturbance as quickly as possible to reduce the heating effect and consequent disruption. The other effect is an electromagnetic phenomenon proportional to rate of change of current $\mathrm{d}I/\mathrm{d}t$.

The process of current interruption usually involves the separation of physical contacts. As the contacts start to separate, the contact pressure gradually decreases and the resistance R of the connection increases. This dissipates power proportional to I^2R. As separation continues, there is a tendency for a stream of hot ionised particles to maintain the arc between the contacts. Insulating gas or vacuum environments reduce this phenomenon. Extinction may only occur when the current passes through a natural cyclic current zero. If the X/R ratio of the network is very high, this will be almost at the time when the voltage is reaching a maximum between the separated contacts, tending to re-strike the arc. It is clear therefore that it is not only the modulus of the current which is significant but its vector relationship to the system voltage.

Embedded generation changes the situation in two ways.

- It introduces a new source of energy at the tail-end of the electricity network system, hence providing an increased fault level to be dealt with by distribution switchgear. Induction machines are known to contribute up to 6 times full-load current as fault level during a close-up 3-phase fault. In the case of a DFIG machine, unless the manufacturer provides a fault level, it must be assumed that it behaves as an induction machine plus a converter. A full speed range machine will behave as the converter behaves. For the most part the fault level contribution will be negligible. It should be noted that, because these machines are generally rated at 690 V, the impedance of the connecting transformer will reduce this contribution greatly when viewed from the HV network. Another factor is that the grounding arrangements of multiple wind turbine transformers need to be considered in the zero sequence network for determination of earth fault currents.
- It changes the X/R ratio because an almost pure inductive source is introduced. This is most significant when the wind farm is electrically close to the distribution switchgear concerned. When the wind farm is at the end of a long line, a different phenomenon occurs, in that the X/R ratio is reduced and the sub-transient and transient part of the fault current are affected. It is important to check whether the switchgear is trying to interrupt during the presence of these currents.

In the United Kingdom, the Embedded Generation Working Party has, through its technical committee, sought to better understand the way forward in addressing the fault level issues on distribution networks. It has looked at a range of solutions.

- Network splitting (which results in lower network security by operating the network with fewer interconnections) which reduces the number of parallel transformers/network paths and hence reduces fault level.
- Fault limiting devices, e.g. fuses and variable impedance links.

There is no uniquely correct way forward as yet, and more work is required.

In relation to wind generators, and DFIG devices in particular, there is a lack of information at the time of writing as to their responses to single-phase faults. They are likely to make some short-term fault contribution, but then trip rapidly as a result of unbalanced phase currents.

4.8 Information

In order to operate a system successfully with high levels of wind penetration, real-time information is required. To manage energy demand, wind farm output and wind speed/direction will need to be reported every few minutes. There will also need to be some instruction and status signals related to whether reserve is being provided and the amount of reserve available. Network RASs may need to be informed as to the output of one or a group of wind farms. Voltage control schemes are likely to require information on the voltage at wind farms and the farm's reactive power generation/absorption in real-time. Instructions for the voltage control system and target may need to be given remotely and alarms sent when out-of-limits control is in operation or an emergency trip is operated. Wind farm status reports need to include, as a minimum:

- down – lack of wind
- down – on instruction or pre-gust control
- down – maintenance
- running free
- running curtailed

How this is best managed is still to be explored. Some experts believe that instruction and information from a wind farm controller is all that is required and that all subsidiary intelligence rests at the wind farm. Others believe it safer to draw information directly from each turbine. Possibly a hybrid will emerge which satisfies both the security and efficiency criteria. Whichever method is adopted will have to be interfaced with a range of supervisory, control and data acquisition (SCADA) and energy management systems (EMS). Protocols differ, but, partly because information exchange has proven difficult during major system incidents, common interfacing packages are emerging.

Performance assessment and model trimming require disturbance recorders with a sampling rate of typically 20 kHz, and, where practicable, remote reading and resetting facilities.

4.9 Protection

4.9.1 System protection

The standard of protection increases with system importance. Thus a simple cartridge fuse may be used to protect an LV circuit whereas dual main protection based upon different protection principles may be used on the super-grid network. The costs and complication of protection rise accordingly and this reflects either the strategic worth of rapid fault clearance to maintain generator stability or the importance of the integrity of a particular part of the system. Many transmission and distribution systems are reclosed automatically after a fault to minimise customer outage. The complications of this are dealt with below.

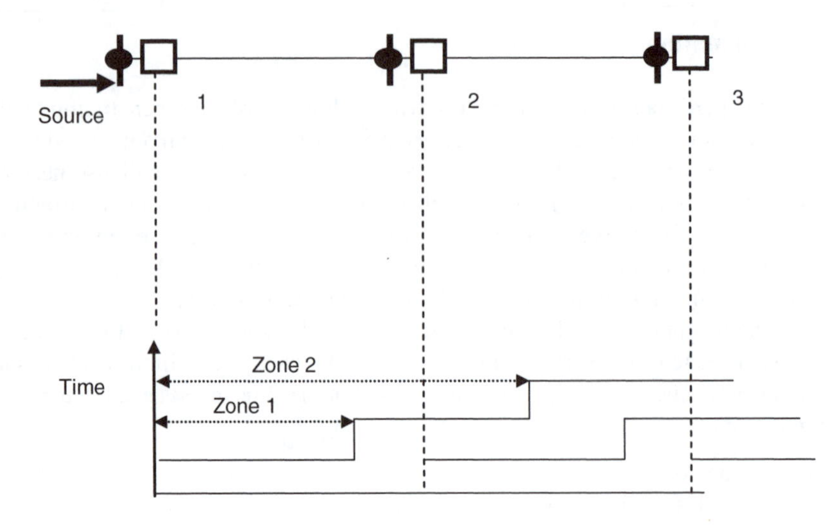

Figure 4.11 *Distance protection scheme*

4.9.2 Transmission connected wind farms

A combination of protection complexity and system security is likely to feature in the debate over how wind farms may be connected. A plain, a switched or a fused 'T' connection may be acceptable at distribution level, but is unlikely to be acceptable at transmission level. Where distance protection (Weedy and Cory, 1998) is one of the systems in use, it is likely that the protection may mal-operate because, by 'seeing' to the end of the zone 1 range necessary to protect the backbone power circuit, the scheme will 'over-reach' into and perhaps beyond the transformer supplying the wind farm. Thus the backbone system may trip for disturbances within the wind farm. The following diagrams explain the problem.

Figure 4.11 shows the normal arrangement for a plain feeder protection scheme. The relays are effectively measuring impedance by relating voltage and fault current. For faults within a small impedance distance (say 80 per cent of the distance to substation 2 – to prevent over-reaching into transformers at substation 2) the fault is cleared quickly by zone 1 protection which has no added time delay. Zone 2 detects faults beyond that range and introduces a time delay. Often faults in the last 20 per cent of the distance leading to substation 2 are detected as a reverse flow at 2 and an acceleration signal is sent to 1 to allow a zone 1 trip time to occur.

Figure 4.12 illustrates the 'over-reach' if the two paths are not of equal impedance length.

As a result, system planners are likely to divide the circuit into two sections using switchgear and locate the wind farm between the two plain feeders. Two or three circuit breakers are required. Figure 4.13 shows a three circuit breaker arrangement and Figure 4.14 a two-breaker arrangement.

This configuration may be preferred over an arrangement with the wind farm positioned between two line breakers, because a failure in the wind farm connection

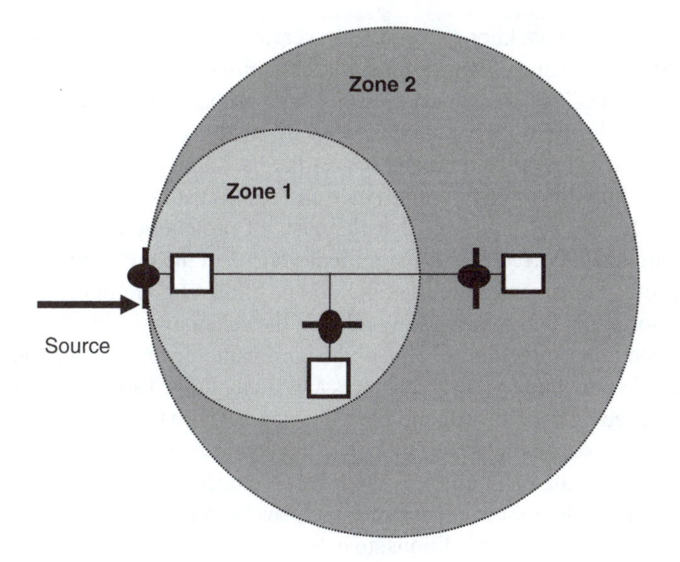

Figure 4.12 Distance protection with over-reach

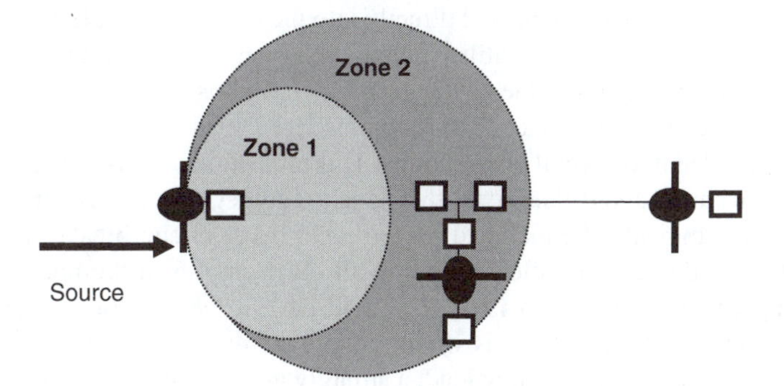

Figure 4.13 Distance protection – three circuit breaker arrangement

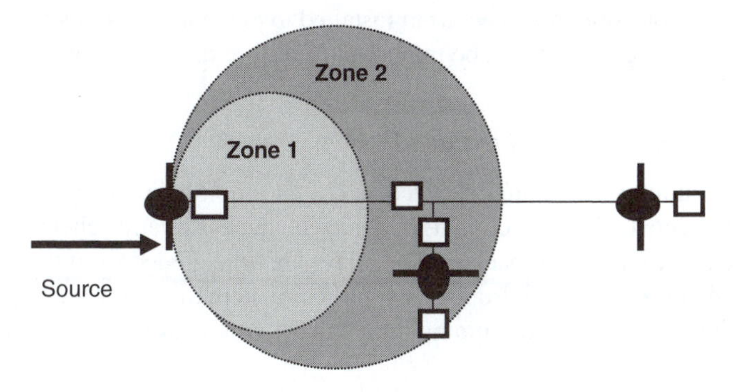

Figure 4.14 Distance protection – two circuit breaker arrangement

does not remove the backbone circuit from service. Allowing for remote operation, earthing, SCADA, battery chargers and security arrangements, an expensive solution is likely to result. The transmission utility will probably seek to retain, as far as possible, the security of its backbone circuit by ensuring that any problem with the T-ed equipment is self-protected and leaves the backbone system intact. Not every transmission circuit is of equal importance and some cost/benefit analysis should be undertaken. It may be, however, that the costs of protecting the more complex 3-ended arrangement correctly are a significant factor in the decision. While 3-ended distance protection schemes are possible, the position and length of the T-ed circuit will be important in determining how reliable the scheme will be. The alternative is a full 3-ended unit protection scheme with significant telecommunications needs.

Automatic post-fault circuit reclosures are often allowed since a large number of faults are caused by transitory phenomena like contact with trees, clashing conductors and birds. This is seen as an important tool in managing customer interruption periods. It will however create difficulties as the level of embedded generation grows. Even where safety considerations prevent auto-reclosure as a routine event, it is often used during lightning and storm conditions, to help reduce the demands on repair teams. As a general rule, auto-reclosures are not allowed on cable or transformer faults because of the low likelihood of transitory phenomena and the high cost of further damage.

Some wind farms are connected directly into the low-voltage side of bulk or grid supply points. It is a matter of utility policy who owns the circuit breaker for such a generation connection, but the switchgear may have bus-zone protection that may need to be extended (or replaced) to accommodate the additional circuit. Strategic nodes are divided into several bus-sections. This protection detects the zone in which a fault has occurred and isolates only that zone. At strategic locations, circuit breaker fail protection may also be included. In the event that a circuit breaker fails to trip, this protection detects the failure, tripping all other circuits at the node to prevent continuing energy supply to a fault. This will have to be modified to account for the new circuit. A further issue is whether the transformers are fitted with reverse power protection. At times of low load, a strongly generating large wind farm might cause the flow through the transformers to be reversed. The issue may only be one of settings, provided the transformers are capable of the level of reverse flow. The reverse flow protection may have been installed to prevent reverse feeding of a fault as shown in Figure 4.15. It will be important to restore this facility in another way.

4.9.3 *Distribution connected wind farms*

Many distribution systems are operated as closed loop or ring networks. Such systems are protected using directional over-current and earth fault schemes with plain over-current and earth fault backup. The relays are time graded from the source substation, and any attempt to introduce a significant alternative source around the ring would result in inappropriate tripping for a fault. A simple example demonstrates the principle.

As can be seen in Figure 4.16, both the over-current and directional over-current protection are time graded from the source. Suppose a fault occurs between

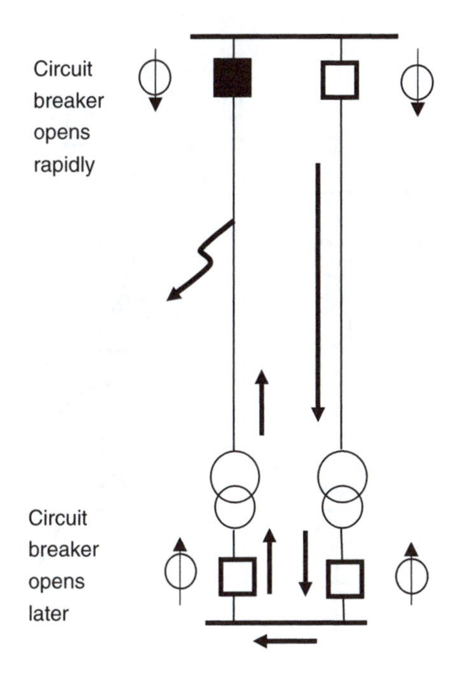

Figure 4.15 Reverse flow protection

substations 3 and 4. Energy flows from the source in both clockwise and anti-clockwise paths. The first relay to activate for anti-clockwise current is the 0.5 s relay at substation 3. The first relay to operate for clockwise current is the 1.3 s relay at substation 4. Hence the line from 3 to 4 is removed from service and substation 3 remains supplied from 2 while substation 4 is supplied only from the source. All customers connected remain supplied.

If a new source is introduced, say at 3, then energy flowing clockwise from this source may cause operation of the 0.1 s relay at substation 1, hence removing customers at substations 1, 2 and 3 from service. There seems little alternative with present commercially available protective equipment but to replace the scheme with either a distance or unit protection scheme.

On distribution systems the long clearance times, which are necessary to allow discrimination (sometimes called fault-grading) with higher network levels, can present a problem. Long clearance times are acceptable on low fault level networks, but as fault level is increased, e.g. by the introduction of local generation, network damage may result if clearance times are not shortened. The same reverse power issue may exist with transformers at lower voltage levels.

Modern distribution systems detect non-permanent faults by attempted circuit reclosure. Another feature is automatic fault sectionalising, using multi-shot reclosing circuit breakers and sectionalisers that count the number of times fault current passes. A further development is automatic change-over. This type of scheme detects loss of one source of radial supply and closes a normally open source automatically within

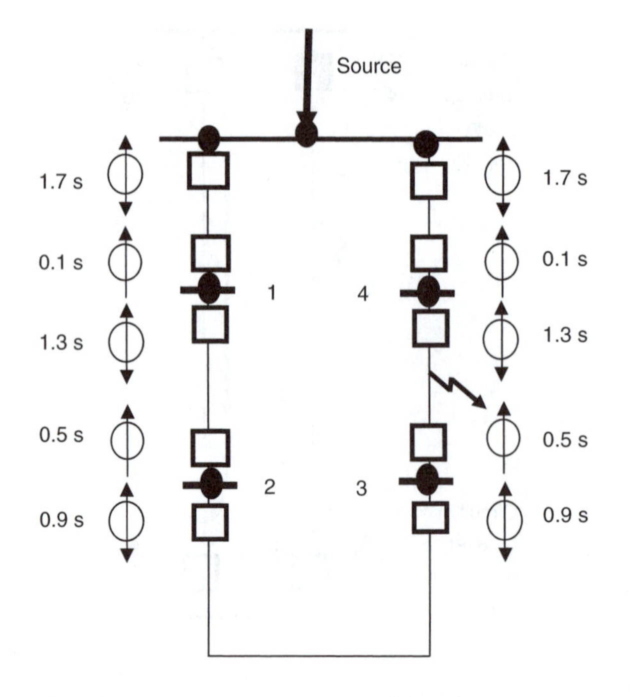

Figure 4.16 Closed loop distribution system with directional over-current and earth fault protection

a few seconds. These systems are designed to minimise both the cost of network development and the number of customer minutes lost due to interruptions. However, this type of system automation represents a significant problem when embedded generation is present beyond the automation. Two issues may arise:

- if the embedded generation trips, as a result of islanding protection, then the transformer voltage control may be on the wrong tap position when the automatic reclosure occurs, giving rise to low voltage to customers
- if the embedded generation is not allowed to trip, there is likely to be a form of mal-synchronisation with the system, resulting in a risk of plant damage

4.9.4 Wind farm protection

Most wind farm protection is a developer's internal matter. However, a prudent network operator will seek assurance that the protection design, installation, commissioning and maintenance are such that the grid is not threatened. One issue remains unresolved.

Embedded generation was protected traditionally by some form of islanding detection. In the UK codes of practice, referred to as G59 or G75, were applied. In essence a view was taken that embedded generation should not remain connected to customer load in the absence of connection to the main grid – a condition known as 'loss of mains'. Disconnection from the grid was detected fundamentally by rate of change of frequency (ROCOF) or rate of change of angle (ROCOA) relay elements (Jenkins *et al.*, 2000). A standard G59 relay emerged which embodied ROCOF or ROCOA and over-voltage, under-voltage, over-frequency, under-frequency and other features. ROCOF tends to be set at about 0.5 Hz/s. The difficulty is that a severe disturbance, especially on a smaller system, may cause ROCOF relays to operate, hence removing all embedded generation. This was not a problem with low levels of penetration. However, as penetration grows, there is the potential for loss of a major system infeed to be accompanied by a further loss of generation due to mal-operation of embedded plant loss of mains protection. The emerging grid codes require wind farms to respond in a manner similar to modern traditional generation, controlling voltage and frequency. If ROCOF or ROCOA cannot be allowed to trip the wind farm, and the other elements have to be set wide enough to allow a satisfactory operational range, there is no obvious way of detecting islanding. Some countries have sought to monitor the immediate feeder circuits using unit protection, but this fails to detect a more remote cause of islanding. The implications for system operation of spurious embedded generation tripping will be discussed further in Chapter 5.

A possible philosophy is that, even though the absolute output of a wind farm is unpredictable, if it is controlling frequency and voltage, it may be re-considered as safe for operation with customers. When the customer load passes outside the range for which it has controllability, it will trip in any case, thus self-islanding. If an emergency trip needs a remote re-set, the wind farm will remain islanded. The more difficult issue is network reclosure. If this is automatic, a reclosure may result in a mal-synchronisation shock to the wind farm and a dramatic transient over-flux in transformers. It would seem that unless network connectivity and wind farm status can be incorporated into a real-time model controlling reclosure, the problem remains unsolved.

Another approach is to continue to trip the wind farms but to use a different principle. Research is on-going to determine whether a measured change in network harmonics would give rise to a way forward. Work is also underway to determine whether tripping the power factor correction capacitors associated with induction turbines would allow discrimination between dynamic events and loss of grid (O'Kane *et al.*, 1999).

Chapter 5
Operation of power systems

5.1 Introduction

The connection of wind farms into distribution and transmission systems will have a *local* effect on voltage levels and reactive power flows, as discussed in the preceding chapters. The primary objective, however, of this network infrastructure is to deliver active power economically and reliably from generation sources to individual loads scattered across a national area. One of the underlying characteristics of electricity supply, which profoundly affects the manner in which it is engineered, is that electrical energy cannot be stored conveniently or economically. The consequence of this statement is that, ignoring any losses incurred in the electricity transmission network, there must be an instantaneous balance between the electrical power generated and the system demand.

This chapter first examines how conventional generating units can be scheduled and operated to cope with variations in the demand under both normal and emergency conditions (Section 5.2). The inclusion of wind generation in this task creates potential problems, as the power output of individual turbines is subject to time varying weather patterns. Section 5.3 begins by presenting a status report on how various countries are addressing the challenges (Section 5.3.1). Then, taking the island of Ireland as an example, the variability of wind output on different geographical scales and operational timeframes is examined in detail (Section 5.3.2). Various key problems of wind power integration on a significant scale, with some current solutions, are discussed in Sections 5.3.3–5.3.8. Wind forecasting is a useful tool, in this respect, and its potential role is discussed briefly here before being examined in detail in Chapter 6.

Finally, a number of alternative *storage* solutions, which can play a role in maintaining the desired energy balance, are examined in Section 5.4, facilitating further wind farm expansion. This section includes a discussion on how demand-side management can provide some of the benefits of energy storage at minimal cost.

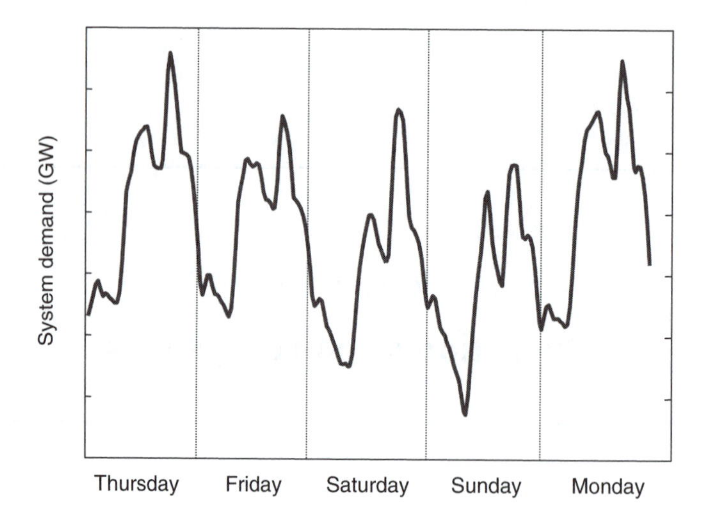

Figure 5.1 Five-day load demand profile

5.2 Load-frequency control

In addition to ensuring a continuous balance between electrical demand and the combined output of the system generation, the system operator must also ensure that network voltage levels and the system frequency remain within statutory limits. Clearly, this is a challenging task, made more difficult by the fact that the total power drawn by consumers of a large utility can fluctuate between wide limits, depending on seasonal weather, time of day, TV schedules, national holidays, major events, and so forth. Figure 5.1 illustrates a typical northwest European demand profile over a five-day period. Superimposed on a daily variation of high demand during daylight hours and low demand at night, it can be seen that the peak demand during weekdays is significantly higher than that at the weekend. Similarly, the late night/early morning minima are significantly lower at the weekend compared to weekdays.

Unfortunately, there can only be minimal control over load behaviour, although appropriate tariff structures, for example, can be used to encourage demand during natural periods of low demand and discourage demand during natural periods of peak demand, so that a flatter, less variable load curve is obtained. One commonly applied caveat to this statement is the use of pumped storage – an upper and lower reservoir are linked by a hydraulic turbine-synchronous machine arrangement which can operate either as a (hydro) generator or a motor (pump). During periods of high demand, water is released from the upper reservoir, rotating the hydraulic turbines and generating electricity, just as in a hydroelectric plant. Thus, the peak demand on the system is effectively reduced, potentially avoiding the need to start up responsive but expensive generation such as diesel engines and open-cycle gas turbines (OCGTs). During periods of low demand (typically at night or at weekends) and when electricity costs are at their lowest the water in the lower reservoir is pumped back to the higher one in readiness for the next day. Hence, the minimum load on the system is increased artificially. For smaller systems, this has the benefit of ensuring that several generating units will always be required to meet the load demand – minimising concern about system reliability and possible system collapse (see Section 5.2.2). Even for larger power systems, with inflexible nuclear generation, pumped storage can be of great benefit in filling in the troughs (and smoothing out the peaks) of electrical demand. However, the overall cycle efficiency is likely to be in the range 60–80 per cent, depending on age and technology, so large-scale energy storage is unlikely to be economic or even advantageous.

Ideally, electrical utilities would like the load demand to be invariant – the cheapest and most efficient generation could then be selected to operate continuously at the desired level. However the demand profile is shaped, sufficient generation plant needs to be connected to the system, at any particular time, to meet the likely variation in system demand. Since it may take several hours, or even days in the case of nuclear generation plant, for an electrical generator to be started up and begin to generate, the demand pattern must be predicted over a suitable period, and generating units scheduled accordingly. Until recent times, when a single, vertically integrated utility would have been responsible for generation, transmission and distribution of electricity, this *unit commitment* task would have been executed daily, with the objective of minimising operating cost against a required level of system reliability. However, as will be seen in Chapter 7 (Wind Power and Electricity Markets), many countries have now evolved to a deregulated environment, where a market may exist for energy trading, along with separate markets for a range of ancillary support services. The peculiarities of a particular market structure will clearly influence the types of generating plant that are deployed, the duties they are required to perform, and any long-term investment decisions that are taken. However, the underlying principles of operating the power system remain the same. Thus, in order to focus on *technical* rather than *economic* issues, a traditional vertically integrated utility, personified as a *system operator*, will be assumed in this chapter to have the sole responsibility for system operation.

As the system demand varies throughout the day, and reaches a different peak from one day to the next, the electric utility must decide in advance which generators to start up and when to connect them to the network. It must also determine the optimal sequence in which the operating units should be shut down and for how long. A power system will consist of many generating units, utilising a wide range of energy sources, employing different technologies and designs, and providing a wide range of generating capacity. As a result, fuel, labour and maintenance costs can vary significantly between generating stations. Even within the same power station the thermal efficiency of individual units will vary with power output. Despite these complexities, a merit order can generally be defined whereby all units are ranked in terms of operational costs. Consequently, the most economic (or inflexible) units are committed to the system first, and so-called *base load* units will typically be required to operate at 100 per cent of their rating for 24 hours per day. At the other extreme, expensive *peak load* units may only be required for brief periods during the day, or year, to meet the system demand maxima. In between these extremes, the

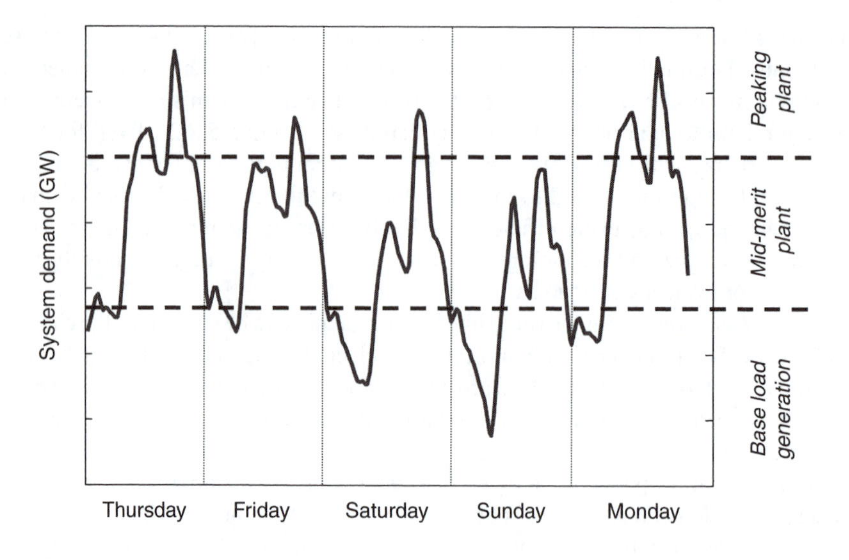

Figure 5.2 Combined merit order/system demand profile

morning rise and evening fall in load (see Fig. 5.1) are tracked by flexible generation plant operating in a *two-shift* mode. Such plant, operational for two shifts each day, and off for the other, are capable of responding fairly quickly to changes in demand. Depending on indigenous resources and the government policy of different countries, individual utilities may operate a significantly different merit order. At present, modern coal-fired plant and combined cycle gas turbines (CCGTs) tend to be the cheapest, while older coal-fired plant and oil-fired units tend to be more expensive. Nuclear plant, while not necessarily the cheapest to run, is comparatively inflexible, and may therefore operate as base load. Hence, countries such as Japan and Sweden have a high penetration of nuclear plant, backed up by flexible pumped storage and hydroelectric generation, respectively. In contrast, combined heat and power (CHP) plant is foremost in Denmark, while CCGTs (and coal-fired generation) are increasingly dominant in Great Britain and Ireland. Combining the demand profile of Figure 5.1, and a likely merit order of generation fuel sources, results in the unit scheduling of Figure 5.2.

A fixed cost can generally be assigned to shutting down a unit, but the start-up cost will depend on how long the unit has been shut down. For a conventional fossil-fuelled power station, the boiler metal temperature and hence the fuel required to restore operating temperature will depend on the duration of cooling. A *hot* plant, which has only recently been shut down, with metal temperatures close to design conditions, may require 2–4 hours to be synchronised and brought to full-load. A *cold* plant, which has been off-line for some time, may instead require 6–10 hours to complete the same process. The start up and shut-down times will vary significantly with generation technology, plant age and operator practice. Since there is a cost (and time) associated with the start up or shut-down of an individual generating unit, then, from an economic viewpoint, the scheduling of units at a particular time cannot be determined without considering the current configuration of committed units, the standby (hot/cold) state of uncommitted plant, and the expected variation in system demand. Consequently, the operating cost at one particular time is related to the units selected for all other periods. Units will be scheduled typically on a day-ahead basis. So, if the 24-hour period is split into a number of stages (say half an hour long), a multi-stage cost minimisation problem results, the objective being to select the combination of generating units at each stage to ensure that the demand is met at minimal cost, while observing the minimum up and down times of each unit. A variety of mathematical techniques are available for solution, including the use of neural networks and genetic algorithms. However, the standard approaches are based on the Lagrangian relaxation technique, whereby a Lagrange function (incorporating the multi-stage cost and weighted loading and unit limit constraints) is optimised subject to the unknown multipliers (Wood and Wollenberg, 1996).

Of course, in order that unit commitment can be achieved successfully, it is essential that accurate forecasts of the system demand for the 24-hour period are available. Many factors affect the system demand, as outlined earlier, but these are reasonably well understood and it is generally possible to forecast the demand with a 1–2 per cent error over the required time horizon. The most critical periods tend to be the morning rise, the daily peaks and the overnight trough. Forecasting is achieved using a combination of data trending and analysis of cyclical variations from historical load profiles, and/or construction of an overall demand profile from a sampling of individual load sectors. Weather forecast data, awareness of major events, including dramatic developments in popular TV programmes and system operator judgement may also have an important role to play in fine-tuning the final predicted demand curve.

So far, against a predicted demand profile, unit commitment has been performed using Lagrangian relaxation techniques, or otherwise, to ensure that sufficient generation plant has been committed to the system at the appropriate time. Since units can only be committed in discrete steps, spare generation capacity will normally be available to cope with errors in the forecasted demand. However, it is clearly not possible to predict the demand perfectly at each instance, so the task remains to ensure that a continuous balance is obtained between demand and generation. As an illustration of the difficulties that can be encountered, Figure 5.3 illustrates the demand profile in Great Britain during the evening of Sunday 9 July 2006.

During a typical evening there is a fall-off in demand as shops and businesses close for the day, domestic (evening meal) cooking activity slows down and people retire to bed. However, examination of the demand trace for 9 July reveals that, superimposed on this behaviour, there are two rising trends in demand beginning at approximately 7.45 and 8.50 p.m. Italy were playing France in the final of the 2006 football World Cup on this day and so the peaks correspond to half-time and full-time in the match – kettles are switched on for a cup of tea, visits to the bathroom (water pumping), etc. The demand profile could thus be considered predictable and generating plant could therefore be scheduled in advance to meet the short-term peaks in demand. However, at the conclusion of normal play the score was tied at 1–1, causing the match to extend into extra time (ending 1–1) before concluding in a penalty shoot-out, which Italy won 5–3. Although the rapid increase in demand after 9.45 p.m. is partially explained

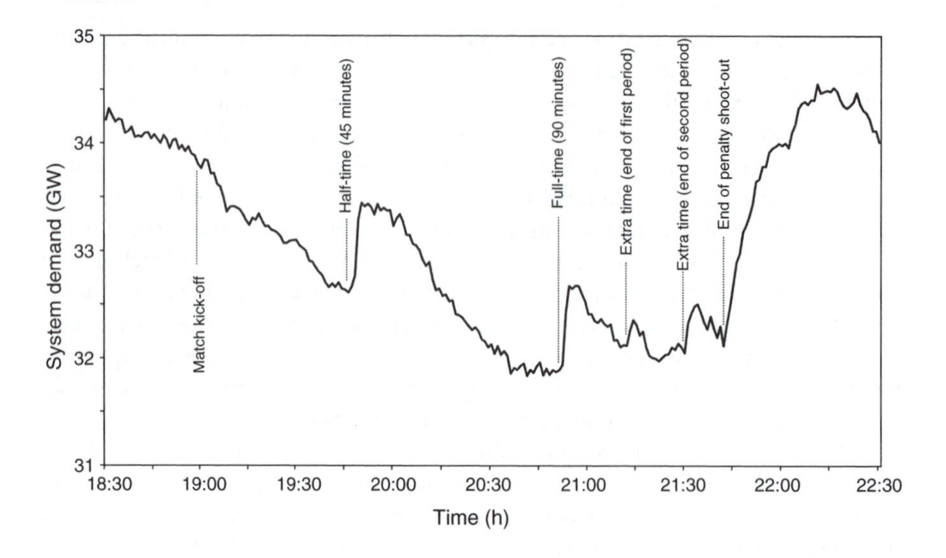

Figure 5.3 World Cup 2006 final – Italy versus France (National Grid)

by lighting load, the utility could not have known in advance that the match would not finish after 90 minutes.

5.2.1 Unit load-frequency control

For most electrical generation a fossil fuel (oil, coal, gas, etc.) is burned in a stream of air producing heat of combustion and a supply of steam at high temperature and pressure. The steam flows through a multi-stage steam turbine, and so drives the rotor of an electrical generator. In a CCGT the exhaust combustion gases also drive a gas turbine, which is coupled to the same, or a different, generator. Similarly, a nuclear reactor may also be considered as a source of steam, which drives a multi-stage turbine and generator. As illustrated in Figure 5.4, the steam flow, and hence ultimately the power output of the thermal power station, can be controlled by adjusting the governor valve feeding the steam turbine. The generator itself will be a synchronous machine (commonly known as an alternator) which has the characteristic that the frequency of the generated voltage is proportional to the rotational speed of the machine.

If the system load increases the electrical generators connected to the system will decelerate, with a consequent drop in rotor speed and system frequency. By similar logic, if the system load decreases the generated frequency will increase. Frequency is thus an appropriate signal for monitoring the balance between electrical supply and demand. Consequently, as seen in Figure 5.4, the turbine-governor adjusts the steam valve position to control the mechanical power output. The governor also monitors the rotor speed, ω, which is used as a feedback signal to control the balance between the mechanical input power and the electrical output power. The traditional speed-sensing device would have been a Watt centrifugal governor. However, with a desire for a high speed of response and accuracy in speed and load control, these

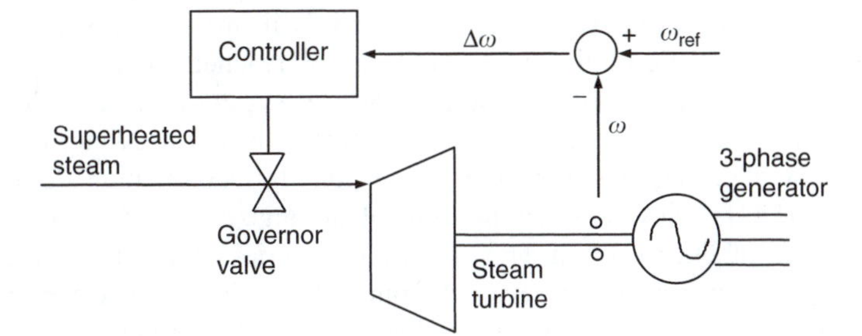

Figure 5.4 Steam turbine-governor control

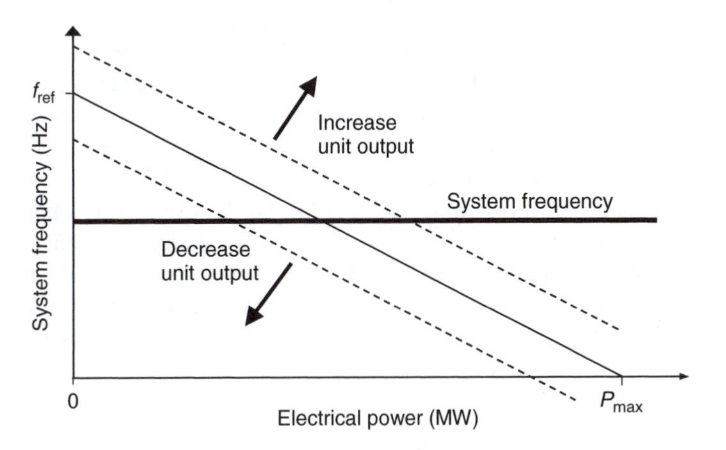

Figure 5.5 Governor droop characteristic

mechanical governors have been replaced by electro-hydraulic equivalents. Each generating unit in the power system will possess its own turbine-governor control system. In order that the units can operate in parallel, the speed versus power output characteristic of each unit requires *droop*, which implies that a decrease in machine speed should accompany an increase in generated output. Figure 5.5 illustrates the idealised governor characteristic of a large steam turbo-alternator.

If f_{ref} is defined as the reference frequency for the unit, that is, the frequency at which the generating unit provides no electrical output, then, as the frequency falls from this level, the unit will begin to generate dependent on the slope, or *droop*, of the characteristic. The droop is defined as the percentage increase in speed (or frequency) following full-load rejection, and a value of 4–5 per cent is typical. An important feature of this arrangement is that the reference frequency, f_{ref}, and hence the main steam valve position, can be changed quite independently of any observed variation in the machine speed. Thus the output of the unit can be controlled precisely, independent of the system frequency.

Under normal conditions, each generating unit will operate under free governor action, whereby an increase/decrease in the system frequency is interpreted as a decrease/increase in the system load. Each unit will move up/down its respective droop characteristic until a new equilibrium is reached, such that generation equals demand. Against a nominal target of 50 Hz, system frequency should normally be maintained within reasonably tight limits – in Great Britain the defined range is 49.8–50.2 Hz, while for the Union pour la Coordination du Transport d'Electricité (UCTE) power system the equivalent range is 49.9–50.1 Hz. The synchronous zone of the UCTE system spans 23 countries, reaching from Portugal in the west to Romania in the east and from Italy in the south to western Denmark in the north. The broader range adopted for Great Britain recognises that the power system is much smaller (and possesses less inertia – see Section 5.2.2) than that of mainland Europe. During significant system disturbances these limits can be widened – perhaps from 49.5 to 50.5 Hz, or wider, depending on the size (inertia) of the actual power system. If all generating units have the same droop characteristic, then each will share the total system demand change in proportion to its own rating (maximum output). Although this is a desirable feature, in terms of sharing the regulation burden across all operational units, a significant change in demand will probably result in an uneconomic dispatch of the load, since the load is no longer distributed according to the merit order ranking of the units. However, an optimal loading of the units can be restored by suitably adjusting the reference frequencies of individual units. Given that the unit commitment process typically defines the required unit trajectory for each scheduling period, economic dispatch can be combined with the load-following requirements of individual units.

5.2.1.1 Automatic generation control

A synchronously operated power system can extend over both regional and national boundaries. Thus, power imbalances in one area will cause regulator action in all other areas, minimising the effect of the original disturbance. However, since the interconnection between subsystems and individual countries may be limited either physically (transmission line rating) or by contractual agreement it is necessary that interchanges between neighbouring areas are monitored and controlled. Germany, for example, is split into four control areas, while France and Austria are operated as one and three regions, respectively. For a particular region, the total error, ΔP_{tie}, between the actual and scheduled tie-line interchanges with all neighbouring regions can be determined by communicating information on the individual tie-line power flows to the central area control. Similarly, by calculating Δf, the error between the reference system frequency and the actual frequency, the error in the regional generation, ΔP_{area}, can be determined if the system *stiffness* is known (Wood and Wollenberg, 1996). The stiffness of an area quantifies the sensitivity of the system frequency to imbalances between demand and generation, and will depend on both the governor droops of operational units and the frequency dependency of the load. The area control error (ACE) can then be defined as

$$\mathrm{ACE} = \Delta P_{\mathrm{tie}} + \Delta P_{\mathrm{area}}\ \mathrm{MW}$$

The area controller (regulator) must then determine the required change in P_{ref}, the regional reference power, and commonly a proportional + integral (PI) strategy is applied,

$$\Delta P_{ref} = \beta \times \text{ACE} + \frac{1}{T}\int \text{ACE}\, dt \text{ MW}$$

with the integral element eliminating the ACE and the proportional element providing improved regulation. β and T are the regulator parameters. Subsequently, $\Delta P_{i(ref)}$, the change in reference power for each unit i, is determined by proportioning the total change in power output with respect to the stiffness (governor gain) of each generating unit. Updated $P_{i(ref)}$ signals are then communicated directly to the individual generating units (in a similar manner to economic dispatch). Identical actions will be performed for each region, with individual generators regulating their output, such that interconnecting tie-line flows are maintained at contract levels.

5.2.2 Emergency frequency control

So far it has been assumed that any change in the system load or generated output of units occurs relatively slowly, over minutes and hours, and that the magnitude of the change is small, compared to the system demand, so that the impact on individual generating units is minimal. If, instead, a significant imbalance occurs, such as a large load being connected to the system, or the loss of an interconnecting tie-line, then there may be difficulty in stabilising the system.

Consider what could happen, *without forward planning*, following the sudden disconnection of a large generating unit from the system. Initially, demand will exceed prime mover power and the system frequency will begin to fall. This will cause the governor control system of each unit to react, in an attempt to negate the imbalance between generation and demand. Within a conventional power station, there are various fans and pumps controlling the flow of air, feedwater, etc. As the frequency falls, the performance of these fans and pumps will deteriorate, but for small deviations in frequency this can be more than accounted for by the governor control system, so that the electrical output of the station can be increased as desired. However, for larger changes in frequency, the former effect will begin to dominate and the electrical output may actually decrease as the frequency falls. Continued operation at reduced frequency will also impose severe vibratory stresses on the steam turbine sections. If the frequency falls too far, power station protection schemes will safeguard the unit by isolating individual generators from the system grid. Clearly, this will cause the system frequency to fall even further, before leading quickly to a system blackout.

For example (Andersson *et al.*, 2005), during the night of 28 September 2003 the Italian power system was importing approximately 6,700 MW (representing 24 per cent of the load demand) from Switzerland, France and other nearby countries (UCTE, 2004). At 3.01 a.m., during storm conditions, a tree hit the 380 kV Mettlen-Lavorgo transmission line, one of the main feeds into Italy, causing it to be isolated. Subsequently, Gestore del Sistema Elettrico (GRTN SpA), the Italian grid operator, lowered the requested power import to minimise overloading on the remaining infeeds. Then, at 3.25 a.m., the second main 380 kV Soazza-Sils interconnector was also hit by a tree, and similarly isolated. The resulting massive generation–demand imbalance within Italy caused the system frequency to plummet and within 10 seconds all connection with the European UCTE grid was lost, due to line overloading and subsequent tripping. As the frequency fell below 49 Hz, generators connected to the distribution grid were tripped (by protection), soon followed by the larger-scale generation. Despite implementing automatic protection measures in the form of pumped storage shedding and load shedding (described later), the frequency had fallen below 47.5 Hz by 3.28 a.m., and a major blackout of Italy was inevitable. Over 50 million people were affected, with some consumers off supply for up to 18 hours.

There are many reasons why a generator may be tripped from the system, and it is sufficiently common that protection plans should be designed to minimise the effect of such a disturbance (Armor, 2003). When units are being committed to a power system consideration must, therefore, be given not only to the predicted demand pattern, but also to the impact of a generator trip. Loss of the largest source of supply (infeed) to the system will generally, but not always, be of greatest concern. At best, it may take several minutes before fast-start units, typically OCGTs, can be brought online to make up the shortfall in generation. Since frequency (and hence system) collapse may occur in seconds, there must be the ability within the existing unit commitment for a number of units to increase their outputs rapidly to curtail the initial fall in frequency. The so-called *spinning reserve* requirement is expensive, since it implies that extra units are committed to the system, and it results in a non-economical sharing of system load. Intuitively, the spinning reserve available for a particular commitment of generators is the difference between the sum of the power ratings of all operating units and their actual loadings. So, if a unit is operating x MW below its upper limit, then it should be able to provide x MW of spinning reserve. As Figure 5.6 illustrates, the actual emergency reserve available from a thermal generating unit is less than this.

Due to time delays present in the power station boiler, mainly the reheating of steam between the high and intermediate pressure stages of turbine steam expansion,

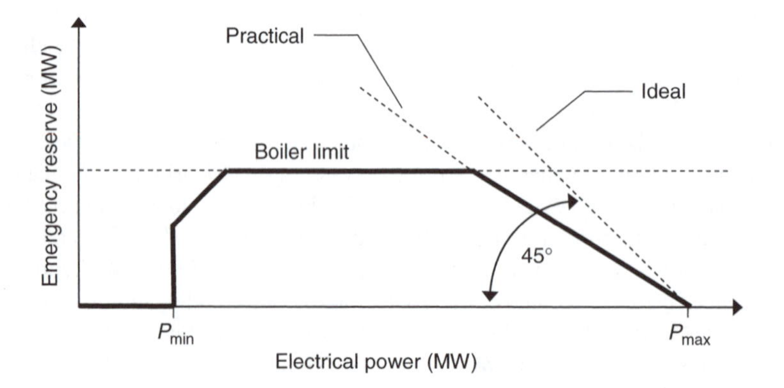

Figure 5.6 Thermal unit spinning reserve characteristic

the immediate reserve available will be reduced. It is worth noting, however, that the addition of the reheater improves the thermal efficiency of the plant significantly. Assuming that a generating unit has a governor droop of 4 per cent, then a 1 per cent fall in frequency would require a $(1/4) \times 100 = 25\%$ short-term increase in boiler output. The boiler is an energy store of limited capacity, and so over such a short time scale, the boiler would not be able to increase, and sustain, such an output. The steam pressure would inevitably fall, leading to droplets of water forming in the steam, which could cause significant damage to the turbine blades. The steam pressure therefore must be controlled, placing a limit on the reserve from each unit, as shown in Figure 5.6. As the frequency falls, fuel and air supply to the boiler will be increased. However, depending on fuel type and the configuration of the plant, it may be tens of seconds before the increased fuel flow appears as a sustained, increased power output.

Since the provision of spinning reserve incurs an operational cost, a balance has to be struck between the cost of the reserve and the likelihood of it being called upon. For small, or synchronously isolated, power systems it may prove too expensive to provide 100 per cent cover following loss of the largest infeed. A decision may be reached that only a certain fraction, say 80 per cent, may be covered. The remaining shortfall of 20 per cent may then be obtained by disconnecting customers, either through selectively disconnecting part of the load, or through the establishment of an interruptible tariff with large industrial consumers. As load shedding is a somewhat drastic control measure, and disruptive to consumers, it is usually implemented in stages, with each stage triggered at a different frequency level, so that the minimum amount of load is lost. In practice, this measure may not be necessary, as the system load tends to be frequency sensitive – as the frequency falls the demand also falls – hence reducing the initial generation–demand imbalance. System load is also sensitive to voltage and hence, amongst other measures, some utilities switch in reactors (using frequency-sensitive relays) at suitable network points following loss of generation to depress the local voltage and hence the locally connected demand. Also, if pumped storage is present and operating in pumping mode, then it can be disengaged (reducing the system demand to its natural level), and then switched to peak output in generating mode. For many power systems, pumped storage may be the most responsive load/generation plant available.

5.2.2.1 Inertial system response

Immediately following a significant generation–demand imbalance, the system frequency will change rapidly. The initial rate of change of frequency (ROCOF) will determine how quickly generation plant must respond before system difficulties begin to worsen. If an individual generator is rotating at a nominal angular speed ω_0, and has inertia I_{gen}, arising from the inertia of the multi-stage turbine and the alternator, then the rotational-stored energy, E_0, is given by $\frac{1}{2} I_{gen} \omega_0^2$ (see Section 4.6.2). As the system frequency varies, and hence the generator rotational speed changes, energy will either be stored or released, tending to negate the original disturbance in the system frequency. A convenient indicator of a generator's inertial response is provided by its inertial constant, H_{gen}, which is defined as the generator-stored energy divided by the rating of the machine (see Section 4.6.2). It can be interpreted as the time that the generator can provide full output power from its own stored kinetic energy, with typical values in the range 2–9 seconds (Grainger and Stevenson, 1994).

So, if there is an initial power imbalance, ΔP, then the accelerating torque, ΔT, is obtained as the total system inertia multiplied by the angular acceleration, α. Consequently, at rotational speed ω, an expression for α can be obtained as

$$\alpha = \frac{d\omega}{dt} = \frac{-\Delta T}{I_{system}} = \frac{-\Delta P/\omega}{2 \times E_{system}/\omega_0^2} \ \text{rad/s}^2$$

where E_{system} encompasses the total stored energy of all generators, and loads connected to the system at the nominal rotational speed ω_0. If f is defined as the system frequency, and it is assumed that $\omega \approx \omega_0$, then the initial ROCOF can be readily calculated as

$$\frac{df}{dt} = \frac{-\Delta P \times f_0}{2 \times E_{system}} \ \text{Hz/s} \tag{5.1}$$

Likewise, the inertial response, ΔP_{gen}, of an individual unit can be determined in relation to its stored energy, E_{gen}, as

$$\Delta P_{gen} = \frac{-2E_{gen}}{f_0} \times \frac{df}{dt} = \frac{-2H_{gen}\,S_{max}}{f_0} \times \frac{df}{dt} \ \text{MW} \tag{5.2}$$

where S_{max} represents the generator rating.

Clearly, the larger the power system, the greater will be the system-stored energy relative to the credible loss of generation, ΔP. Hence, the fall in frequency will be slower, giving generators more time to respond. It follows that the loss of a generator in a small, synchronously isolated power system, for example, the islands of Ireland or Crete, will be more significant than for a large interconnected power system (e.g. mainland Europe). For small power systems, the initial ROCOF will tend to be higher, requiring highly responsive generating units, perhaps less than 5 seconds. This burden can be particularly onerous during periods of low demand when the system inertia is less. The spinning reserve duty is then spread across fewer units, and individual generators are likely to provide a greater proportion of the total system load. In contrast, for a relatively large power system, such as the England and Wales system, the most significant scenario normally considered is not the loss of one generator, but two (or the connecting transmission line), namely the 1,320 MW (2 × 660 MW) nuclear power station of Sizewell B. Within the UCTE system, 3,000 MW of regulating reserve is required within 30 seconds following major contingencies.

5.2.2.2 System restoration

Following a loss of generation event, or other major power imbalance, the system inertial response will determine the initial rate of fall of frequency and the maximum response time required of operational generating units. The subsequent actions that

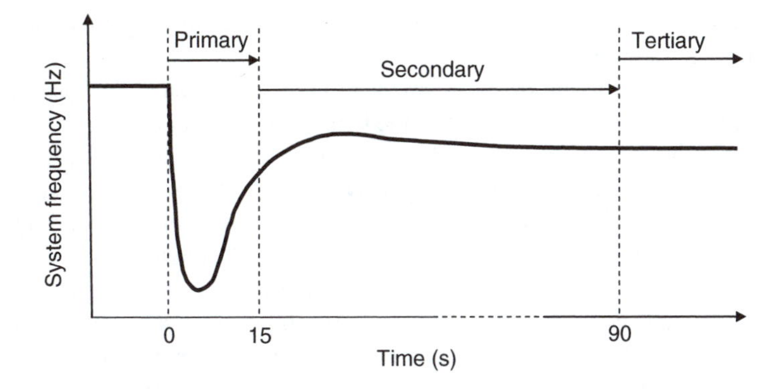

Figure 5.7 Power system frequency transient

a particular power system operator undertakes will depend on the resources and facilities available, but are likely to include the following options:

- Currently operational thermal units will increase their steady-state output.
- Quick-start units (e.g. OCGT) will be brought online.
- Replacement (and economic) generation units will be requested to come online.
- Pumped storage units will be switched to generating mode, and hydroelectric plant rescheduled.
- Previously shed load will be restored, and shunt reactors removed.

As illustrated in Figure 5.7, distinct categories of reserve (e.g. primary, secondary, tertiary and replacement) can be defined which must be supplied on different time scales. Primary reserve is the most critical, especially for small (low inertia) systems, and is required in the period immediately following an incident. The Ireland grid code specifies a primary response between 5 and 15 seconds after an event, for example, while in the much larger UCTE system primary reserve is specified over a 15–30 second period. Primary reserve is provided mainly through the governor response of individual generating units. Often included implicitly in the primary reserve is the inertial response of the synchronised generation plant. Primary reserve ensures that the power system *survives* the event and that the frequency never falls below an *acceptable* minimum frequency (nadir).

In the following minutes, the power system is likely to be in a stable, but insecure, state, that is, generation and demand are in balance, but at a reduced system frequency, possibly with some load shedding, and with depleted levels of spinning reserve. Secondary reserve is required to return the system frequency towards 50 Hz, replacing some of the *consumed* primary reserve, and is provided mainly by part-loaded generating units, which can ramp-up their output in a matter of minutes, and *spinning* pumped storage plant. In Ireland, secondary reserve is specified from 15–90 seconds. For much larger power systems the time horizons can be much longer, for example 10–15 minutes in mainland Europe. Hence fast-starting plant such as diesel engines, OCGTs and hydroelectric generation may be categorised as secondary reserve. During this phase, area regulators may apply automatic generation control to coordinate system response.

The role of tertiary and/or replacement reserve is to replace the lost unit, that is, generation output should be maintained (minutes to hours) until such time as replacement plant can be brought online. Although OCGTs can be started quickly (5+ minutes), they are comparatively expensive to run, so that long-term replacement reserve plant will also be requested to start as soon as possible. Depending on the temperature of such plant, the delay may range from 30 minutes (hot) to 4+ hours (cold). Customer loads, previously shed, will also be progressively reconnected as the stability of the system permits. The final objective is to restore the system to its original state, whereby an economic combination of units is operational, and sufficient spinning reserve is available to cope with plausible fault scenarios. How long this task takes will depend on the severity of the original incident, but will also be affected by the time of day at which the disturbance occurs. Depending on whether the system load is decreasing (evening time) or increasing (early morning), and hence whether generation plant has been scheduled *a priori* to go off-line/online, system restoration may be completed in a matter of minutes or may instead take several hours.

5.3 System operation with wind power

5.3.1 Overview of system operational challenges of wind power

Power systems generally consist of a large number of fossil-fired power stations, able to maintain their output indefinitely at specified levels, to follow defined loading/unloading schedules, or to vary their output in accordance with system demand. The introduction of significant wind generation presents an energy source not amenable to central control, and which may not be available to generate when required due to lack of wind. It is important therefore to consider the variability of wind generation over various time scales, and to consider the behaviour of distributed wind farms from a system perspective. The load-following implications for the remaining generation and the response to generation loss incidents can then be examined.

In the normal operation of thermal power systems, generating units are scheduled to meet the predicted load demand profile and spinning reserve requirements at minimum operating cost, given a specified reliability criterion. Units committed to the system must be capable of changing their output to match changes in system demand over the scheduling period. This should include the ability to follow normal daily load changes (particularly the morning rise and evening fall), participate in frequency regulation and supply replacement power following a loss of generation event, or other major contingency. Those units required to *load follow* are capable of meeting normal load fluctuations as well as sudden, unexpected changes in demand.

Increasing concern about global warming has, however, led to greater interest in, and exploitation of, renewable resources for electrical generation. Individual countries have approached this challenge in different ways and with varying levels of priority. The European Union has been perhaps the most forward thinking, particularly with regard to its available wind resource. The EU Renewables Electricity Directive (2001) requires that the amount of electricity supplied from renewables should increase from 14 per cent in 1997 to 22 per cent by 2010 (European Commission, 2001). As part of the directive, member states must ensure that transmission system operators (TSOs) and distribution system operators (DSOs) guarantee the transmission and distribution of electricity produced from renewable energy sources, subject to the reliability and safety of the grid. TSOs are also required to give priority dispatch to renewable energy sources, insofar as operation of the power system permits. The costs associated with grid connection and grid reinforcement should also be transparent and non-discriminatory, while charges associated with use of the transmission and distribution network should not discriminate against renewable sources.

So, if significant wind generation is now introduced into the plant mix, displacing conventional thermal plant, the task of load-frequency control will be affected. Generator loading levels and ramping requirements, availability of spinning reserve, amongst other issues, also need to be addressed. However, it is worth remembering that, irrespective of current and future levels of wind generation, power systems are already required to cope with significant variability and intermittency concerns. For example, on Wednesday 11 August 1999, following a mid-morning solar eclipse over the United Kingdom, and as people resumed their normal daily activity, the system demand in England and Wales rose from 33.2 to 36.2 GW (a jump of 3,000 MW) in just 5 minutes. In comparison, the wind farm capacity connected to the National Grid (England and Wales) network in 2005, six years later, was 679 MW.

Furthermore, generator forced outage rates of 4–5 per cent can be considered typical, although highly dependent on a variety of factors including plant age, implemented technology and operator practice (Armor, 2003). Thus, unit commitment is undertaken daily with the *expectation*, rather than the *possibility*, that unplanned outages of the conventional generation plant may occur, leading to the need for various categories of reserve (see Section 5.2). In general these outages are not predictable, resulting in a rapid reduction from full output to zero output. In order to reduce the immediate impact, some utilities, such as Electricity Supply Board National Grid or ESBNG (Ireland), practise a slow wind-down, where possible, of the affected generator. Clearly, the abrupt loss of a wind infeed, perhaps due to a problem internal to the wind farm, or as a result of external network problems, could cause similar problems to the loss of a thermal unit – the frequency would fall, and the remaining synchronised units would increase their output to restore a demand-generation balance. The fact that a wind farm consists of multiple turbines, whose individual operation is largely independent of each other, significantly increases the overall availability of the wind farm. For economic and possibly technical reasons (see Section 5.3.6), the wind farm will probably not be scheduled to provide spinning reserve. Hence the loss of a wind farm should be less significant than the loss of a thermal generator of similar rating. So, assuming that the largest wind farm block, likely to be an offshore wind farm, does not represent the largest infeed, then the short-term emergency reserve targets for the system will be unaffected.

Indeed, it has been argued that a power system's *natural* ability to cope with unexpected events and existing load variations, through the provision of spinning reserve and the part loading of units, ensures that *low* levels of wind generation can be safely accommodated without undue cost or difficulty. In Germany, a study was undertaken to investigate the long-term (2020) consequences of an increased contribution from renewable energy sources, and in particular wind generation (DENA, 2005). It was adjudged that wind generation capacity could grow from 18 GW (2005) to 36 GW (2015) and 48 GW (2020), representing 14 per cent of energy production by 2015. Despite the concentration of wind farms in northern Germany (an area of relatively low demand) and offshore, required expansion/reinforcement of the grid would be limited. Additional *balancing* power stations to combat wind variability would not be required, and consumers should see at worst a marginal increase only in electricity price. Similarly, a study for the New York state system investigated the impact of a 10 per cent capacity wind penetration scenario (projected for 2008). It was determined that there was no credible single contingency that would lead to the loss of all capacity (3,300 MW) distributed across 30 locations (Smith, 2005). With the system already designed to cope with a 1,200 MW loss, it was considered that there was no need to revise that planning criterion.

Increasingly, developers are considering the siting of wind farms offshore. By the end of 2005, 20+ offshore sites, with a capacity of about 600 MW, had been installed, mainly in the shallow waters off the European coast. The advantages include reduced visibility and noise problems, low wind shear and higher wind speeds. Capacity factors for offshore sites tend therefore to be in the range 40–60 per cent, noticeably higher than those onshore. Since the sea–air temperature difference tends to be less than the land–air temperature difference, turbulence is also less, which reduces mechanical fatigue and increases equipment life. However, offshore sites are more expensive to construct and operate. In addition to costs associated with offshore turbine installation, grid connection distances will be increased, including the need for submarine cabling, while access by boat for maintenance may be restricted by poor weather conditions. The above factors have encouraged existing, and proposed, offshore wind farms to be much larger (100–1,000 MW) than those on land. The European Wind Energy Association (EWEA) estimates that, from a predicted total of 75 GW installed wind farm capacity, 10 GW will be sited offshore by 2010 (EWEA, 2005). By 2020 this will rise further to 70 GW offshore, from a total of 180 GW. Germany, in particular, has ambitious plans for future offshore growth in the North Sea and Baltic Sea. The Deutsche Energie-Agentur (DENA) study assumed 10 GW offshore (and 26 GW onshore) by 2015, and 20 GW offshore (and 28 GW onshore) by 2020 (DENA, 2005).

In some cases, transmission constraints impose limits on the amount of wind generation that can be accepted into a power system, but generally the limiting factor is the response of conventional generation: minimum load limits, connected capacity required to maintain sufficient reserve (and inertial response), load-following flexibility of part-loaded plant, and start-up times of conventional and peaking generation to meet wind forecast uncertainties. The growth in large-scale offshore wind farms may also cause operational concerns. Formerly, tens and hundreds of turbines may have been distributed over a national area, offering the advantages of wind diversity and distributed reliability (see Section 5.3.1). Offshore installations may instead involve large (5+ MW) turbines concentrated within a small geographical area. It is probable that offshore wind conditions will be more uniform across the wind farm, leading to greater variability of output, while network, or other, faults may lead to the loss of more significant blocks of power.

Once these issues were the concern of island systems and/or those of small capacity only. The same difficulties now face much larger power systems, or those forming an interconnected grid system such as that in mainland Europe. For example, Denmark currently has a wind energy penetration level exceeding 20 per cent, although proposals exist for a 35 per cent contribution by 2015, and a 50 per cent contribution by 2030 (Sorensen *et al*., 2004). Spain has recently set a target of 20 GW by 2011 (up from a former target of 13 GW by 2011), equivalent to 15 per cent penetration, up from 6.5 per cent in 2004. In the Navarre region of northern Spain wind generation already provides 50 per cent of electrical production. In Germany, the federal government has declared that renewable sources shall provide 12.5 per cent of electricity needs by 2010 and 20 per cent by 2020. Similarly, in California a standard was passed in 2002 requiring energy from renewable sources to increase from 10 to 20 per cent by 2017.

It remains clear that beyond a certain level, significant levels of wind generation will impact on system dynamics and, therefore, must affect the manner in which generating units are scheduled and spinning reserve targets are set and apportioned across the system. In response to this changing environment many TSOs and DSOs are introducing new guidelines for the connection of wind power (EWEA, 2005; Jauch *et al*., 2005; Johnson and Tleis, 2005). In order that safe and reliable operation of the electricity network can be assured, *all* users of that network are required to fulfil the requirements of a grid code (see Chapter 4). The grid code assigns responsibilities to parties and regulates the rights of all those involved in generation and supply. Some grid codes provide unified requirements for all generators, while others treat wind generation, for example, as a special case. The general tone of these documents is that wind turbines should behave in a manner akin to conventional generation, for example, predictable and controllable real and reactive power output. Chapter 4 examined wind generation's potential support for reactive power and voltage control, required operation under network fault conditions (fault ride-through), and contribution to transient stability. Here, issues relating to load-frequency control, reserve provision and longer-term system balancing will be addressed.

The following section summarises the Irish system and current wind generation. This will provide useful background for the more generic approach of later sections, which will draw on examples based on the Irish experience of wind power integration to date.

5.3.2 Wind power in Ireland

The island of Ireland, comprising Northern Ireland and the Republic of Ireland, provides an interesting case study to quantify wind variability. Ireland is unusual in having an excellent wind resource, but limited interconnection with other power systems. Unlike Denmark or regions of Germany and Spain, with higher wind penetrations at present, Ireland cannot conveniently exploit external links for energy balancing and system support. The Irish TSOs therefore have to face, and solve, a number of operational and planning problems latent in other parts of the world.

The Northern Ireland power system, operated by Northern Ireland Electricity (NIE) in coordination with System Operator Northern Ireland (SONI), is a medium-sized network comprising 700,000 customers, with a system demand ranging from a summer minimum of 600 MW to a winter peak of 1,700 MW. The Republic of Ireland power system is owned by Electricity Supply Board (ESB) and operated by ESBNG, and the system demand for the 1,900,000 customers ranges from 1,700 MW in the summer to 4,850 MW in the winter. The main energy sources within the island are fossil-fuel generation: gas-fired plant, mainly CCGTs, represent approximately 48 per cent of island capacity, coal-fired plant accounts for 27 per cent, oil-fired plant 11 per cent, and peat-fired plant 4 per cent. The remainder includes one pumped storage station, providing 292 MW, and smaller-scale hydroelectric plant, CHP and wind generation (Barry and Smith, 2005).

Ireland is well exposed to southwest winds and storms blowing across the north Atlantic. Low-pressure cyclones (depressions) regularly pass over or near the country, with a period of 4–6 days, resulting in large variations in both wind speed and direction. Storm activity is more intense and frequent during the winter months. The available wind resource for Northern Ireland has been estimated at 106 TWh/year. However, considering physical/environmental constraints and economic viability, the *accessible* resource is just 8.6 TWh/year, equivalent to a capacity of about 3,000 MW (Persaud *et al.*, 2000). Likewise, in the Republic of Ireland, the *feasible* resource has been estimated at 344 TWh/year (ESBI and ETSU, 1997). Against this background, the EU directive for the Promotion of Electricity from Renewable Energy proposed renewable *energy* targets for electricity supply for individual countries within the European Union (European Commission, 2001). As part of the United Kingdom, which has an EU target of 10 per cent renewables penetration, the government target for Northern Ireland is 12 per cent by 2012. The comparable target for the Republic of Ireland is 13.2 per cent by 2010, corresponding to an installed capacity approaching 1,300 MW, of which approximately 1,000 MW could be obtained from wind generation.

For the period 1 January 2004 to 31 December 2004 metered data has been collected for the 46 operational wind farms on the island. Within the Republic of Ireland, metered data is available every 15 minutes, while for Northern Ireland the equivalent 30-minute data has been interpolated to generate 15-minute data. During the year, wind farm capacity expanded from 248 to 394 MW. Figure 5.8 depicts the geographical distribution of the wind farms in Ireland in 2004, expressed as the percentage contribution from each county. It can be seen that the wind farms, located generally in hilly areas, are well dispersed, although more concentrated in the northwest of the island. Also included is the Arklow Banks 25 MW (of a potential 520) offshore project, 10 km off the east coast of Ireland, and close to the major load centre of Dublin. Most of the existing wind farms have been connected to the distribution system, although connection at transmission level is becoming more common. The 2004 operational figure of 394 MW corresponds to a wind *energy* penetration level of about

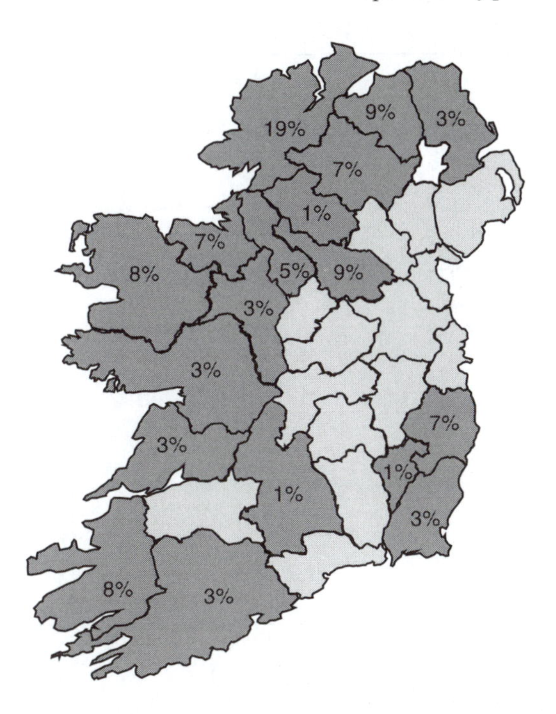

Figure 5.8 2004 county distribution of installed wind farm capacity

4 per cent for the island. 205 MW of capacity was installed in 2005, giving a total of 599 MW, and a further 450 MW with signed connection agreements. Applications to connect to the system exceed 2,000 MW, suggesting a potential island capacity approaching 3 GW (SEI, 2004).

5.3.2.1 System capacity factor

A convenient, and often used, measure for the available wind resource for a particular wind farm or region is the capacity factor. The capacity factor is calculated as the actual energy production over a given period divided by the maximum potential production over the same period. The capacity factor has been calculated for each quarter of the year (Fig. 5.9). Higher energy production is achieved during the autumn and winter months, with the capacity factor reaching a maximum value of 60 per cent during February (not shown), and lower output during spring and summer. Over a complete year, the average capacity factor for the island is 36 per cent, which for most countries would be considered high. In Germany and Denmark, onshore capacity factors in the range 25–30 per cent would be more typical. The existing wind farm locations have been selected, amongst other reasons, for their good wind regime and hence high capacity factor. However, it would be expected that newer wind farms will be located in less favourable locations, so that over time the average capacity factor will decrease. A counter argument is that future wind farms will be of

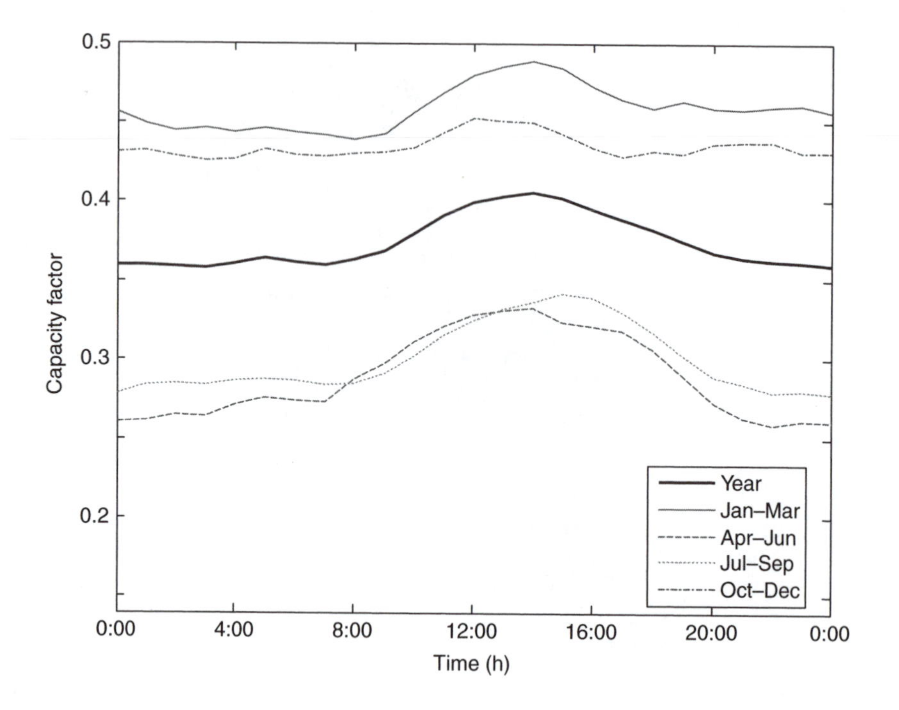

Figure 5.9 Wind daily capacity factor

improved aerodynamic design, increased hub height and (sometimes) sited offshore, all leading potentially to higher capacity factors.

The seasonal variation in wind power production is similar in shape to the seasonal load demand profile for Ireland (Fig. 5.10), such that wind generation tends to be high during periods of high demand (winter), and low during periods of low demand (summer). The peaks in the neighbouring systems occur at similar times, but differences in national holidays cause the summer minima to occur a few weeks apart. Similarly, the daily variation in capacity factor for the different seasons (Fig. 5.9) is characterised by a fairly constant, but low, capacity factor at night, rising to a mid-afternoon peak. Superimposed on this trend is the seasonal variation, which suggests greater variability during the spring and summer months. The correlation between the wind profile and system demand is, however, relatively weak, so that, for a particular day, it does not follow that high system demand will coincide with high wind generation. In general, though, wind farm production does tend to support generation needs at peak periods, leading to the possibility of a capacity credit for wind farms (Section 5.3.5). High wind generation at periods of low demand is also unlikely, lessening the need for wind curtailment (Section 5.3.6). This beneficial relationship is true for Ireland (and northwestern Europe), but is not always the case. For example, peak wind generation in New York State tends to occur at night (when demand is low), while during daylight hours wind generation tends to be less (Piwko *et al.*, 2005). Similar behaviour is seen in southern California, where wind production tends to fall

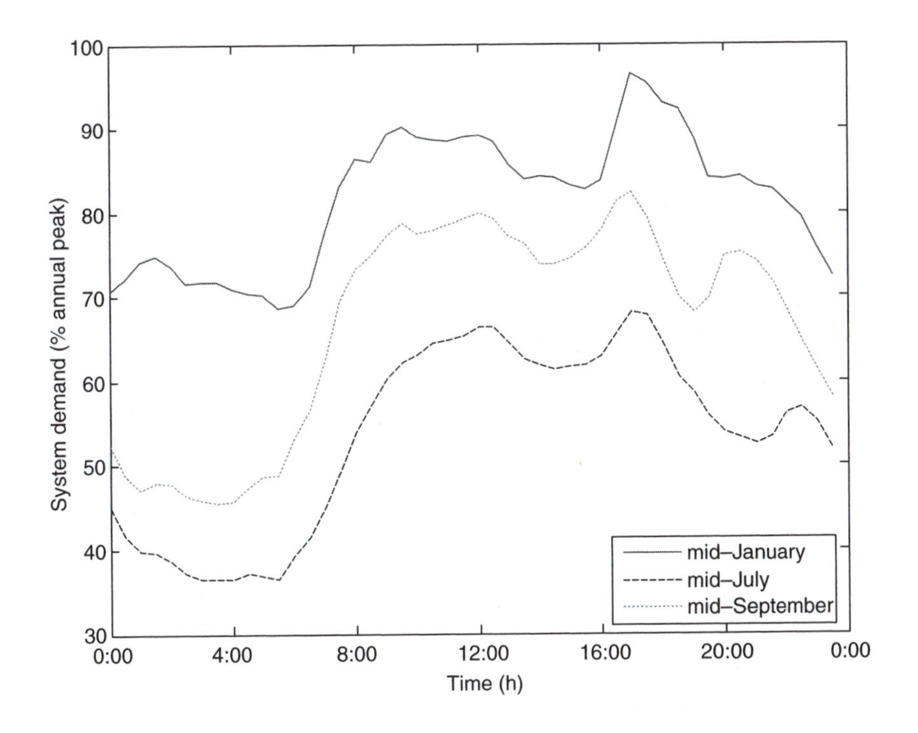

Figure 5.10 System demand seasonal profile

sharply during the morning before rising in the afternoon (Kahn, 2004). In contrast, in south Australia electrical demand and wind generation are almost uncorrelated, so that peak wind production is equally likely at times of high and low demand (AGO, 2003).

5.3.2.2 Individual wind farm variability

The energy production from wind farms varies on the time scales of seconds and minutes to months and years. Hence, an understanding of these variations and their predictability is essential for optimal integration. Figure 5.11 depicts the normalised output for an individual wind farm, utilising fixed-speed turbines, over the period of 1 minute. Local topography and weather patterns are dominant factors in determining wind variability, and for an individual turbine will affect both instantaneous wind speed and direction. In addition to the slow drift downwards in electrical output shown here, superimposed on the power output trace is a low-frequency oscillation, due to the variation in wind speed seen by the turbine blades. A combination of factors is involved: tower shadowing, wind shear and turbulence. The frequency of the oscillation will thus depend on the rotor rotational speed, and the number of blades, but is typically in the range 1–2 Hz. For fixed-speed turbines, these variations in mechanical input (wind speed) are directly translated into electrical output oscillations, which can reach 20 per cent of the average power production. In weaker

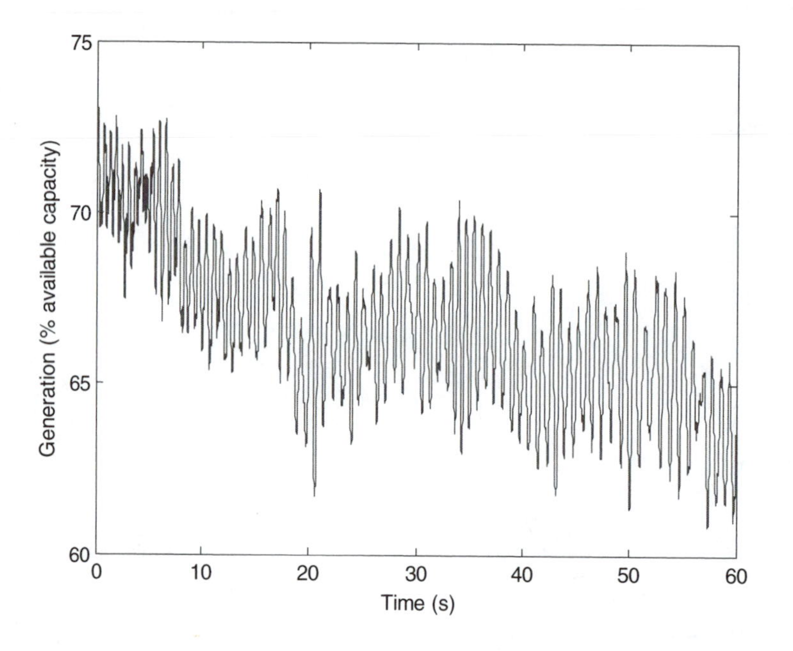

Figure 5.11 Single wind farm variability – 1 minute

networks, the power fluctuations can also cause voltage fluctuations, which may be perceived as light flicker. This is much less of a problem for variable-speed machines, since short-term variations in wind speed (gusting) can be buffered as changes in rotor speed through the rotating blade inertia. Short-term variations in wind speed will also introduce oscillations in power output, arising from the limited bandwidth of the pitch-regulation mechanism (Larsson, 2002).

Now looking over the period of one week (in June 2004), Figure 5.12 illustrates the variation in output of three *neighbouring* wind farms. Periods can be seen of almost zero output, over 100 per cent rated output, and often fairly rapid excursions between the two extremes. Although each characteristic is distinct, there is clearly a similarity in the observed power output from the three wind farms.

An understanding of the variability of production over this period can be gleaned from European weather maps for the period 22–25 June 2004 (Fig. 5.13). A low-pressure cyclone, beginning in Sole/Fitzroy, can be seen moving in a north-easterly direction towards Ireland on 22/23 June, before heading towards Scotland and Scandinavia on 24/25 June. Within the cyclone, the wind blows in a counter-clockwise direction around a region of low pressure, with stronger winds close to the centre, except for the central region itself where wind strength generally falls off. Figure 5.12 (wind farm production), taken with Figure 5.13 (weather map) on the morning of 22 June, shows low wind power output before the pressure system arrives. Over the next 24 hours, production increases, as the outer regions of the depression reach the wind farms. On the evening of 23 June, wind power temporarily decreases as the central region of the cyclone passes, before rising to a second peak as the trailing

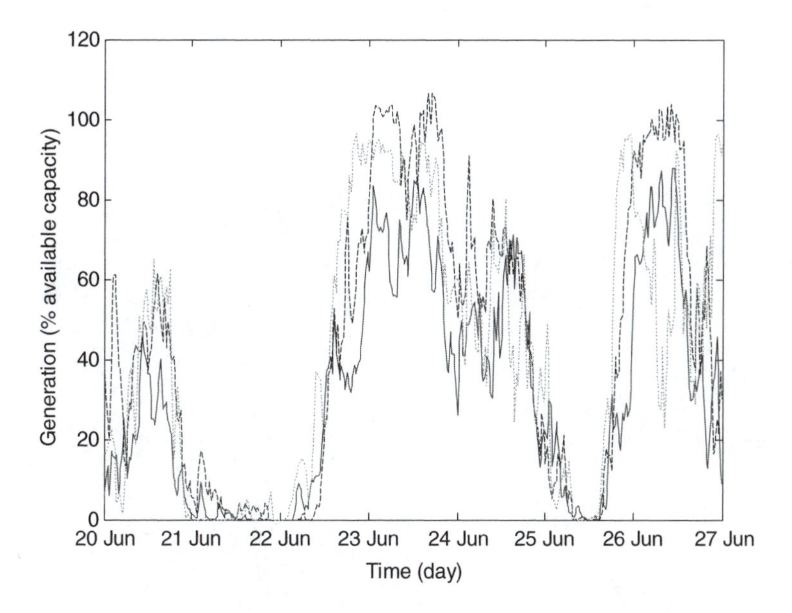

Figure 5.12 Single wind farm variability – one week (three wind farms)

edges of the cyclone begin to leave Ireland behind. There is a further lull in wind farm output before the next low-pressure region, in the mid-Atlantic ocean, reaches Ireland.

Wind variability, over different time scales, can often be of greater interest than the actual magnitude of the wind energy production, as this can place an additional load-following burden on conventional plant. Figure 5.14 illustrates the frequency, or probability, distribution of the observed power fluctuations for a single wind farm, over periods ranging from 15 minutes to 12 hours, using logarithmic scales for clarity. As discussed in Section 5.2 (and later in this chapter), the horizons considered reflect power system operational time frames. Up to 1–2 hours, wind variability can deplete/replace secondary and tertiary regulating reserve. The start-up and loading time for conventional plant depends on the recent loading duty of the unit, but is typically in the range 3–8 hours. Thus, wind variability 4, 8 and 12 hours into the future may affect scheduling decisions. This is in advance of unit commitment decisions, typically performed daily. The various distributions in Figure 5.14 are approximately Normal (Gaussian) with zero mean, and with a standard deviation that is a function of the time horizon. With increasing time horizons, the probability of 0 per cent variation in output is reduced, as would be expected, while the likelihood of variations in production exceeding 20 per cent of the available capacity are more prevalent beyond 1 hour.

The same information can be expressed much more informatively as a cumulative distribution graph (Fig. 5.15). For a lead time of 15 minutes, the magnitude of the wind power fluctuations is likely (>95 per cent probability) to be less than 20 per cent of wind farm capacity, and unlikely (>99 per cent probability) to exceed 30 per cent. A similar pattern is seen for 30 minutes, 1- and 2-hour lead times, except that the

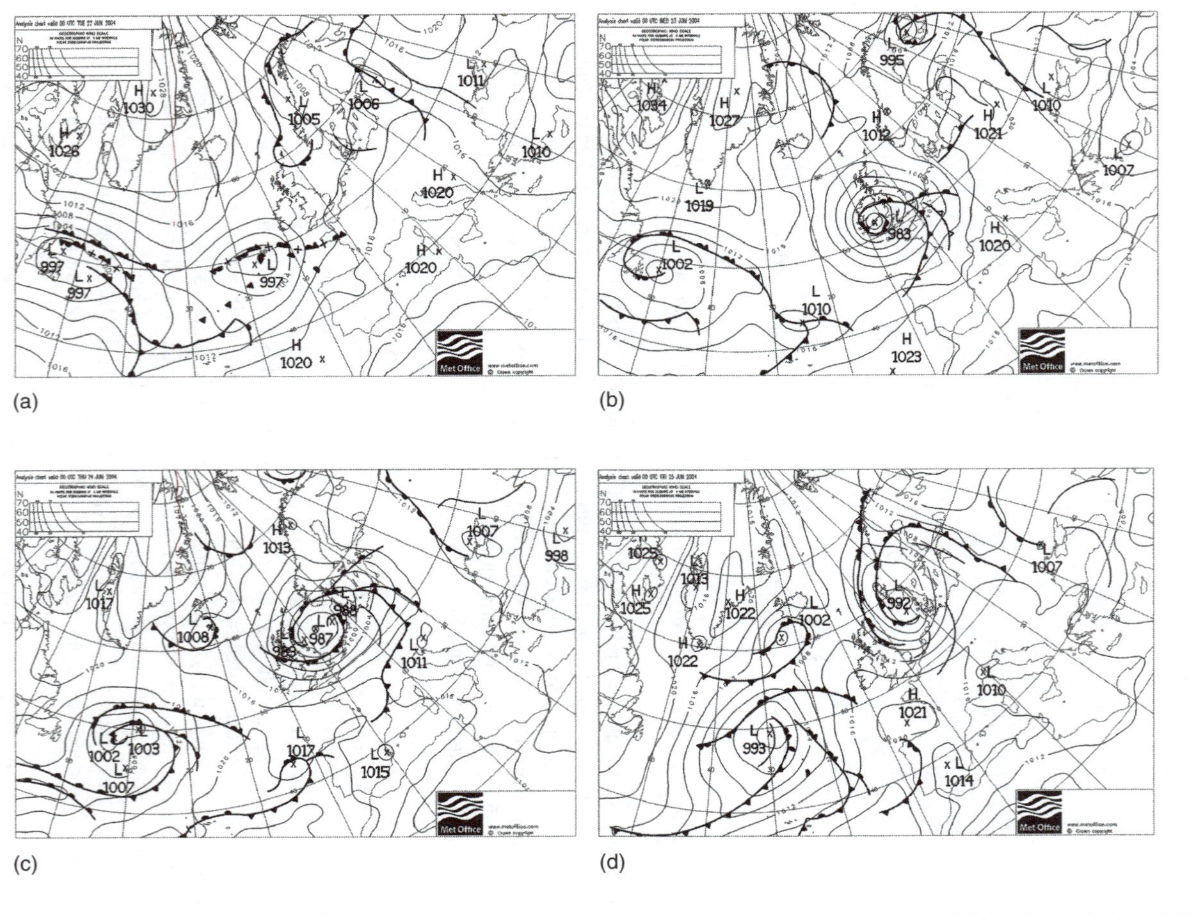

Figure 5.13 European weather maps 22–25 June 2004 (a) 00:00 UTC 22 June 2004 (b) 00:00 UTC 23 June 2004 (c) 00:00 UTC 24 June 2004 and (d) 00:00 UTC 25 June 2004

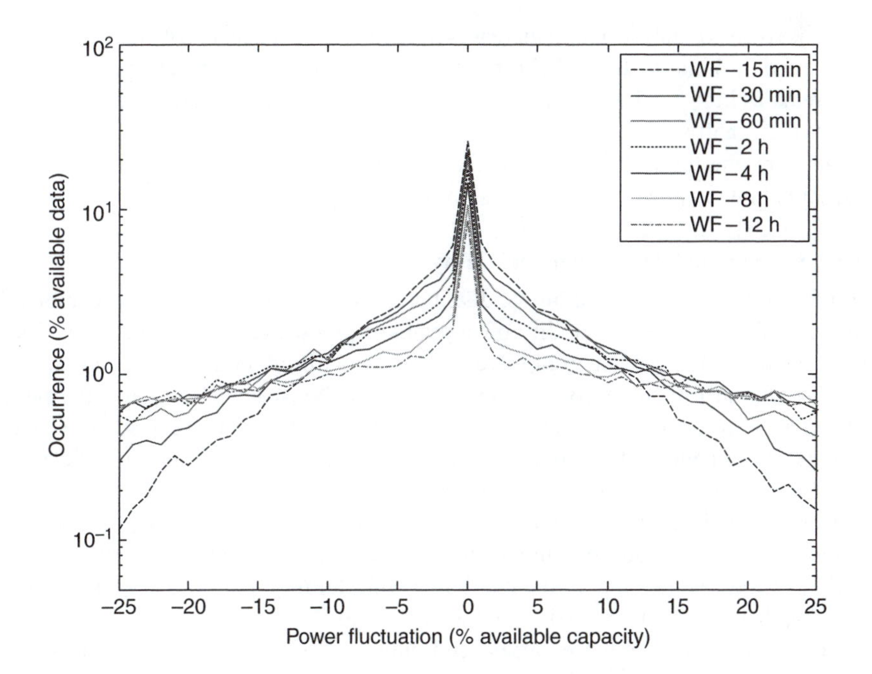

Figure 5.14 Single wind farm power fluctuations – variability

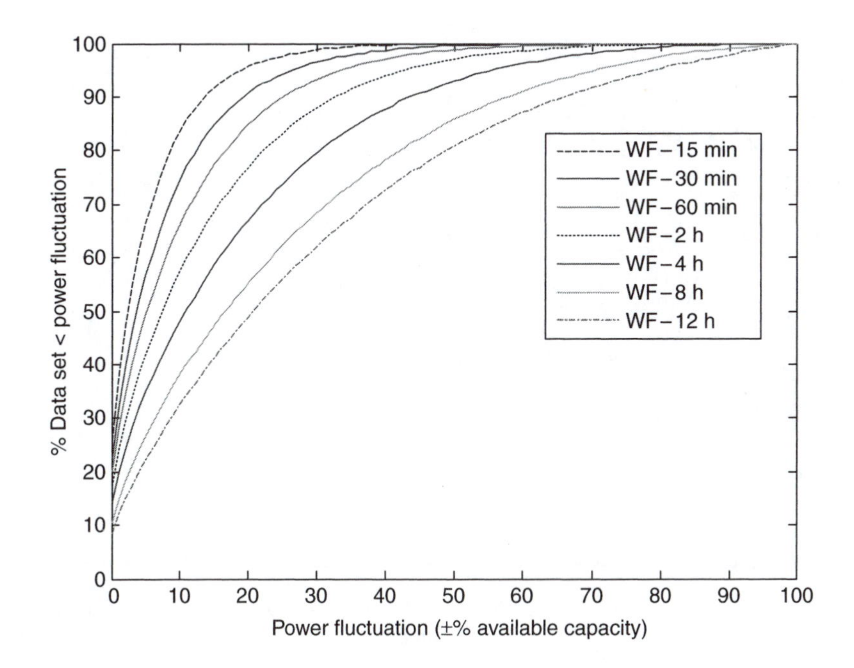

Figure 5.15 Single wind farm power fluctuations – cumulative distribution

most likely maximum variation increases from 26 to 34 per cent and 44 per cent. Over the period of 4 hours, the probability of a 20 per cent maximum variation in output still remains high at 67 per cent, but the likelihood of greater variations is noticeably increased. Finally, over a 12-hour period, variations in output of less than 10 per cent occur only one third of the time, while the probability of a 50 per cent or greater variation in wind power output almost exceeds 20 per cent.

5.3.2.3 Regional wind farm variability

If the output of several wind farms, distributed over a wide area, are now combined together, the observed variability of normalised production should decrease. Consider the case of a weather front travelling across uniform terrain at a speed of, say, 8 m/s (29 km/h). If a disturbance were to occur in the weather front, then the same disturbance would be felt 1 hour later 29 km away. So, by dispersing wind farms over a large area a degree of smoothing is likely to occur. Of course, the time delay in the disturbance will be reduced at higher wind speeds, and increased at lower wind speeds, but the benefit still remains. Similarly, within a wind farm, the physical separation between each turbine implies that they each *see* a slightly different (or time shifted) wind regime, and so the combined wind farm output is smoother and less variable than that from any individual wind turbine.

Figure 5.16 compares the normalised output of five equally sized wind farms within the Northern Ireland region with the averaged output (solid line) of all the wind farms. The distance between the wind farm locations ranges from 40 to 120 km,

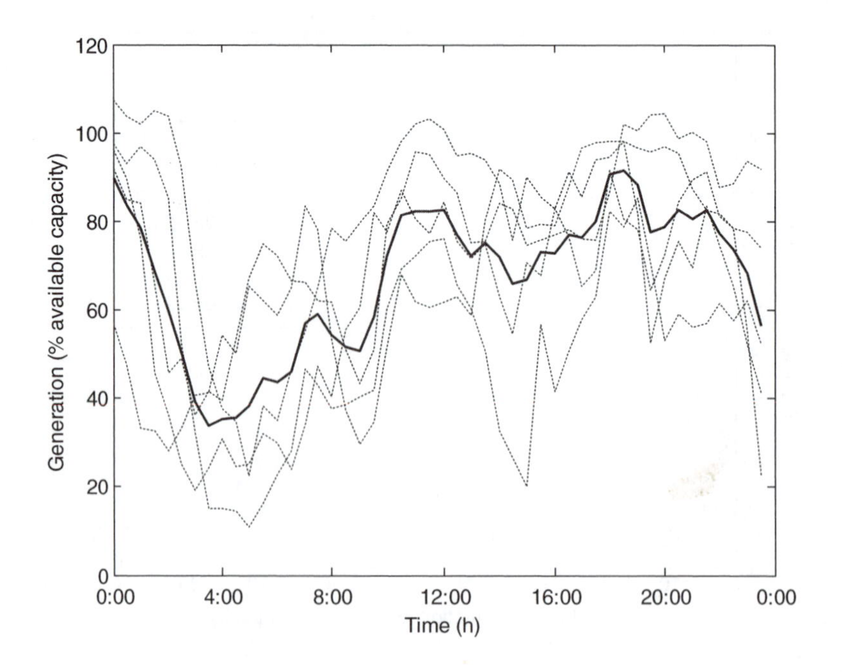

Figure 5.16 Five wind farms' variability – 24 hours

with an average inter-site distance of approximately 65 km. It can be seen that the combined output is much smoother than that of any individual wind farm. The standard deviation of the 30-minute variations for the individual wind farms ranges from 5.5 to 8.4 per cent, expressed as a percentage of the wind farm capacity. For the aggregated wind farm outputs, the standard deviation drops to 3.2 per cent, confirming the benefits of diversity. A similar study using measurements from Danish onshore and coastal wind farms investigated the wind power gradients (15-minute variation) for different topologies (Pantaleo *et al.*, 2003). For a 1,000 MW capacity, distributed at three sites, the variability dropped by 50 per cent, while assuming perfect geographical dispersion caused the ramping gradient to fall by a further 20 per cent.

Considering all the wind farms in the Northern Ireland zone, a cumulative distribution plot can be constructed, similar to Figure 5.15. By increasing the area of interest, short-term and local wind fluctuations will not be correlated and, therefore, should largely balance out. Consequently, the maximum amplitude of wind power fluctuations as seen by the power system should be reduced (Fig. 5.17). For a time delay of 30 minutes, the magnitude of the wind power fluctuations is most likely to be less than 7 per cent of wind farm capacity, and unlikely to exceed 11 per cent. Both figures are noticeably improved on the variability of a single wind farm. Similarly, over the period of 1 hour, 2 hours and 4 hours the most likely maximum variation within Northern Ireland is 13, 20 and 32 per cent of wind farm capacity. Even over a 12-hour period, the most likely maximum variation is 56 per cent of wind farm

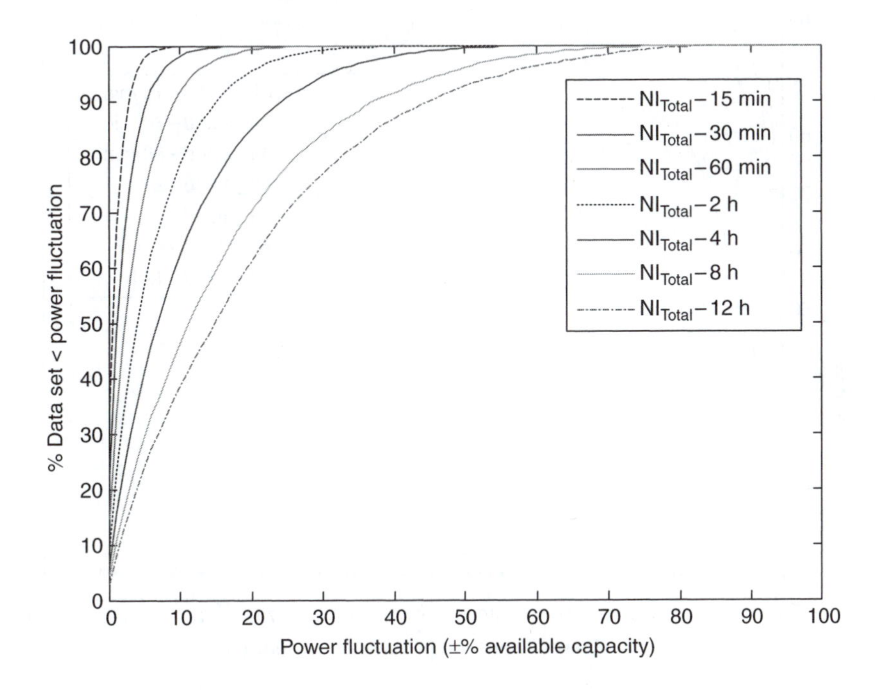

Figure 5.17 Northern Ireland wind power fluctuations – cumulative distribution

capacity, and the probability of a 30 per cent or greater variation in production is only 23 per cent.

5.3.2.4 All island variability

A cumulative distribution plot can now be created for the entire island (see Fig. 5.18). This looks similar in shape to Figure 5.17 for the Northern Ireland region, and Figure 5.15 for an individual wind farm, except that the normalised variability is further reduced. For a time delay of 60 minutes, the magnitude of the wind power fluctuations is most likely to be less than 8 per cent of wind farm capacity and unlikely to exceed 11 per cent, both comparable with the 30-minute variations for the Northern Ireland region. Over 2- and 4-hour periods the most likely maximum variation is 15 and 25 per cent of wind farm capacity, with the combined wind farm output unlikely to ever vary more than 20 and 32 per cent over the same periods. Even over a 12-hour period, fluctuations exceeding 50 per cent occur less than 5 per cent of the time, while two thirds of the time, the same variations are less than 20 per cent of wind farm capacity.

These results confirm that a sudden collapse (or indeed rise) in wind power generation across the island is most unlikely. One exception to this rule could be an advancing storm front that could cause turbine high wind speed protection to activate, disabling production across the network. Within Ireland excessive wind speeds are very rare and

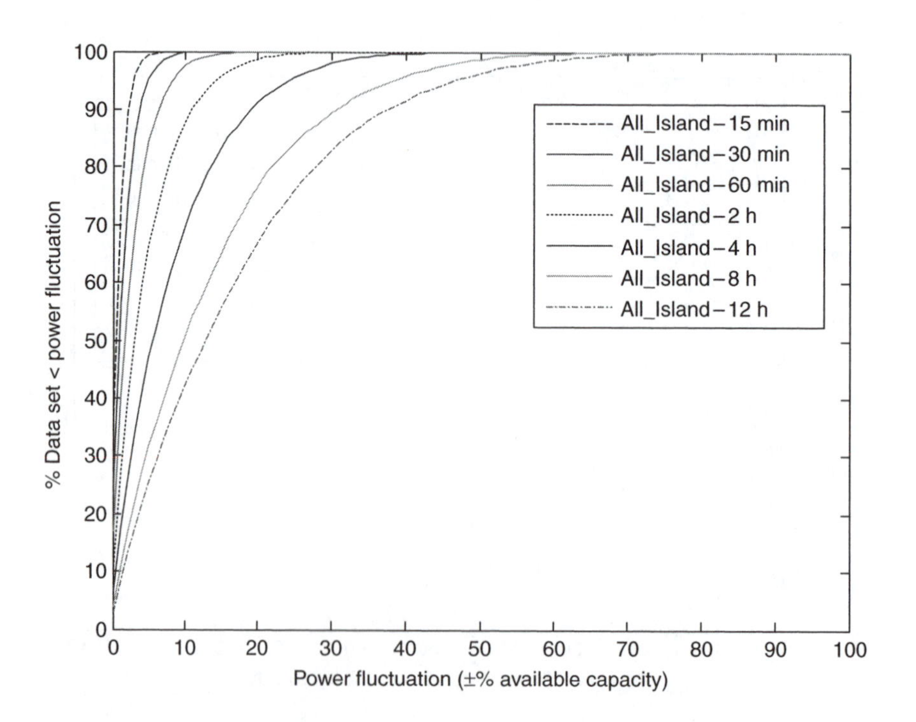

Figure 5.18 Ireland wind power fluctuations – cumulative distribution

likely to be localised, so that such a scenario is inconceivable. However, as discussed below, various measures are available to limit short-term wind variability.

The variability benefits in the transition from the Northern Ireland region (14,000 km^2) to the entire island (84,000 km^2) are less dramatic than might be expected, given the increased land mass. This is mainly due to the concentration of wind farm sites in the northwest of the island, a region which is partly in Northern Ireland and partly in the Republic of Ireland (Fig. 5.8). A contrary factor, however, is that most wind generation in Northern Ireland is of the older fixed-speed induction machine type, sensitive to changes in wind speed, while in the Republic of Ireland DFIG (doubly fed induction generator) machines are more common. Newer wind farms are likely to be clustered near existing ones, so the benefits of geographical diversity will begin to wane. For Ireland, it has been estimated that most of the benefits of diversity will have been achieved with an installed wind farm capacity of 850 MW (SEI, 2004). One factor which could help push this threshold higher is a greater penetration of offshore generation – in general, the correlation in output between two onshore sites is higher than that between an onshore and offshore wind farm. Where appropriate, similar benefits could be achieved in siting wind farms in a variety of terrain and subject to differing weather patterns (e.g. hills, coasts, deserts).

An alternative method of quantifying wind variability is to determine the frequency distribution of partial wind power production. Figure 5.19 shows the load duration curve for a single wind farm, the Northern Ireland region and the entire island. All three curves are broadly similar in shape, and only differ in the low and high extremes. Considering first a single wind farm, there is a quantifiable period

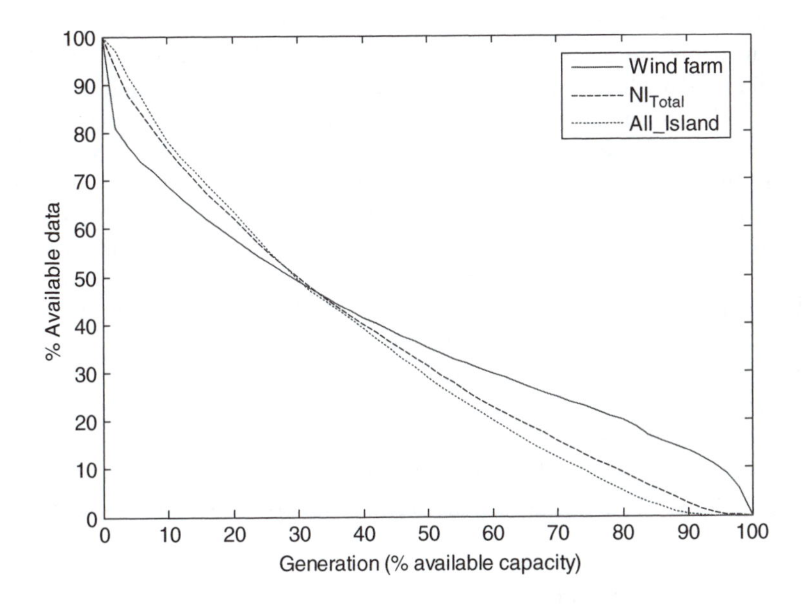

Figure 5.19 Load duration characteristic – single wind farm, Northern Ireland region, all island

of time, approximately 20 per cent, when the total production is almost zero, either due to maintenance outages or becalmed conditions. Similarly, for 2 per cent of the time the wind farm is operating at maximum output, since above rated wind speed the electrical output is effectively curtailed. The addition of further wind farms has the effect of flattening the load duration curve and removing the extremes. Total wind production equalling 100 per cent of available capacity implies that all wind farms must be operating at 100 per cent output. Thus, considering the Northern Ireland region, the wind farms occupy 32 per cent of their time at 50 per cent or more output, and no time above 95 per cent output. For the entire island the trend continues, so wind production exceeds 50 per cent of available capacity for 29 per cent of the time, and never exceeds 92 per cent. At the same time, total wind production across the island exceeds 10 per cent of capacity for 78 per cent of the time, while the equivalent figures for Northern Ireland and a single wind farm decrease to 76 and 70 per cent, respectively.

The system demand, like wind power production, is also a highly variable quantity, and it is equally informative to plot the cumulative distribution of variations over different time scales, ranging from 15 minutes to 12 hours, akin to Figure 5.18. In this case, the variation in load, over a particular time horizon, has been determined as a fraction of the average system demand over that time. As expected, increasing the time horizon of interest results in a greater likelihood of increased variation. For example, over 15- and 30-minute periods fluctuations in demand tend to be random and uncorrelated – 99 per cent of variation lies within 6 and 11 per cent of system demand, respectively. Particularly over the time scale of 4–8 hours the probability of significant variation is quite high, with 30 per cent variation in system demand occurring 17 (4 hour) and 42 per cent (8 hour) of the time. However, this variation is due mainly to the daily morning rise and evening fall in demand pattern. System demand can be predicted with high confidence (1–2 per cent error), and, as discussed in Section 5.2, generating units are scheduled to cope with this variation.

Comparing Figure 5.18 (wind power fluctuations) and Figure 5.20 (system demand fluctuations) suggests initially that large-scale variation in wind production across the island is much more frequent than large-scale variation in system demand. However, assuming an average annual system demand of 3,600 MW and an average installed wind farm capacity of 300 MW over the year, enables the relative (%) variations to be approximated to absolute (MW) variations. Over 15 minutes, 90 per cent of system demand variation is less than 120 MW, and 90 per cent of wind variation is less than 6 MW. Similarly, over 60 minutes, 4 and 12 hours, 90 per cent of the expected variation in system demand/wind production are 390/20, 1,200/57 and 1,950/120 MW, respectively. Clearly, the system demand variations dominate. In particular, short-term (15–30 minutes) variations in wind production can deplete regulating reserve, which will be increasingly evident with increased wind penetration levels.

5.3.2.5 Regional variability management

Interconnection between the Northern Ireland and Republic of Ireland systems is limited, consisting of two 275 kV parallel circuits rated at 600 MVA and two additional

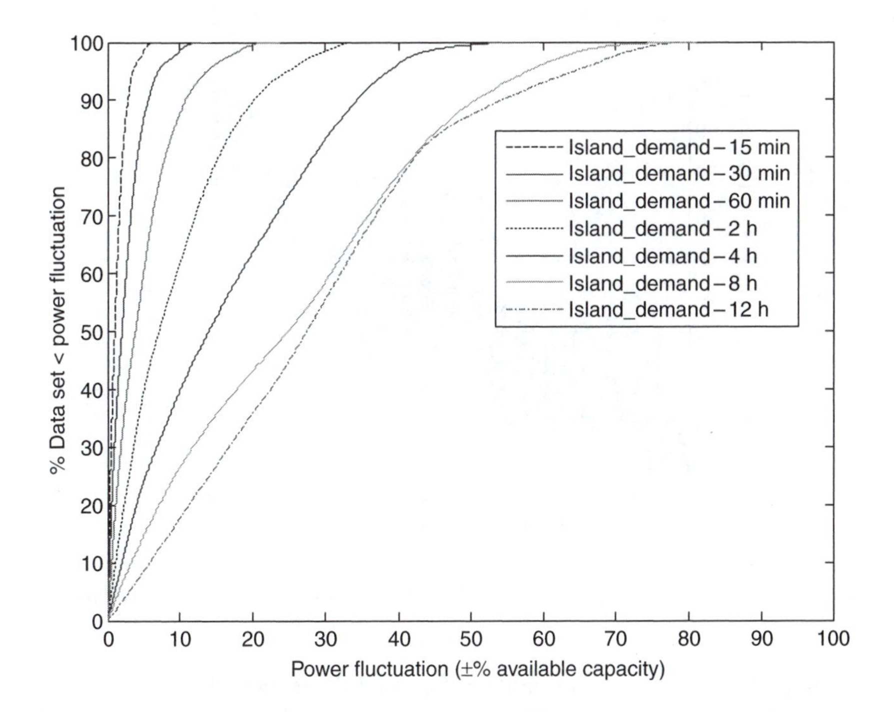

Figure 5.20 Ireland system demand fluctuations – cumulative distribution

110 kV lines, providing a combined rating of 240 MVA. A 500 MW high-voltage direct current (HVDC) interconnector also ties the Northern Ireland power system to Scotland, although the link is primarily configured to provide energy rather than ancillary services. It is, therefore, of interest to assess whether cross-border *wind* transfers can reduce the variability seen within the respective regions. Figure 5.21 shows a scatter plot of the Northern Ireland wind production plotted against the Republic of Ireland wind production. It can be seen that there is a strong correlation in output (92 per cent at a lag time of 15 minutes) between the two regions, suggesting that interconnection provides minimal benefit here. This is probably because of the high concentration of wind farms in the northwest of Ireland, accounting for 50 per cent of the wind penetration. Part of this area is located in the Republic of Ireland and the remainder in Northern Ireland, hence accounting for the high overall correlation and minimal time lag between the two regions. Greatest diversity in output between the two regions is seen at mid-range output, coinciding with mid-range wind speeds, and in accord with the sensitivity of the turbine power curve (Fig. 3.3).

The benefits, or otherwise, of interconnection as a wind variability balancing mechanism can be further assessed by determining the potential net energy transfer between the two regions. Consider that wind output in the Northern Ireland region increases by 10 MW during a 15-minute period, and that wind production in the Republic of Ireland decreases by 6 MW over the same time period. If the two regions were not electrically connected the total variability would be $10 + 6 = 16$ MW.

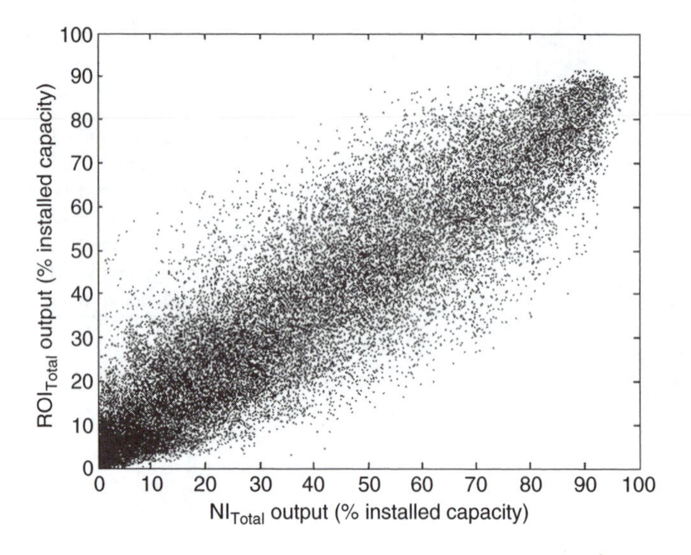

Figure 5.21 *Northern Ireland versus Republic of Ireland wind power variability – scatter plot*

With interconnection, a north-south transfer of 6 MW reduces the net variability to only 4 MW. Figure 5.22 shows the necessary power transfers to minimise variability across the island, for time horizons of 15 minutes, 4 hours and 12 hours. Over 15 minutes, power transfers of only 2–3 MW in either direction are necessary, which further increases to 7–10 MW up to 12 hours ahead.

The cumulative distribution of these variations is shown in Figure 5.23. Looking 15 minutes ahead, interconnection provides a 50 per cent reduction in variability for approximately 20 per cent of the time and a 90 per cent reduction for only 4 per cent of the time. Similarly, over a 12-hour period, a 50 per cent reduction in variability is achievable 8 per cent of the time, and a 90 per cent reduction for 2 per cent of the time. Although, in this case, there is limited potential for interconnection to reduce island variability, an alternative perspective is that reserving, say, 1 per cent of interconnector capacity would provide much of the available benefit. Direct current (DC) interconnection already exists between Northern Ireland and Scotland, while a similar connection between the Republic of Ireland and Wales has been proposed. However, even though, conceivably, a larger geographical area could be defined uniting Ireland and Great Britain, initial analysis suggests that the benefits in terms of reduced wind variability are limited – when wind production in Ireland was less than 10 per cent of available capacity, wind production in Great Britain was also less than 10 per cent of capacity for 94 per cent of the time (Gardner *et al.*, 2003).

5.3.3 System operation and wind variability

As distinct from wind forecasting, and the desire to predict wind profiles accurately over a desired time period, wind output will also exhibit variability. Even if the

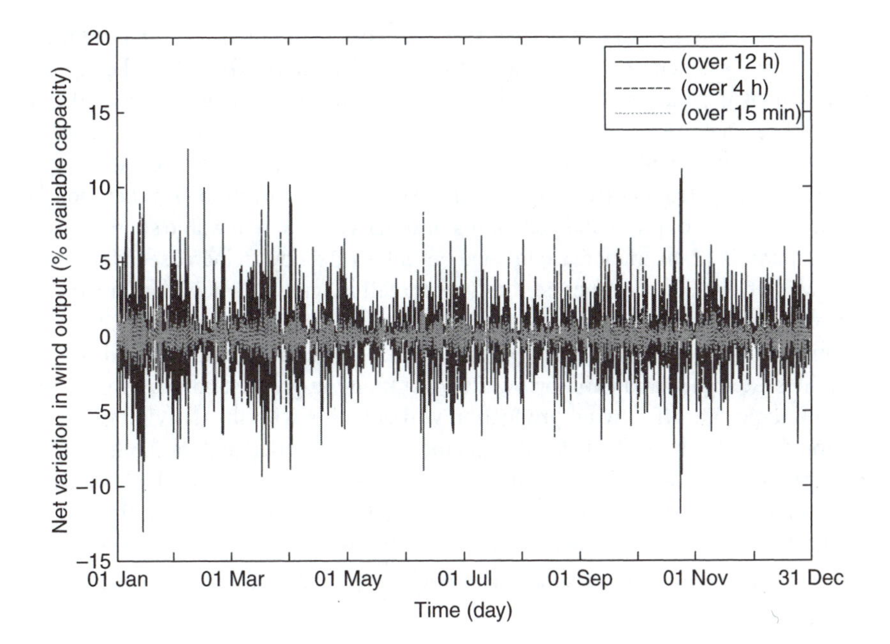

Figure 5.22 Northern Ireland/Republic of Ireland net wind power transfers

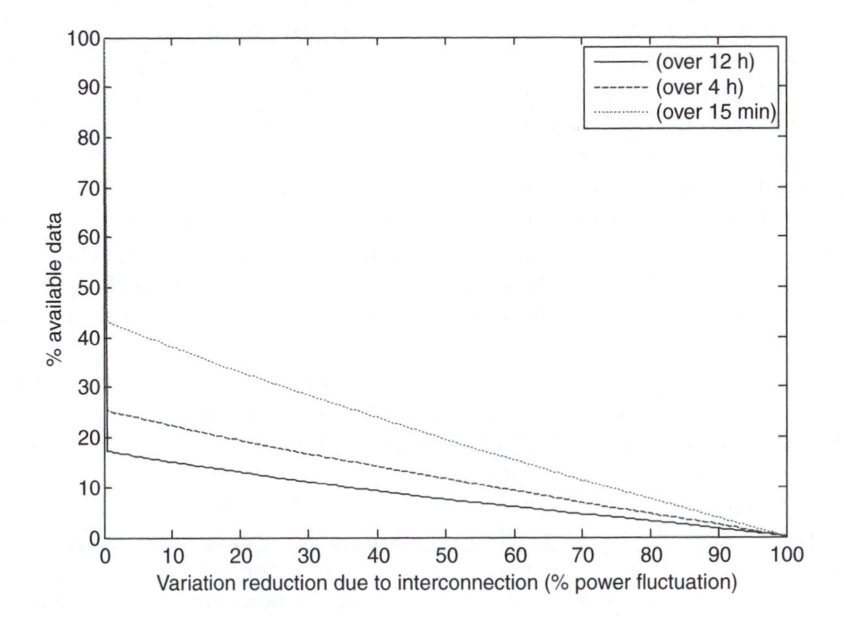

Figure 5.23 Northern Ireland/Republic of Ireland net wind power transfers – cumulative distribution

wind output could be forecast perfectly, the task of operating the power system with a time varying generation source would remain. This may affect the scheduling of conventional generation units and the requirements for spinning reserve and other ancillary services. At a system level, it is probably of greater interest to consider the net demand variability rather than the wind variability. The net demand may be defined as the system demand requirement to be met by conventional generation, the rest being provided from wind generators or other variable renewable sources. There may be periods when an increase in wind output is offset by an increase in the system load, while there may be other periods when a fall in wind output is reinforced by an increase in demand. Analysis of data from western Denmark suggests that most of the time (52 per cent) wind power production rises with the system demand and falls with the system demand (Milborrow, 2004). Often the greatest concern for a utility is that a large fall-off in wind production will coincide with the daily morning rise. Section 5.3.2 illustrates that, for the Ireland system, net demand variability is very similar to demand variability for current wind penetration levels. Similar results have been obtained for other power systems (Hudson *et al.*, 2001). This follows, since the short-term variance of the net demand variations, $\sigma^2_{\text{net demand}}$, is given as the sum of the variances of the wind time series and system demand time series, σ^2_{wind} and $\sigma^2_{\text{system demand}}$,

$$\sigma^2_{\text{net demand}} = \sigma^2_{\text{wind}} + \sigma^2_{\text{system demand}}$$

This assumes that they are statistically independent processes. Since the system demand is dominated by the morning rise and evening fall-off in load, the net demand variance will be dominated by the system demand term until wind penetration reaches significant levels. However, while features such as the weekday morning rise introduce high variability, they are also largely predictable. For example, in Ireland between 6 and 9 a.m. the morning rise is approximately 1,200 MW in winter and 1,100 MW during the summer, representing almost 50 per cent rise in demand (Fig. 5.10). Recognising this daily pattern, generating units can be scheduled to come online at appropriate times during the day, and later switched off in the evening as the demand falls. Despite continuing improvements in forecasting techniques, the same prediction confidence is lacking for wind power production, as will be seen in Chapter 6.

Before integrating wind generation into power system operation it is important to understand the likely variability of production. Section 5.3.2 provided a brief analysis of wind variability on the Ireland power system. Considered over a wide enough area, wind power does not suddenly *appear* or *disappear*, even during storms or severe weather conditions. Instead, studies for many countries, and groups of countries, around the world have shown that (probabilistic) limits can be placed on the likely variability and maximum variability over different time horizons. The larger the area considered, the more gradual will these transitions in wind production be, and the smaller the impact on system operation.

Variation in output on the time scale of tens of seconds up to tens of minutes will tend to be small, due to the averaging effect of individual turbines and individual wind farms *seeing* slightly different wind regimes. The greater the network area under consideration and the larger the inherent load fluctuations, then the less is the impact (if any) on demand variability (Dany, 2001). In Denmark and Germany the maximum wind gradient per 1,000 MW installed wind capacity is 4 MW/minute increasing and 6.5 MW/minute decreasing (Milborrow, 2004). In comparison with typical load variations, these ramp rates are usually insignificant, for example, in extreme conditions 6.5 MW of regulating reserve per 100 MW installed capacity would be required within Germany to maintain system balance, resulting from a reduction in wind production. Significant depletion of reserves could occur if the maximum ramp rate was sustained (30 minutes × 6.5 MW/minute = 195 MW), but the probability of this occurring is negligible. Secondary reserve, in the form of part-loaded thermal generation, may be called upon to replace *consumed* primary reserve. Instead, during this time period, primary reserve requirements are dominated by the need for *fast* reserve following the loss of a conventional generator (see Section 5.2). A number of studies have shown that wind generation has minimal impact on these short-term reserve requirements (DENA, 2005; SEI, 2004).

In the time period 15–60 minutes, wind variability becomes more significant, and, as discussed in Chapter 6, wind variations in this time frame are not easy to predict – persistence methods are generally more successful than meteorological approaches. The greatest variation in output is seen with turbines providing between 20 and 80 per cent of rated power, as illustrated by Figure 5.24, modified from Figure 3.3, operating on the steep part of the power curve. If the wind speed is *low*, and below the cut-in speed for most wind turbines, then a small change in wind speed will have no appreciable effect on wind power production. Similarly, if the wind speed is *high*, with most wind turbines operating close to rated output, then a small variation in wind speed will also cause minimal change in wind power production, since, depending on design, pitch- or stall-regulation of the turbine blades will ensure that equipment ratings are not exceeded. However, at *mid-range* wind speeds, then the cubic relationship between wind speed and power production implies that a small variation in wind speed will cause a relatively large change in electrical output.

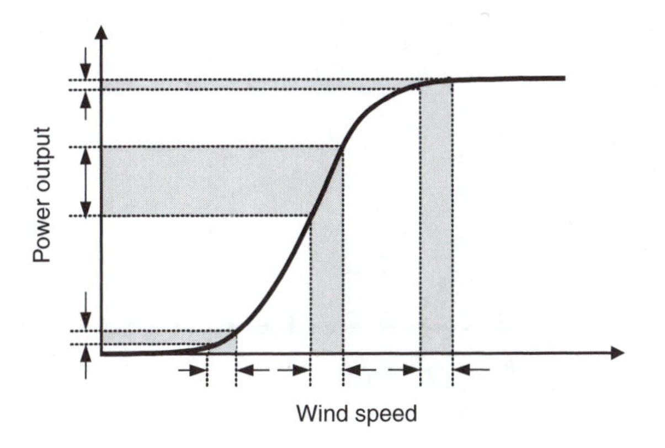

Figure 5.24 Wind turbine power curve

Geographical dispersion again provides a smoothing effect, so, for example, in Ireland only a maximum 8 per cent variation in output could be expected within a 1-hour period (Fig. 5.18). Similarly, in the United Kingdom, variations less than 2.5 per cent of the total installed wind capacity are most probable from 1 hour to the next – 20 per cent changes in hourly output are only likely to happen once per year (ECI, 2005). As the area under investigation further increases in size, the variability tends to decrease. In Germany, for example, the maximum hourly variation in output rarely (<0.01 per cent) exceeds 20 per cent of the wind farm capacity, with a standard deviation for these deviations of only 3 per cent (Gardner *et al.*, 2003). For the four Nordic countries (Norway, Sweden, Finland and eastern Denmark – forming part of the Nordpool system) the largest hourly changes will decrease to approximately 11 per cent of the installed capacity, although for 98 per cent of the time the variation decreases to 5 per cent of installed capacity (Holttinen, 2005a).

Should wind power penetration reach 5–10 per cent, the wind variations become comparable with random, short-term demand variations (see Figure 5.20). Concern may arise not only from the magnitude of the variability, but also the rate of change, and hence the dynamic requirements placed upon the conventional generation (Fig. 5.25). There will thus be a requirement for extra regulating/secondary reserve – typically somewhere between 2 and 10 per cent of the installed wind power capacity for a 10 per cent wind penetration (Holttinen, 2005b; ILEX Energy Consulting and Strbac, 2002; Parsons *et al.*, 2004; SEI, 2004). For example, in the United Kingdom, assuming a 10 per cent wind penetration, reserves equivalent to 3–6 per cent of the wind capacity were deemed necessary (ILEX Energy Consulting and Strbac, 2002). This figure rose to 4–8 per cent of the wind capacity, assuming a 20 per cent wind penetration. Similarly, for the Ireland system the secondary reserve target should increase by 5–6 per cent, assuming 1,300 MW installed wind capacity

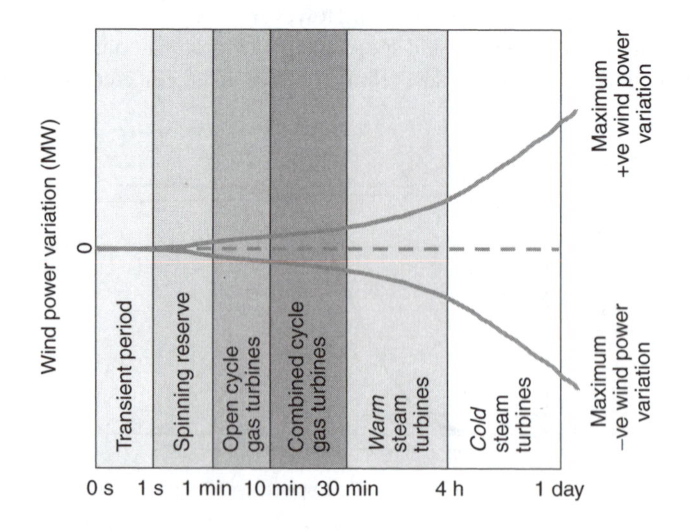

Figure 5.25 Wind variability versus power system time scales

($\approx$10 per cent wind penetration), and by 9–10 per cent for 1,950 MW wind capacity (SEI, 2004). The wide variation in reserve requirements is dependent on the inherent flexibility of the plant mix, and the interaction between wind-forecasting time horizons and electricity market operation (see Section 5.3.4 and Chapter 7). Clearly, the regulating impact and cost will be less for a hydro-based power system than for a (part-loaded) fossil-fuelled/nuclear-based power system. Market arrangements may also range from hourly predictions to day-ahead forecasting, greatly affecting the errors associated with the forecast, and hence the need for both secondary and tertiary/replacement reserve. At low wind penetration levels hourly wind variability can often be handled by the existing plant commitment (no extra costs). Increased part-loaded plant and dedicated reserves may be required at higher penetration levels, leading to extra costs, but this should be obtainable from the existing generation. For example, the DENA study in Germany, projecting forward to 2020, envisaged no need for additional *balancing* plant to regulate wind variability (DENA, 2005). Assuming day-ahead wind forecasting, the required day-ahead reserves, available in the form of intra-hourly primary and secondary reserve, were estimated for a 2015 scenario with 14 per cent wind penetration. The required *positive* regulation was on average 9 per cent of the installed wind farm capacity, with a maximum value of 19 per cent. The corresponding *negative* regulation was 8 per cent on average, with a maximum requirement of 15 per cent. This compares with 2003 average requirements of $+9/-5$ per cent of installed wind farm capacity, and maximum requirements of ±14 per cent, when wind penetration was 5.5 per cent. For the Nordic power system, the hourly reserve requirements were estimated conservatively to increase by only 2 and 4 per cent for wind penetration levels of 10 and 20 per cent, respectively (Holttinen, 2005b). Clearly, the Nordic power system is more widespread than that of Germany (reducing variability), while market operation enables party intervention 1 hour ahead only, mitigating forecasting errors.

For isolated power systems, such as Ireland, which are small in terms of both geographical area (and hence high variability) and electrical demand requirements (diminished system flexibility), there *may* be encouragement for greater exploitation of fast-starting OCGTs and diesel generating sets. Even with the advent of 40 per cent efficient OCGTs, the economic and environmental arguments are not straightforward. Wind curtailment and ramping limits are alternative options.

Variations in wind production 1–12 hours ahead are significant because they can affect the scheduling of conventional generation. Wind variability should be compared with the cold-start times of conventional generation, which may require 6–10 hours before operating at full capacity. In Ireland, over 4-hour and 12-hour periods, the maximum variations are 64 and 100 per cent of wind farm capacity (see Fig. 5.19). In Germany, the maximum variation 4 hours ahead is 50 per cent of installed wind capacity, rising to 85 per cent looking 12 hours ahead. In the Nordic area, even for the longer time horizon, the maximum variation will only be 50 per cent (Holttinen, 2005a). Here, however, wind-forecasting tools can be of assistance, although the uncertainty of the prediction tends to increase with the time horizon. Uncertainty can also be associated with the demand prediction, but again the magnitude of this is likely to be much less. Hence it is not the variability of the wind over such time horizons that causes scheduling difficulties, but the errors associated with the wind forecasts.

A scenario that has concerned system operators is the passage of a storm front, where wind speeds exceed the cut-out speed, resulting in a shutdown from full output to zero within minutes. In practice, this is only ever likely to occur for small geographical areas. The distributed nature of wind generation implies that shutdown will take place over hours and not instantaneously. In the United Kingdom, for example, high wind speeds (>25 m/s) are extremely rare, with a probability less than 0.1 per cent at most sites. There has never been an occasion when the entire United Kingdom experienced high winds at the same time (ECI, 2005). In fact, the windiest hour since 1970 affected around 43 per cent of the United Kingdom – an event expected to occur around 1 hour every 10 years. Similarly, on 8 January 2005, a storm all over Denmark resulted in wind speeds over 25 m/s, and a reduction in wind generation approaching 2,000 MW, but over a period of roughly 6 hours (Bach, 2005). Of course, there are parts of the world where the wind resource is sufficiently good to introduce its own problems. In New Zealand, for example, sites with average wind speeds in the range 10–12 m/s are common, and wind speeds can regularly exceed 25 m/s every three to four days on average (Dawber and Drinkwater, 1996). More generally, the expected growth of offshore wind generation also suggests that the existing benefits of dispersed (onshore) generation may be degraded. Without intervening control actions, discussed below, generation from large-scale offshore sites may be lost within about an hour during severe storm conditions.

5.3.3.1 System interconnection

A convenient method of reducing regional wind variability is through system interconnection. Of immediate interest, synchronised interconnection permits aggregation of loads and generation over a wide area, increases significantly the rotating inertia of the system during severe transients and reduces the reserve burden that individual regions must carry. It is, therefore, believed to follow that large-scale integration of wind power will be easier than for isolated, or asynchronously connected, power systems. The logic proceeds as follows: wind variability decreases as the area of interest increases and the outputs of individual wind farms are aggregated together. So, while Ireland may possess significant variability in wind power production, consideration along with Great Britain reduces the effective variability. Applying the same process further, for example to mainland Europe, results in a resource of increasing time invariance – the wind always blows somewhere! Hence it is asserted that, with more interconnection, wind power can be *wheeled* from areas of high production to areas of low production.

Denmark is a prime example of the benefits, being a member of the UCTE European grid network. Transmission links are in place with neighbouring countries such as Germany, Norway and Sweden, which have enabled Denmark to achieve 20 per cent wind penetration (Lund, 2005). In *western* Denmark, wind penetration is approximately 25 per cent, and instantaneous wind generation has reached 70 per cent for the entire country. About 60 per cent of the remaining generation is provided by

CHP plant, much of which operates according to the *heat* demand and time-of-day tariffs, rather than *electrical* demand requirements. It therefore falls upon the small fraction of conventional generation to provide grid stabilisation and reserve duties. As a result, Denmark uses its external links to both export its increasing surplus of electricity production and import spinning reserve capability (Bach, 2005).

For a number of reasons, however, an expansion of system interconnection cannot solve the issues surrounding wind variability. Considering the example of the European UCTE network, power trading occurs between national grids, as defined by contracts agreed over 24 hours in advance. Transmission capacity will be reserved well in advance to meet these agreements. Automatic generation control (see Section 5.2) is applied within control areas to ensure that cross-boundary power flows are as specified. The temptation may exist to exploit the interconnection capacity to *spill* excess wind generation or *fill* wind generation shortfalls. Assuming that scarce interconnector capacity would even be made available to accommodate wind power imbalances, penalties are likely to be imposed for not complying with agreed power exchanges. It will probably be more economic to spill excess wind power when transmissions networks are congested, rather than construct additional interconnection capacity to access the small amount of additional energy involved. In Denmark, for example, it is realised increasingly that relying on external sources for regulating power can be risky and expensive, while limiting the export transfer capability. Domestic sources of reserve have been examined, with the focus on centralised CHP units to decrease electrical production during periods of high wind, and use of small-scale distributed CHP plant for load-following duties (Lund, 2005).

Interconnection between neighbouring countries is often limited, and the expansion in wind farm sites in recent years has not generally been matched by an increase in interconnection capacity. Northern Germany, for example, already has a high concentration of wind farms – the state of Schleswig-Holstein supplies 30 per cent of its needs from 1,800 MW of wind installations. This will only increase further with the adoption of Germany's ambitious plans for offshore sites (DENA, 2005). Thus, interconnectors to the Netherlands, France, western Denmark and Poland can often be overloaded. Similar difficulties exist between Scotland, with its rich wind resources, and England. Here export southwards can be limited by the existing Scotland-England interconnector and the north of England transmission network. Ireland presents a contrasting example. Due to the concentration of wind farms in the northwest, interconnection provides minimal opportunity for *wind* power flows. Of course, island systems can also be adversely affected by limited interconnection. Sardinia, located off the west coast of Italy, but electrically joined via Corsica to the mainland by a 300 MVA submarine cable, is expected to experience significant growth in wind generation (Pantaleo *et al.*, 2003). The existing base-loaded, tar-fired plant have minimal regulating capability. In the future, when load-following requirements are high (morning rise and evening fall), and/or local demand is low, wind curtailment may be required.

Figure 5.26 compares the installed wind power capacity, relative to both their population size and export capability, for a number of European countries (Rodríguez *et al.*, 2005). The former measure provides an indirect indication of installed wind

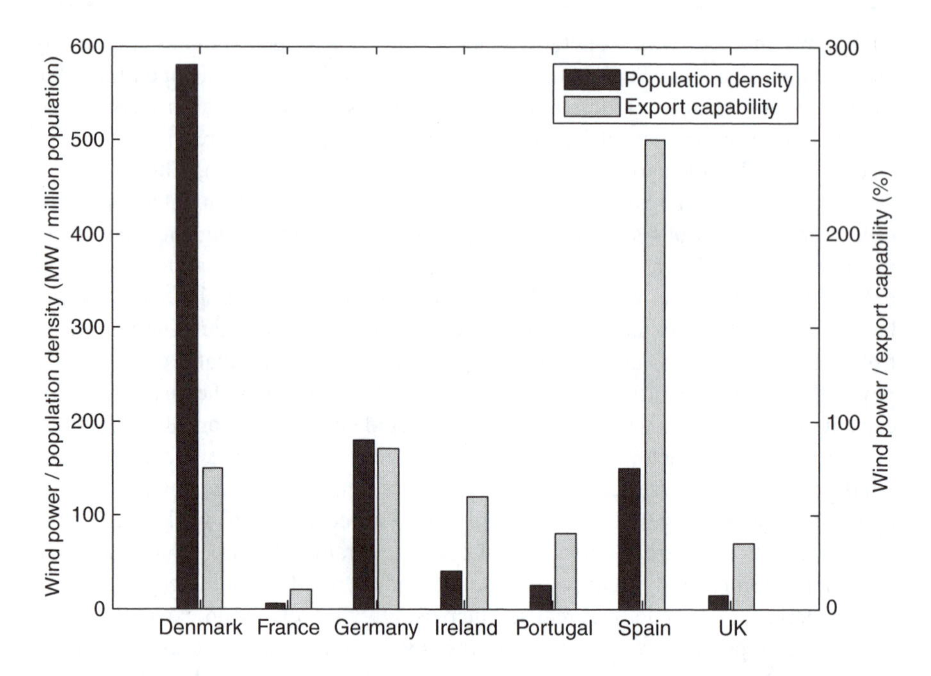

Figure 5.26 Interconnection capability of selected European countries

power capacity relative to power system size. For most of the countries shown the export capability is high relative to current wind penetration levels, suggesting that there is scope for external regulation. However, it is the countries of Denmark, Germany and Spain which are of greater interest. Clearly, Denmark has the highest wind farm density, but this represents about 75 per cent of the interconnection capacity. At the end of 2004, Germany and Spain had installed wind capacities of 16.6 GW and 8.3 GW respectively. However, while this translates into a relative export capability of approximately 85 per cent for Germany, the equivalent figure for Spain is 250 per cent. In other words, Spain has reduced opportunity for transient support from neighbouring countries. In 2005, wind power capacity in Spain grew by a further 1,764 MW, second only in Europe to Germany (1,808 MW). The Portuguese and Spanish TSOs have carried out joint studies into the effects of a sudden loss of wind generation following a system disturbance. The increase in power flow from neighbouring areas can cause overloading of the limited interconnection capacity, including ties from Spain to France, leading to a requirement for wind curtailment under certain operating conditions (Peças Lopes, 2005). Ireland also presents an interesting example. Although connected by a submarine HVDC cable to Scotland, market arrangements have militated against frequency control and reserve provision being enabled (SEI, 2004). So, except under emergency conditions, Ireland is *dynamically* isolated from Great Britain.

The low capacity factor (typically between 30 and 40 per cent) of wind generation also counts against further interconnection. Excepting countries such as Denmark and

Scotland where annual energy production exceeds local demand, periods of excess wind production can occur during low demand periods when electricity prices are low. The potential value of large-scale wind power import/export is therefore unlikely to justify investment in transmission. So, given the economic and environmental difficulties of building additional transmission capacity to facilitate occasional large-scale wind imbalances, it is likely that individual countries with ambitious wind production targets must manage the uncertainty and variability of the ensuing generation largely within their own networks. However, there is a potential caveat to this statement. The European Commission has identified a lack of system interconnection as one of the main barriers to effective competition in European power markets. In 12 of the 14 EU-15 member states the top three utilities control more than two thirds of the market, while the average capacity share of the generation market of the top three utilities is 76 per cent (EWEA, 2005). Amongst various solutions proposed by the European Commission and European Council is that the volume of interconnection capacity should equate to 10 per cent of the installed production capacity. Increased import should, of course, reduce dominant positions, and, almost as a side effect facilitate increased wheeling of wind power from Spain to northern Germany say, or vice versa.

5.3.3.2 Wind turbine ramping control

Grid codes normally specify a minimum ramp rate for conventional generation. In Ireland, the requirement is set at 1.5 per cent of registered capacity per minute, or 90 per cent per hour, while in Holland (TenneT), forming part of the interconnected UCTE system, the equivalent target is 7 per cent of registered capacity per minute, ensuring full plant availability after 15 minutes. The ramping requirements will have been based upon the known variability of the system demand, and the relative size of the generators to that of the power system itself. However, consider a case during the evening period when the load falls off naturally and conventional generators reduce their output accordingly. If wind power is also increasing at the same time, are the conventional generation plant sufficiently flexible to balance the system load? Should wind output be constrained in advance? Should conventional generation be decommitted early? Similar questions can be posed during the morning rise in demand. In general, conventional generation *is* able to cope with worst-case ramp rate scenarios, over time scales up to 1 hour. (Over longer time horizons, stand-by generation may be required.) Prudence would suggest that, as wind penetration increases, strict limits must be placed on the rate of change of wind power production. Such limits should apply under all conditions (i.e. turbine start-up, normal operation and shutdown). Power variability from an individual wind farm may also result in local voltage problems, as was seen in Chapter 4.

Staggered connection prevents several turbines starting at the same time, and hence a reduction in the initial loading rate (see also Section 4.2). Some grid codes, for example ESBNG (Ireland) and E.ON (Germany), require that ramping rate requirements are complied with during start-up, in addition to normal running. It is also advisable that wind turbines should not be permitted to start if the frequency is

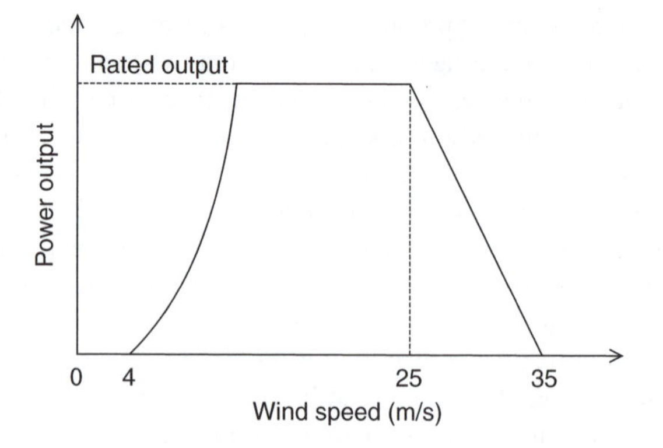

Figure 5.27 Modified wind turbine power curve

high – indicative of excessive generation supply. In Ireland, a ramp frequency controller is set to prevent ramping upwards when the frequency is 50.2 Hz or higher, implicitly including turbine start-up.

Staggered shutdown is the natural complement to staggered start-up, although not quite so straightforward to introduce. Turbines are likely to be shut down because of high wind speeds, and hence any delay in doing so implies increased maintenance costs. A typical cut-out speed for modern turbines is 25 m/s. Above this speed the turbine shuts down and stops producing energy. A hysteresis loop and a programmable delay are usually introduced in the turbine control system, such that small changes in wind speed around the cut-off threshold do not require the turbines to persistently stop and start. Restart of the turbine may require a (hysteresis) drop in wind speed of 3–4 m/s. Recent turbine designs have focussed on continuous operation during such high wind speed conditions, with electrical output gradually curtailed as wind speed approaches 35 m/s, as shown in Figure 5.27. Such an approach can mitigate wind variability, while increasing energy capture under extreme conditions.

A TSO may require that individual turbines within a wind farm have distinct (but similar) cut-out speeds, providing a gradual, rather than sudden, *wind-down* of wind farm production. 2500 MW of wind generation was lost in the German E.ON grid on 26 February 2002 due to high wind speed protection over several hours. As a result, Energinet (Denmark) and Svenska Kraftnät – SvK (Sweden), for example, both require that high wind speed must not cause simultaneous cut-out of all wind turbines. There may also be benefits in signalling early warning to the regional control centre of a potential over-speed condition. At a higher level, prediction methods and weather warnings can provide indication of imminent high wind and/or storm conditions. A phased shutdown over 30 minutes, a requirement of the former Scottish grid code, would ensure that the downward ramp rate is not excessive. However, just as for the case of staggered shutdown in *low* wind conditions, any advance in this action implies a loss of energy production.

When the turbines are operational, the *positive* ramp rate can be controlled easily by adjusting the rotor pitch angle (see later, Fig. 5.30(a)). This operation can be implemented independently for each turbine or coordinated across the entire wind farm. In contrast, the output of stall-controlled (passive) wind turbines cannot be readily controlled. However, above rated wind speed, and depending on the configuration and design of a particular pitch control system, stall-controlled wind turbines commonly present power fluctuations of lower amplitude. The German (E.ON) maximum ramping rate specification is 10 per cent of turbine rating per minute, while in Ireland two settings are specified – ramp rate per minute and ramp rate over 10 minutes. The one-minute ramp rate is set currently at 8 per cent of registered capacity per minute (not less than 4 MW/minute and not higher than 12 MW/minute) while the 10-minute ramp rate is 4 per cent of registered capacity per minute (not less than 1 MW/minute and not higher than 6 MW/minute). In Great Britain, the ramping requirements are defined by the size of the wind farm – no limit for wind farms up to 300 MW capacity, 50 MW/minute between 300 and 1,000 MW capacity, and 40 MW/minute beyond 1,000 MW in size. With sufficient notice the ramp rate should be adjustable by the TSO, with increasing wind penetration. In Ireland, for example, both settings (per minute and per 10 minutes) should be independently variable over the range 1–30 MW/minute. In Energinet (Denmark), the ramp rate should be adjustable within the range of 10–100 per cent turbine rating per minute.

This can become a useful tool for the TSO following the daily peak demand period, for example, when the remaining generating plant are required to track the general fall-off in demand. At higher wind penetration levels, there may also be concern should wind variability approach the maximum ramping rates of conventional generation. Finally, limiting the *negative* ramp rate must be considered more challenging, as it requires a degree of forecasting. Due to the possibility of a sudden drop in wind speed, it would be difficult to constrain to a maximum negative ramp rate on the time scale of minutes. However, in theory this should be more achievable for longer time periods (several hours), so that the magnitude of more extreme fluctuations, such as widespread turbine shutdown preceding an approaching storm front, could be reduced.

5.3.4 System operational modes

Wind variability, unpredicted or even if predicted, occurring on the time scale of tens of minutes to hours, will have a significant impact on economic and reliable power system operation. The main concern is that a continuous balance must be maintained between generation and demand, whilst ensuring an adequate reserve capability. Extensive experience exists in predicting demand behaviour, and credible fluctuations in demand can be covered at minimal cost (see Section 5.2). The introduction of large-scale, distributed, but variable power sources will clearly impact on the scheduling of conventional generation and the operational procedures implemented to ensure sufficient generation reserves. A number of challenging scenarios can easily be imagined. For example, a large increase in load during the morning rise, coupled with an unpredicted fall-off in wind generation, could deplete the system's reserve margin, as well as strain the ramping capability of conventional plant. Alternatively, at times of low demand and unexpected high wind output, conventional generators could be wound down towards their (less efficient) minimum output. Subsequently, should excess wind power be curtailed, leading to an uneconomic, and possibly *high emission* configuration? Or, should some thermal units be de-committed, assuming that wind output will be sustained? Solutions are required which make minimal use of fast-start, but expensive, OCGT back-up generation, or the need for wind and/or load curtailment.

For simplicity, two extremes of system operation are proposed, namely fuel-saver mode and wind-forecast mode. Fuel-saver mode is essentially an extension of current practice, with wind turbines treated as negative load devices (i.e. they provide energy). They thus reduce the effective system load, but do not provide any ancillary support services. The alternative to fuel-saver mode is to include wind forecasts directly within the unit commitment process. Assuming here for expediency that wind curtailment will not be required, and that wind generation will not provide any ancillary services, then unit commitment will be based on the net demand forecast (equal to the demand forecast less the forecast wind generation). The advantage is that a forecast of significant wind power could reduce the required commitment of conventional generation, leading to a more cost-effective mode of operation with lower emissions.

5.3.4.1 Fuel-saver operating mode

Assuming a vertically integrated utility structure, as in Section 5.2, the system operator will complete unit commitment, typically for the following day, based solely on the demand forecast, and ignore any potential contribution from wind generation. The subsequent intention is that individual generating units will start up/shut down in sympathy with the expected system demand pattern. In real time, however, assuming that wind farms are operational and thus contributing to the generation–demand balance, conventional plant will reduce their output in sympathy with the variation in wind production, and so *fuel is saved*. Wind generation thus appears as *negative* load, causing an over-prediction of system demand. Further fuel could be saved if one or more of the generating units were de-committed, but this action is not taken for fear of a wind power lull. The result is a very secure power system, with high levels of reserve, enhanced load-following capability (due to the number of part-loaded plant) and a solution that is easily implemented. However, there are a number of notable disadvantages.

There will at times be an over-commitment of conventional plant, with wind production having to be curtailed to ensure that the minimum output restrictions of generators are not violated. This issue can be particularly significant for smaller systems, where at times of low demand, and ignoring the likelihood of wind generation, a relatively small number of large (high efficiency) generators would probably be employed. Wind curtailment may be required should there be significant wind power import. A more expensive solution, but offering greater load-following flexibility, would be to operate a larger number of small (lower efficiency) conventional units. Furthermore, many generators will now be part-loaded, causing an increase in unit heat rate (which relates to the fuel consumption rate), and so CO_2 savings (from fuel saved) may not be fully realised. For example, a 5–10 per cent absolute reduction in thermal efficiency for fossil-fired plant could be considered typical (Leonhard and Müller, 2002). Similarly, CCGTs offer the advantage of high efficiency and low emissions when operating under full-load conditions, but both can drop off significantly if units are part-loaded. At higher outputs, fuel and air can be pre-mixed *before* combustion. However, when running at part-load (<50–60 per cent) this is no longer possible, in order to maintain flame stability, and oxides of nitrogen (NO_x) emissions increase dramatically (CIGRE, 2003).

Looking at the longer term, the fuel-saver mode effectively assumes a zero capacity credit for wind. It follows that there will be no reduction in (conventional plant) capital costs to meet anticipated growth in demand. The fuel-saver approach is thus simple to implement, but potentially expensive. However, despite all the drawbacks, such conservative unit scheduling has been the natural response of most system operators – it is probably the best approach when wind energy penetration is low, say less than 5 per cent of annual energy supplied.

5.3.4.2 Wind-forecasting mode

A wind energy penetration figure of 10 per cent is often quoted as a threshold figure after which the fuel and emissions cost of part-loading fossil-fuel generation compels integration of wind production into the daily scheduling process. Allowing for wind-forecast errors and wind variability would suggest, however, that system reserve levels would have to be increased slightly if *no-wind* system reliability levels are to be maintained. Unit scheduling following a wind-forecasting approach therefore tends to favour smaller, more flexible generation plant which may be required to start up and shut down more quickly compared with the fuel-saver mode.

For this mode to be fully effective, state of the art, meteorologically based wind-forecasting tools are required, with a prediction horizon of at least 4 hours and more acceptably 24–48 hours. Even longer time horizons can be beneficial for managing limited hydro reservoir reserves. Significant errors in longer-term forecasts, say 12 hours hence, can be corrected, at a cost, by additional unit start-ups. Beyond this time horizon, forecasts can assist in maintenance scheduling. For shorter time horizons, an overly optimistic wind power forecast could lead to a shortfall of online generation to meet the current demand and/or seriously undermine the system's operating reserve. Additional load-following requirements may be placed on the scheduled conventional plant, impacting on unit maintenance costs and plant life expectations. Generally speaking, an under-prediction of wind generation will not cause great concern to the TSO since scheduled plant can be backed-off (tending towards fuel-saver mode), with positive ramping rates applied to sufficient wind farms if required.

It may not be possible to redeem a *shortfall* situation by starting up additional conventional plant: drum boiler steam plant can typically be brought from start-up to full-load in 2–4 hours (for hot plant) and 6–10 hours (for cold plant), while the equivalent cold-start figures for once-through plant (3–4 hours) and CCGTs (5–6 hours) are noticeably reduced. OCGTs and diesel engines offer much shorter start-up times, but at the expense of high operating cost and high CO_2 emissions. Conventional generation also tends to have many scheduling constraints, often embodied within connection agreements, which are linked to the cost of hot, warm and cold starts. Severe limits can be placed on both the minimum run time of individual units and the number of start-ups allowed during a given period (e.g. two hot starts per day or up to 200 per year), and up to 50/10 warm/cold starts per year.

All of these factors can place restrictions on the operational flexibility of the units, and consequently the generation mix will determine the required wind-forecast prediction horizon – the more flexible the units, the later unit commitment decisions can be delayed. For example, a power system with a large capacity of quick-start plant (such as OCGTs, hydroelectric generation, pumped storage schemes, etc.) can more readily cope with large forecasting errors than a power system comprised mainly of less responsive plant (such as nuclear power stations or CHP schemes). An optimum level of scheduling aggression can be envisaged, whereby the more accurate the wind power forecast and the higher the associated confidence level of the predictions (both of which may be subject to weather patterns), the more confidently the system operator can de-commit conventional plant in readiness for wind generation. So, for example, during periods of lower prediction confidence (typically associated with the passage of storm fronts when there may be rapid and large changes in wind output) upper limits could be placed on the amount of wind generation that can be accepted, and/or tighter restrictions imposed on the rate of change of wind power.

As discussed further in Chapter 6 (Wind Power Forecasting), for up to 3–4 hours ahead simple persistence forecasting methods, whereby it is assumed that wind generation output will *persist* at its current level, perform reasonably well. Alternatively, statistical regression techniques, which follow the recent trend in wind farm output, can improve on persistence methods for similar time periods. However, for time periods beyond 3–4 hours, approaches employing forecast data from national meteorological office numerical weather prediction (NWP) models offer significant improvements in forecasting ability. The NWP output often feeds a separate model that generates site-specific wind speed and power estimates. Two approaches are commonly adopted here: physical equations and relationships are utilised to estimate wind farm power output; or statistical models, sometimes backed up with online measurements for short-term predictions, are applied. Subsequent up-scaling of these results provides a prediction of the production for an entire region. State of the art approaches are able to achieve an *average* prediction error of 8–10 per cent, relative to installed capacity, over a 12-hour period, with some degradation in performance for longer time horizons. In comparison, average load forecasting errors are of the order of 1–2 per cent, and are much less dependent on the time horizon. For system operation it is the *combined* wind and load forecasting error that is important, since this will determine the additional regulating requirements placed on the power system as a whole.

Since the wind-forecasting errors should be largely independent of the demand forecasting errors, the total error should be less than the sum of the individual errors. Figure 5.28 illustrates the 36-hour ahead wind speed forecast for a particular site in Ireland in hourly steps for three distinct days, allowing comparison between the actual and predicted values. In general, the wind speed predictions are seen to be

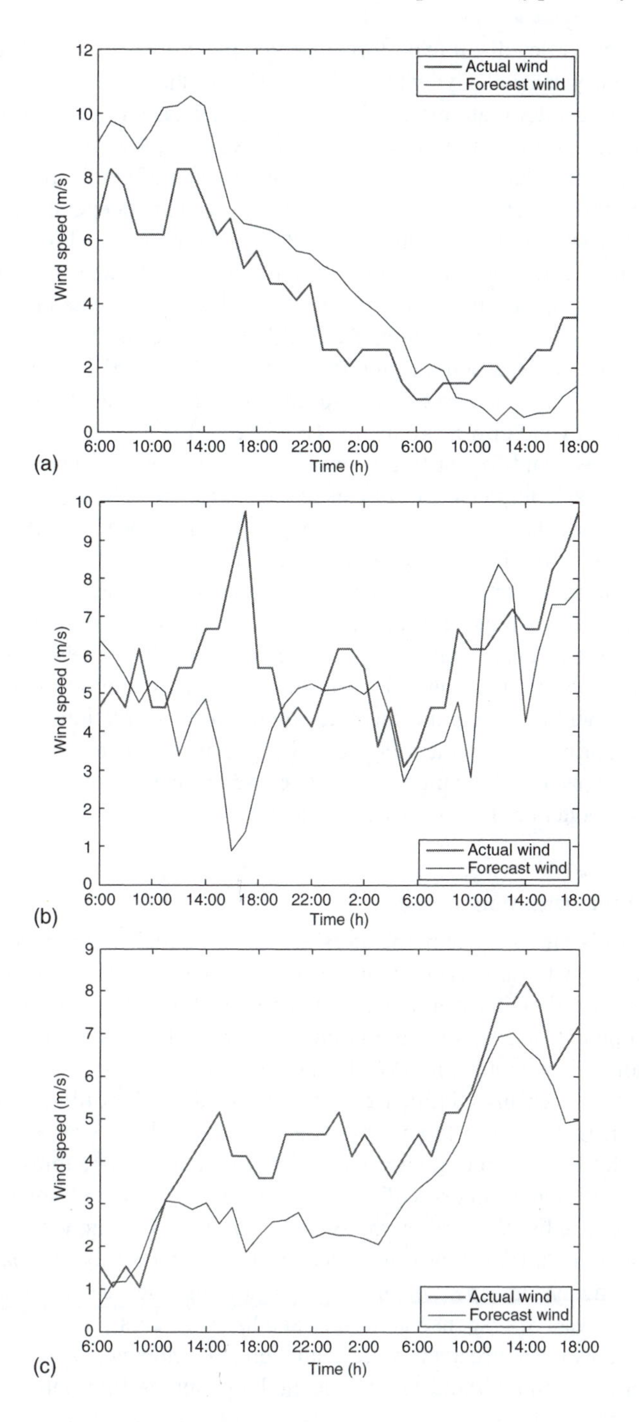

Figure 5.28 36-hour ahead wind speed prediction (a–c)

reasonable, with little fall-off in performance for longer time horizons. Figure 5.28(b) illustrates a fairly unusual day: when looking 10 hours ahead the actual wind resource is significantly greater than predicted. For the Ireland case, this has probably arisen due to the inherent difficulties associated with predicting low-pressure systems. So, for example, wind behaviour at the forecast site can be profoundly affected by minor deviations in the expected path of the low-pressure system, deepening or shallowing of the weather system and/or unpredicted changes in its speed. Figure 5.12 earlier illustrated the effects of a low-pressure system crossing Ireland, so errors associated with the direction and timing of its passage can be directly translated into large errors in forecast wind power production. Such difficulties are likely to be common. In Denmark, for example, assuming an installed base of 2,400 MW of wind generation, a deviation of ±1 m/s in average wind speed has been estimated to result in a ±320 MW variation in wind power production (Bach, 2005).

In most cases, wind-forecasting errors arise from timing significant weather fronts incorrectly: a 4–6 hour phasing error can be seen in Figure 5.28(a). Since the passing of such fronts can be associated with changes in wind speed, large power errors can occur. Over time these *phasing* errors tend to cancel out, so that wind *energy* forecasting can be very good, compared with wind *power* forecasting. As discussed in Section 5.4, this should encourage greater use of energy storage and demand-side management. Consequently, a system operator may be concerned about a large forecast variation in wind power later in the day. Uncertainty about the timing of such an event will affect commitment/de-commitment instructions to conventional plant. Anticipation of extreme storm conditions will also cause uncertainty about whether wind speeds will be high enough to cause turbine protection to activate, and a subsequent reduction in wind power production.

5.3.4.3 Implementation options

It is clearly sensible for system operators to follow the fuel-saver approach initially, since, at low penetration levels, wind generation can be accommodated economically and technically without too much forethought. However, with a growth in wind farm development, and increased confidence in wind-forecasting technology, system operators will undoubtedly *drift* towards the latter wind-forecasting approach, reflecting ever increasing confidence in the predicted wind profiles. It is worth noting, however, that demand forecasting is also a probabilistic process and assumes that, although the behaviour of an individual consumer cannot be foreseen, the *likely* behaviour of different categories of customer can be predicted with some confidence. It is therefore simplistic to present the switch from fuel-saver mode to wind-forecast mode as involving a transition from a deterministic problem to a probabilistic one. At all times the system operator must still ensure that the demand is met and that sufficient fast reserve response and other ancillary services are available. Clearly, the wind-forecasting approach is more challenging to implement and will require the system operator to modify and expand his thinking, but the environmental and other benefits are obviously significant.

In order to support understanding of the wind profile at a particular operational time, TSOs require that regular data be supplied increasingly for individual wind farms using supervisory control and data acquisition (SCADA) systems. For example, a requirement for real and reactive power production is common (Energinet, ESBNG, E.ON, SvK). Meteorological data is required by some TSOs, including wind speed (Energinet, ESBNG), in addition to wind direction, ambient temperature and ambient pressure (ESBNG). Control status information (SvK, Energinet, ESBNG) may include available capacity, curtailment setpoint (on/off), regulation capability (on/off) (SvK), percentage shutdown due to high wind speed, and islanding detection. Prior to its merger with the new electricity trading arrangements (NETA) market, the requirements in Scotland were perhaps the most stringent, with data required on most of the above, and also frequency control status (on/off) and power system stabiliser (PSS) functionality.

The data, and in particular any meteorological information, can be used to inform the output of wind-forecasting tools. Wind power prediction tool (WPPT) is employed in Denmark, with updated forecasts provided every 6 hours for the next 48 hours in 1-hour steps for the 2.5 GW installed capacity. Wind speed, wind direction, air temperature and power output from reference wind farms in each sub-area are provided as inputs to the system, to enhance the accuracy of the forecasts (Holttinen, 2005c). Advanced wind power prediction tool (AWPT) also uses a statistical model, in conjunction with data from 25 wind farms and measurements from 100 single turbines. For a 15 GW capacity in the E.ON area, AWPT achieves typically a 6 per cent average (root mean square or RMS) error for 6 hours ahead and a 10 per cent error between 24–48 hours ahead (Ensslin *et al.*, 2003). For the future, Zephyr has been proposed as a merger of Prediktor (physical) and WPPT, both outlined in Chapter 6, using data from *all* Danish wind farms rather than representative sites (Nielsen *et al.*, 2002). Although online data improves forecasting, particularly for short-term horizons (2–3 hours), this should be balanced against the additional costs involved. In Ireland the MoreCARE evaluation programme has been running since 2003 (Barry and Smith, 2005). Measurements are obtained from 11 geographically dispersed wind farms, which are then used to create a countrywide, 48-hour forecast in 1-hour steps. Regular HIRLAM (high-resolution limited area model) data are provided by the national meteorological office, but this introduces a 4.5 hour (3.5 hours in winter) time lag. In Spain, systems such as Prediktor, Casandra and Sipreolico have been applied (Giebel *et al.*, 2003), although from January 2006 there has also been a requirement for all wind farms to provide a 24-hour forecast of production, 30 hours in advance of the day beginning. At present, the vast majority of existing wind farms are located onshore. With expansion offshore, wind forecasting will also be required to include these new installations. It is generally assumed that offshore forecasting should be easier than onshore, due to the flatter *terrain*, although wind/wave interactions may complicate matters, and the benefits of error aggregation will be largely lost. The pan-European ANEMOS project, for example, has an aim of providing a short-term forecast of wind power production up to two days ahead, including the impact of offshore developments (Kariniotakis *et al.*, 2006a).

Wind forecasting clearly brings benefits to a power system, in lessening the impact of high wind penetration, and hence the associated cost, while increasing the penetration limit at which wind generation can be safely tolerated. This needs to be balanced against the financial outlay for the forecasting system itself, the associated staff training and software maintenance costs, and the requirement to collect and store production data from operational wind farms. The question then arises as to who should pay for this wind-forecasting service, and over what time horizon predictions should be made. The answer depends on the particular electricity market arrangement. In California, for example, it has been agreed that wind developers will create a (2 hour) wind-forecasting system but allow the independent system operator (ISO) to operate it. If the wind farm owners pay a forecasting fee, and schedule according to the forecast, then all generation imbalance penalties are cancelled (Asmus, 2003). In Germany it is the sole responsibility of the TSOs (and DSOs) to balance wind power, and so they provide their own forecasting tools. The day-ahead market closes at 3 p.m. on the preceding day, requiring a prediction horizon of 33 hours. Eastern Denmark participates in the Nordpool Elspot day-ahead market, which closes at noon on the preceding day, requiring a prediction horizon of 36 hours (Holttinen, 2005c). The England and Wales NETA spot price market originally had a *gate closure* time of 3.5 hours before time of delivery, later reduced to 1 hour. Deviations from the agreed power schedule resulted in penalties being imposed through the mechanism of the balancing market (Johnson and Tleis, 2005). It was, therefore, in the wind farm operators' best interest to invest in a forecasting tool. In April 2005 NETA was extended to include Scotland, forming BETTA (British electricity trading and transmission arrangements). Nordpool also operates a 1-hour market, Elbas, which can be used by parties from Sweden and Finland, and Denmark since 2004.

It follows from the above that the requirements for secondary (and tertiary) control depend greatly on the market closure time. An increasing delay between the wind forecast (and demand forecast) and the actual time of production will inevitably lead to greater forecast uncertainty, which must be matched by higher reserve levels. This can be translated into additional unit start-ups, part loading of thermal plant, and increased capability for ramping up and down. As outlined in Section 5.2, forecasting of system demand can generally be achieved with a 1–2 per cent error of peak demand over a 24-hour period. This is in contrast to average wind-forecasting errors of 8–10 per cent of wind capacity, with the forecast generally becoming less valid for longer time horizons. As wind penetration levels increase it is inevitable that increased uncertainties will arise in the forecast of *net* demand, that is, that part of the system demand met by conventional generation. If the system operator is uncertain exactly how much wind generation will be available at a particular time then increased operating reserve must be carried. It should be noted, however, that the primary reserve requirement will be largely unaffected even at relatively high wind penetration levels (see Section 5.3.3). Primary reserve is intended to cover the sudden loss of a large infeed to the system – it is probable that this will always be an existing conventional generator, with smaller wind farms scattered around the distribution and transmission networks. Large-scale reduction in wind output is most likely to occur during storm conditions, and even here may take place over several hours and should be largely predictable.

In operational practice, utilities will perform the unit commitment task according to the available wind forecast, acting effectively as a reduction in the system demand. The unit commitment may be modified/tweaked to ensure that the generation profile remains viable and economic within high and low confidence boundaries of the wind prediction. Reserve levels must be increased to combat wind variations on the time scales of tens of minutes to hours. As seen in Section 5.3.2 for Ireland and in Section 5.3.3 more generally, although the wind output cannot be predicted accurately, bounds can be placed on the likely variation over extending periods of time. For example, taking Figure 5.18 representing wind variability for Ireland from 1 hour ahead to 12 hours ahead, it can be seen that, looking 1 hour ahead, a maximum variation of 11 per cent could be expected. Variability increases to 32 and 50 per cent when looking 4 and 8 hours ahead, respectively. Advance action plans will also be defined for extreme wind scenarios, in addition to those normally created to ensure system *survival*.

5.3.5 Capacity credit

On 28 February 2005 the Spanish electricity system was near collapse, due to a combination of cold weather (increasing electrical demand) and a shortage of conventional generation. Four nuclear power stations were out of service, and hydroelectric reserves were largely depleted. Five gas-fired units were also unavailable due to their fuel supply being interrupted for domestic purposes (increased heating load), while tankers carrying a back-up liquefied natural gas (LNG) supply could not dock at port due to the poor weather conditions (Ford, 2005). Over the peak evening period, 4,000 MW of wind production prevented emergency actions being implemented. The following day only 900 MW was provided from wind generation, resulting in exports to France and Portugal being reduced, and large industrial consumers with interruptible tariffs being disconnected. On both days, the availability of wind generation meant that power cuts were not required, and a partial blackout was avoided.

At some time during the year, the peak demand on a power system will occur. In northwestern Europe this is likely to occur on a winter weekday evening. In warmer climates, the peak demand is more likely to happen during the summer time, coincident with high air-conditioning load. Since electrical energy cannot be conveniently stored, it follows that there must be sufficient generation plant capacity installed and available to meet the peak annual demand. Under-utilised plant is a natural consequence, and system capacity factors (calculated as the ratio of the average system demand to the peak system demand) of 55–60 per cent are typical for many utilities. The additional need for spinning and back-up reserve, and the possibility that individual units may be out of service when the peak occurs, leads to a requirement for additional capacity beyond the forecast peak.

Against this background, there is a continuous increase in demand for electrical power – world growth continues at about 7 per cent per year, while in the UK consumption has increased by less than 2 per cent per year for a number of years. Given that it may take several years to obtain planning permission and complete plant construction, the electrical utility must predict the peak system demand 10–20 years into the future and ensure that sufficient plant capacity exists when required. A question that then arises is – how should the (anticipated) growth of wind farms be factored into determining the required future conventional plant capacity? In other words, does wind energy have a capacity credit?

With a fuel-saver strategy, the task is straightforward – wind forecasts do not form part of the unit commitment process, and so the question is redundant. If a wind forecast strategy is adopted, however, the question is more challenging. In parts of the world where peak annual load occurs during the summer (driven by air-conditioning load) the capacity credit of wind generation is likely to be reduced (Piwko *et al.*, 2005). However, in locations where maximum demand falls during the winter (increased heating demand) the capacity benefits of wind generation should be more apparent. Taking Ireland as an example, and by inference northwest Europe, it was shown in Section 5.3.2 (Figs. 5.9 and 5.10) that there is a weak correlation between both the daily and annual variation in system demand and wind generation. Thus, it tends to be windier during the day rather than at night (more noticeably in the summer), and during the winter rather than the summer. Consequently, wind generation will tend to cause a reduction in both the system peak and the minimum load. In the United Kingdom, a positive correlation has been observed between the average hourly capacity factor and the electrical demand (ECI, 2005), where the average energy provided from wind farms during peak demand periods (winter evenings) is around 2.5 times that produced at minimum demand (summer nights). However, it does not automatically follow that it will be windy during the periods of peak demand on the system, although a wind chill factor could impact on the heating load component of the system demand and hence influence when the peak actually occurs (Hor *et al.*, 2005). Recognition that wind energy has a capacity credit should ultimately result in the avoidance, or at least delay, in the construction of additional conventional generation. Unlike hydroelectric generation, which can be subject to significant annual variation in available production, long-term analysis of wind speed records suggests that the inter-annual variability is noticeably reduced – in Europe the distribution tends to be normal with a standard deviation of 6 per cent (EWEA, 2005).

Actually determining a figure for the capacity credit of wind generation (i.e. the ability to displace an equivalent amount of 100 per cent firm capacity) has proven to be controversial and subject to interpretation. Anecdotal evidence has indicated that wind output can be low during periods of high system demand, although others have suggested that across a sufficiently large region (Great Britain/Europe) it will always be windy somewhere. Over a 30-year period, based on wind speed measurements from over 60 locations across the United Kingdom, no single hour was identified during which the wind speed at every location fell below 4 m/s (turbine cut-in speed). On average, there was only 1 hour per year when over 90 per cent of the United Kingdom experienced low wind speeds, and only 1 hour in every five years when this would have occurred during the winter higher demand period (ECI, 2005).

A variety of approaches have been proposed to calculate the capacity credit, based mainly on loss-of-load expectation (LOLE) and loss-of-load probability (LOLP) methods, with wind generation alternatively modelled statistically or using time series data. For a particular area of interest, a wide range of capacity credit figures can often be quoted, dependent on assumptions of system reliability, seasonal wind regime, distribution of onshore/offshore sites, degree of system interconnection, etc. Time series approaches can be particularly affected by coincident weather/demand patterns, particularly during peak demand periods. Sensitivity analysis, in the form of time shifting wind production relative to the demand in 24-hour steps, improves robustness.

Following a probabilistic approach, the availability of each generating unit is determined, based on past operational performance. Such an approach cannot be extended directly to individual wind turbines, since power production will be affected by the wind regime, which in turn results in the output of neighbouring turbines being to some extent correlated. Consequently, geographical dispersion and the smoothing effects of aggregation need to be recognised. Some approaches define a dispersion coefficient (Voorspools and d'Haeseleer, 2006), which ranges between a value of 1 (the outputs of all turbines are perfectly correlated – no dispersion) and 0 (total wind power output is constant – infinite dispersion) dependent on the wind penetration level. For Ireland, as an example, a dispersion coefficient of 0.33 has been adopted, resulting in a capacity credit of approximately 30 per cent at a wind penetration level of 10 per cent (ESBNG, 2004b). Also considering the Ireland system, Figure 5.29 illustrates the estimated capacity credit of wind generation for a 2020 scenario (Doherty *et al.*, 2006). It can be seen that for low wind penetrations the capacity credit is approximately 40 per cent, exceeding the assumed capacity factor of 35 per cent. This follows from the weak correlation of wind power with system demand in Ireland, such that it tends to be windier during peak demand periods (see Figures 5.9 and 5.10). As would be expected intuitively, both reduced dispersion of wind farm sites and increased wind penetration cause the capacity credit to fall. So, for example, with an installed capacity of 3,500 MW, wind's capacity credit falls to approximately 20 per cent.

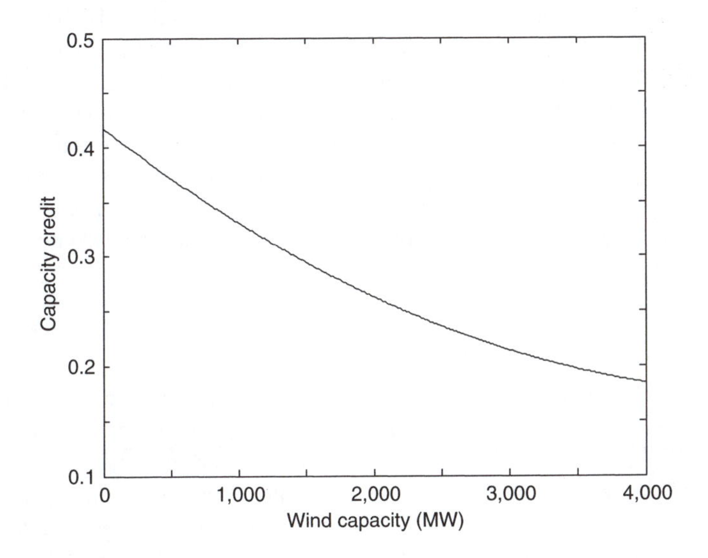

Figure 5.29 Wind capacity credit for 2020 Ireland system

A significant number of studies for *northwest* Europe have concluded that wind does have a capacity credit, roughly equal to the capacity factor of wind energy during the winter period (Milborrow, 2004). The validity of this approach is likely to lessen with increased wind penetration. A more conservative calculation would be based upon the annual, rather than winter period, capacity factor. Hence, for example, considering 1,000 MW of wind generation, and assuming an annual capacity factor of 0.30, the capacity credit would be 300 MW, which could be offset against equivalent conventional generation. Again, these results are only suitable for low wind penetration levels. Alternatively, the capacity credit can be approximated as the wind capacity factor divided by the average reliability of the remaining power system. So, for example, choosing a system reliability of 90 per cent would indicate a wind capacity credit, in this case, of 330 MW (assuming a significantly large power system). At higher wind penetrations (greater than 1 per cent of peak load), the capacity credit decreases asymptotically to a level dependent on the system reliability.

In the United Kingdom, the *annual* capacity factor for onshore wind has varied between 24 and 31 per cent since 1992, with a long-term average of 27 per cent (DTI, 2005). During the top 10 per cent periods of demand the capacity credit exceeds 36 per cent (ECI, 2005). With increased development of offshore sites, and improving turbine technology, these figures may rise higher. In the United States, various utilities, regional transmission organisations, public utility commissions, etc. have estimated annual capacity factors ranging from 2 up to 40 per cent. The observed variation arises from a multitude of analytical/ad hoc techniques, varying confidence in wind availability, and, of course, consideration of differing (climatic) regions of the United States with greatly differing penetrations of wind generation (Milligan and Porter, 2006). Quoted annual capacity factors for Denmark and Germany are around 20 and 15 per cent, respectively. The lower values can be explained largely by lower average wind speeds, and stronger requirements for system reliability (DENA, 2005). An investigative study of mainland Europe, assuming an installed wind capacity of 40 GW (EU 2010 white book target), suggested that the capacity credit for wind could vary between 5 and 35 per cent, with an average value of approximately 19 per cent (Giebel, 2000). The variation in figures quoted was achieved by examining the effect of time shifting a time series wind profile with respect to the system demand pattern. The low figure (5 per cent) is partially explained by the working assumption that conventional plant must operate above 50 per cent of maximum output and hence wind curtailment can be required.

5.3.6 *Ancillary service provision*

Conventional generation, in addition to providing energy, may also be required to provide a number of additional ancillary services – spinning reserve, load following/frequency regulation, voltage/reactive power support and black start capability. Traditionally, wind turbine generators have not provided any of these services, and indeed fixed-speed machines, in particular, may introduce high/low local voltage distribution network problems (Dinic *et al.*, 2006; Romanowitz *et al.*, 2004). As wind penetration levels increase, however, there are likely to be significant operational difficulties if the provision of these services becomes depleted. The majority of existing wind farms have been installed during the last 15–20 years. Sited largely onshore, and scattered across distribution networks, they present an evolution of many different types of technology and provide significant variation in controllability. At one extreme is the passive stall, fixed-speed wind turbine, introduced in the 1980s, in the 10–100 kW range and most widespread in Denmark. From a grid perspective, the output of such machines is uncontrollable (apart from stopping/starting) and they are generally unresponsive to system needs – as discussed later in Section 5.3.7 they do provide an inertial response. At the other extreme are modern, pitch-regulated, full- or partial-variable-speed wind turbines, providing real-time active and reactive power control, and potentially capable of contributing to dynamic stability, voltage support and network control.

The wind turbine market has grown significantly in the last five to ten years, so that most installed wind farms are controllable to some degree. The system benefits of retrofitting secure communications for centralised monitoring and control, in addition to modifying the existing control systems, are likely to be minimal for older wind turbine designs, considering the low capacity of the installed plant, and balanced against the cost of implementation and the remaining equipment life. Instead, newer installations, likely to be of much greater capacity (particularly if sited offshore), can be specified to include communication networking and advanced control systems at a small fraction of the total project cost. A modern SCADA permits comprehensive information to be collected from individual turbines, including meteorological data, which enables optimum setpoints to be defined, along with external system operator commands.

TSOs and DSOs, through the development of their associated grid codes, have tended not to impose requirements that are retrospective, that is, those wind farms causing a critical further reduction in ancillary service provision will be required to replace the shortfall. Increasingly, grid codes are being developed to be *future-proof*, with a range of functionalities specified, although not all are active. It is probable that different issues will become pressing at increased levels of wind penetration, and so individual requirements will have to be implemented only when the total capacity exceeds a defined limit. The objective, clearly, is to evolve a power system that, from a system perspective, differs little from that today, with only the original energy source(s) being changed. Issues relating to voltage and reactive power support and the contribution of wind farms to power quality management, system stability and transient performance, etc. were examined in Chapter 4. Here, the focus is on the ability of wind turbine generators to provide spinning reserve and continuous load-frequency control.

5.3.6.1 Power-frequency characteristic

Utility grid codes will generally specify that individual generating units must be able to maintain continuous operation within certain frequency bounds, and maintain short-term operation over slightly wider frequency extremes (see Table 4.3). Conventional generation is required to maintain 100 per cent of its real power output within a defined band of the nominal system frequency. In Great Britain, for example, the defined range is 49.5–50.4 Hz, while in Germany (E.ON) the comparable boundaries are 49.5 and 50.5 Hz. This ensures that any fluctuations in frequency (arising from generation – demand imbalance) are not exacerbated by subsequent variation (excluding governor action) in generator output. At lower frequencies, notably during emergency conditions, some licence is normally given for a reduction in generator output (see Section 5.1). In Great Britain, generators are required to operate continuously between 47.5 and 52 Hz and for a period of 20 seconds between 47 and 47.5 Hz. Any reduction in generator output below 49.5 Hz should be proportional to the change in system frequency, that is, at 47 Hz (a 5 per cent reduction in frequency) the power output should be at least 95 per cent of that available at 49.5 Hz. In Germany, since it is synchronously interconnected to the rest of the UCTE grid, the requirements are less stringent – generator output may fall by 20 per cent of rated output when the frequency reaches 47.5 Hz.

Since all generating units in a synchronous power system will *see* essentially the same system frequency, it is extremely beneficial that all units are capable of providing the same power-frequency response. Consequently, grid codes do not in general differentiate between the requirements for conventional generation, as considered above, and those of renewable generation. The grid codes for Germany (E.ON) and Great Britain represent good examples.

5.3.6.2 Frequency regulation and spinning reserve provision

Stall-regulated turbines comprise rotor blades that are aerodynamically shaped such that the power output is naturally curtailed to a required maximum as wind speed increases (see Section 3.4). In contrast, pitch-regulated wind turbines have the ability to adjust the rotor blade angle in real time, and thus the amount of power that is extracted from the wind can be controlled. For fixed-speed wind turbines, operating above rated power, the blade pitch angle is increased in order to restrict the power output to the rating of the electrical generator. Available wind energy is therefore *spilled*. Below rated output, in contrast, the blade pitch angle is set at an optimum angle to maximise energy capture. The story is similar for variable-speed wind turbines operating above rated conditions, so that the blade pitch angle is increasingly feathered with wind speed in order to maintain rated output. Beneath rated speed (not necessarily coincident with rated power), however, it is conventional for the rotor rotational speed to be adjusted in sympathy with wind speed, with the blade pitch angle fixed at optimum, so that energy capture is maximised (see Section 3.6). For some variable-speed wind turbine designs a combination of speed control and pitch-angle regulation is employed in light wind conditions.

For both fixed- and variable-speed wind turbines, load-frequency control can be obtained by slightly increasing the nominal blade pitch angle, deloading the wind turbine by a corresponding amount (Fig. 5.30(b)). Fixed-speed machines tend to be more sensitive to changes in pitch angle compared with variable-speed machines due to their constant speed operation. Thus, the wind turbine output can be adjusted

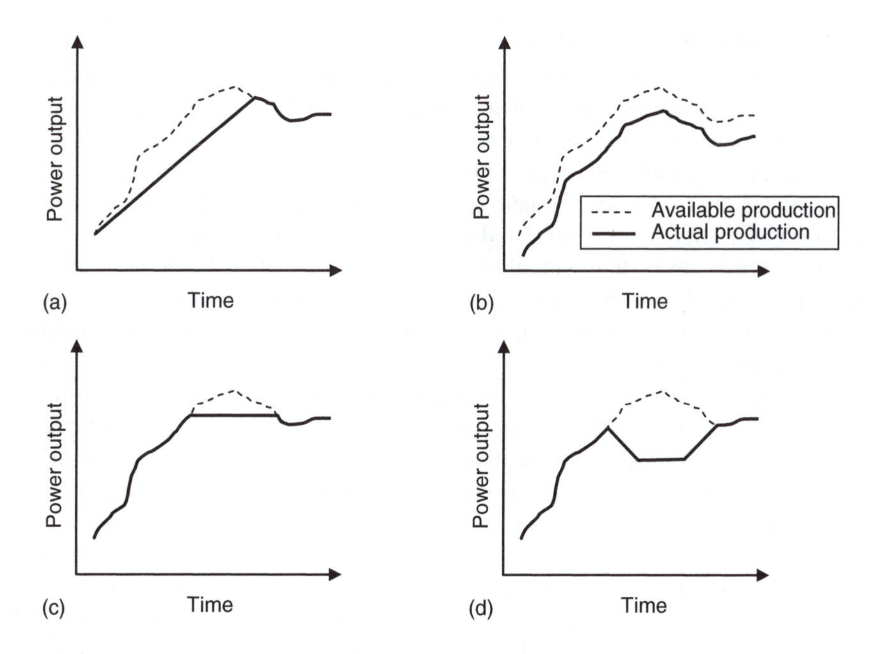

Figure 5.30 Wind turbine regulation capability (a) ramp rate limit (b) frequency (delta) control (c) maximum power constraint and (d) area balance control

in sympathy with frequency variations, akin to governor control on a conventional generator. Here, a fall in frequency (demand exceeds generation) causes a decrease in pitch angle and hence an increase in electrical output. Conversely, an increase in frequency (generation exceeds demand) causes an increase in pitch angle and a decrease in electrical output. Such an approach is feasible with conventional pitch control. Above rated output the power reference setting can be lowered – thus restricting the maximum power output (Fig. 5.30(c)). At lower turbine outputs the minimum pitch angle can be increased, again pushing wind turbine operation away from the optimum (Holdsworth *et al.*, 2004).

For variable-speed wind turbines, power output regulation can also be achieved by varying the rotor speed – a small decrease in rotational speed away from the optimum tip-speed ratio will cause a reduction in electrical output. The majority of variable-speed wind turbines are based currently on DFIGs, offering only limited variable-speed capability (70–110 per cent synchronous speed). In contrast to pitch regulation, the ability to decrease output is somewhat restricted. However, since speed control is ultimately achieved using power electronics (adjusting the injected rotor quadrature voltage – Section 3.6.1) no *moving parts* are required. This clearly contrasts with pitch regulation, and hence speed control may be well suited for continuous, *fine* frequency regulation. Blade pitch control can provide fast-acting, *coarse* control both for frequency regulation as well as emergency spinning reserve.

5.3.6.3 Implementation options

As conventional generation plant is displaced increasingly by wind generation, the task of frequency regulation will naturally fall on the remaining generating units. If wind variability is combined with existing demand variation, then at times the flexibility of the conventional plant may be insufficient to regulate wind-induced variations and maintain the frequency constant. It follows, therefore, that wind farms will have to contribute to frequency control, either by maintaining a fixed load profile, or contributing directly to system-wide frequency regulation. This is likely to be an issue not only for small/isolated power systems, but for all power systems. A simulation study, based on the E.ON German network (interconnected to the wider European network), considered the addition of 3.5 GW of offshore wind, against an existing background of 3.5 GW of onshore wind and a system capacity of approximately 33 GW (Koch *et al.*, 2003). It was proposed that the planned offshore generation provide frequency regulation capability, equivalent to an arbitrary 3 per cent of the nominal wind capacity, enabling the regulating contribution from conventional plant to be reduced from 28 to 11 per cent of the primary reserve requirement.

Wind turbine generators have good potential to provide frequency regulation and spinning reserve, and, indeed, pitch-regulated turbines are likely to be much more responsive than conventional plant. For example, in Denmark, under network fault conditions, individual turbines are required to reduce the input power (by pitch regulation) below 20 per cent of turbine rating within 2 seconds – an indicator of turbine response. Provision of a conventional governor droop response implies that an increase in power output should follow a fall in the system frequency, and hence the wind turbines must operate below the potential output level for the current wind conditions. Similarly, reserve can only be provided if the wind turbines have been deloaded by the required amount. So, for example, the turbine pitch angle could be adjusted for partial output, while maintaining a reserve margin of, say, 1–3 per cent of rated output (delta control). SCADA systems enable *turbine* wind speed to be measured, providing an estimate of potential power capture, with the turbine output limited to a defined fraction of this value. The pitch angle can then be adjusted to provide continuous frequency regulation or occasional spinning reserve.

The Horns Rev offshore wind farm, located on a submerged sandbank off the west coast of Denmark, provides the first true example of what is achievable. Arranged in five clusters over a 20 km^2 area, the wind farm consists of (5 $\times$ 16) 2 MW, pitch-controlled OptiSpeed turbines, providing a total capacity of 160 MW. Installed in 2002, Horns Rev was intended as a demonstration project of both large-scale offshore technology and wind farm/individual turbine control functionality. The control system is integrated with the network SCADA system enabling bi-directional communication with the regional dispatch centre (Elsam) and the TSO (Energinet). This allows a number of control strategies to be implemented (Christiansen, 2003): production may be constrained to a set reference (absolute production limiter; Fig. 5.30(c)). Alternatively, the wind farm may be required to participate in regional balance control (automatic generation control; Fig. 5.30(d)), such that power output is reduced at a defined ramp rate, and later increased at a defined ramp rate (see Section 5.2.1).

Ramping control can also be applied to limit short-term variations in production, arising from rapid changes in wind speed (Fig. 5.30(a)). Output fluctuations of up to 100 MW have been seen over 5 minutes. The above functions inherently imply wind curtailment, and hence the possibility of regulating/spinning reserve provision. Alternatively, using delta control, the power production is reduced by a defined setpoint (e.g. 25–50 MW; Figure 5.30(b)). In combination with balance control, the Horns Rev wind farm can thus provide system regulation, similar to conventional generation, but much faster. Each of the turbines also transmits an indicator of *available* production, such that the extra *reserve* can be monitored by the TSO. Since all these control functions may be active at the same time, integration of turbine/wind farm control introduces communications and coordination challenges (Kristoffersen and Christiansen, 2003). It should be noted that Horns Rev has had its problems. Faults, originally with the transformers and later the generators, discovered in 2004, required each of the nacelle assemblies to be brought ashore and retrofitted to better handle sea conditions.

Given that the wind source is effectively free, energy is being wasted needlessly by providing regulation – economic efficiency would suggest that this service could be better provided by conventional generation. Wind farm developers would also have concern about a loss of revenue from deloading their output. However, taking Great Britain as an example, wind generators are able to specify the price at which they are willing to be deloaded (Johnson and Tleis, 2005). The generator then receives both holding and response energy payments. There may be short periods during the year when such an approach is suitable – for example, when a light-loaded conventional unit is running mainly to provide reserve. Or, during periods of high wind output and low system demand, when the operational alternative is to constrain off entire wind farms, it may be more economic to request that wind turbines reduce output and provide frequency regulation and reserve capability.

A general framework for the desired response from a wind turbine generator, based upon the ESBNG (Ireland) grid code requirements, is illustrated in Figure 5.31. Under normal conditions, the turbine operates within the deadband region between points B and C (i.e. there is no contribution to load-frequency control). Alternatively, points B and C could be coincident such that a contribution is provided. Within this deadband region, the wind turbine output is limited to a defined fraction of the available wind resource, following the methods described earlier. Hence, should the frequency fall below point B, the wind turbine is required to increase output (a low-frequency reserve response) until the frequency falls to point A, whereupon 100 per cent of available power output is provided. At system frequencies above point C, the power output should gradually be reduced (a high-frequency reserve response) until at point D (and beyond) the power output falls to zero. For frequencies beyond f_D the wind turbines are allowed to disconnect.

At present, for many utilities, points A–D are set at 100 per cent of actual power, that is, neither low- or high-frequency reserve is provided and the wind turbine does not participate in load-frequency regulation. Instead, only the *capability* of providing a frequency response is required. In the future, however, as wind penetration levels increase, one or indeed all of the three ancillary services will be required. In Great

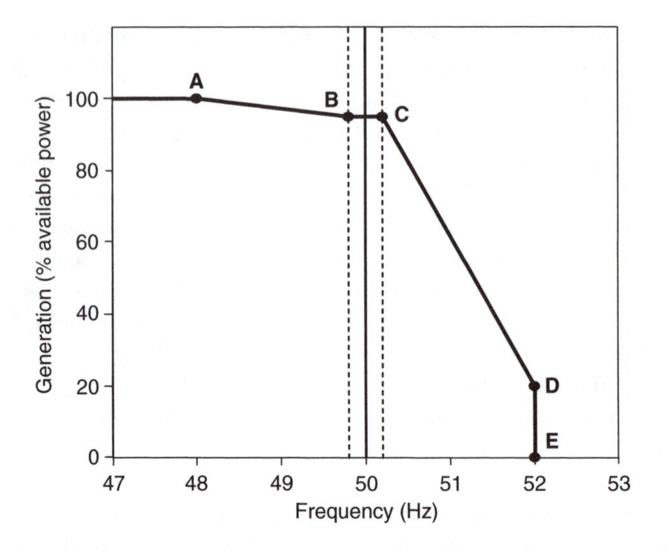

Figure 5.31 Wind turbine power-frequency capability chart

Table 5.1 ESBNG load-frequency control requirements

	System frequency (Hz)		Power output (% available power)	
			WF > 10 MW	5 MW < WF ≤ 10 MW
f_A	47–51	P_A	50–100	100
f_B	49.5–51	P_B	50–100	100
f_C	49.5–51	P_C	50–100	100
f_D	50.5–52	P_D	20–100	20–100
f_E		P_E	0	0

Britain, from January 2006, all new wind turbines must be capable of contributing to frequency control. Similarly, in Ireland, while newer wind turbines are required to fulfil the defined ranges for $f_A - f_E$ and $P_A - P_E$, as listed in Table 5.1, the current requirements are that $P_A = P_B = P_C = 100\%, P_D = 20\%$, while $f_C = 50.25$ Hz and $f_D = 52$ Hz (ESBNG, 2004a). In other words, only a high-frequency reserve response is required beyond 50.25 Hz, for wind farms exceeding 5 MW in size, with an equivalent governor droop of 4.4 per cent. In Germany, wind turbines are required to provide a high-frequency response (region C–D on Figure 5.31). In E.ON, above a system frequency of 50.5–51.5 Hz, turbines should possess a 40 per cent droop characteristic (see Figure 5.5 and Table 4.4). The capability should also exist to reduce output at a rate of 5 per cent of rated output per second above 50.5 Hz. When the frequency has stabilised, the restoration ramp rate must not exceed 10 per cent of the rated output per minute.

Individual TSOs will define the locations of points A–D differently, and their exact position may depend on wind farm location and/or system conditions (e.g. high wind/low demand versus low wind/high demand). For example, NIE (Northern Ireland) utilises a similar capability chart (Fig. 5.31) to that for ESBNG, with two sets of settings defined off-line, switchable within 1 minute. Under normal conditions, a 4 per cent droop characteristic is required above 50.25 Hz, with a high-frequency trip setting of 51.5 Hz. In constrained mode, however, turbine output should be restricted to 85 per cent of available output at 50 Hz, while operating on a 4 per cent droop characteristic above and below this frequency. Constrained mode is likely to be requested by the TSO during summer nights when the system demand is low, and a reduced number of governor-controlled conventional generation units may be online.

5.3.6.4 Wind curtailment

When a power system is stressed, one option to help alleviate existing problems is to curtail wind generation output. Curtailment may be required to reduce local voltage, provide enhanced system security at times of minimum load (by avoiding switching off a conventional generator), limit the ramp rates on conventional generation (during the daily morning rise) or avoid transmission system overloading (following network faults or other failures). More dramatically, wind generation may be reduced gradually in advance of a forecast storm front – before wind farm protection trips out individual turbines due to the high wind speeds. Additionally, considering mainland Europe as an example, uneven growth in wind generation could lead to fluctuating power exchanges across international tie-lines. Excessive export of power can be curtailed by taking conventional generators off-line, at the expense of reduced dynamic response and system fault level. With wind farms of increasing size, rapid curtailment of individual sites becomes feasible, rather than communicating with a distributed network of small turbines.

For most power systems wind curtailment is a backup *control* option being considered for the future when wind penetration levels are likely to be much higher. However, in Crete, wind curtailment has been required for stability reasons for some years at times of high wind output and low system demand, in order to maintain a minimum number of committed conventional units. Fortunately, high winds tend to coincide with the high demand summer period, implying that curtailment is mainly required during the winter. In 2001 curtailment was already 6 per cent of annual wind production, rising to 11 per cent by 2002 (Papazoglou, 2002). During winter (November to January) curtailment is accentuated, rising to almost one quarter of wind power production. The concentration of wind farms on the east of the island, and hence a lack of diversification, accentuates the difficulties. With nearly 400 MW of new wind farm proposals, and a system capacity of 581 MW, these figures are expected to grow considerably from the installed wind capacity of approximately 70 MW (Kaldellis *et al.*, 2004). In California, wind production tends to be high during the late spring months. At the same time, melting mountain snow also causes a peak in hydroelectric generation (Piwko *et al.*, 2005). During the night, when the demand is low, there can be an excess of generating capacity, on occasion requiring the curtailment of wind generation.

Even in mainland Europe, a requirement for wind curtailment is not that uncommon. In Spain, concerns about transient stability, particularly during periods of minimum demand, has led to wind being curtailed several times to maintain sufficient online conventional generation. Denmark presents an equally challenging problem of meeting heating demand and electrical demand requirements, particularly during cold, windy periods. Within the Energinet system, 1,500 MW of CHP plant, in addition to 2,400 MW of wind power, feeds the distribution network, providing almost 50 per cent of the annual energy production (Bach, 2005). The distributed CHP plant are not dispatchable, instead following heating tariffs and local supply contracts. This can cause difficulties when wind production is high, while heating demand is also high but the electrical demand is low. Again, wind production must be curtailed so that sufficient centralised generation can remain online to provide frequency regulation and other ancillary services. Since January 2005, however, legislation has required distributed CHP units (>10 MW) to respond to (electricity) market signals. In conjunction with measures such as electric water heating and flexible demand, discussed later in Section 5.4, it has been suggested that further wind curtailment could be avoided until its penetration reaches 50 per cent (Lund, 2005).

Before wind curtailment can be implemented, individual wind farms will require integration with a centralised SCADA-type system. Turbine control systems should be capable of receiving an external power setpoint from the TSO, and the ability, under extreme conditions, to be constrained off remotely. Grid codes are requiring these facilities for new wind installations. In E.ON, for example, a wind farm, upon instruction from the TSO, should be able to reduce its output to a reference value at a ramp rate exceeding 10 per cent of the rated capacity per minute without tripping. Similarly, wind farms in Ireland should be capable of receiving an external setpoint, curtailing output. Denmark and Sweden also require that it should be possible to reduce output to less than 20 per cent of registered capacity within 2 seconds (Energinet) and 5 seconds (SvK) – this facility is thought to be required mainly under fault conditions and prevents rotor over-speed. E.ON and Energinet also require that wind turbines can be connected and disconnected remotely.

Future wind farms may also be sized with the probability, rather than the possibility, of curtailment. Network investment can often lag behind wind farm expansion, even though areas with strong wind regimes may suggest high wind farm concentration. Consequently, expensive upgrading of weak networks should be balanced against the probability of distribution, or even transmission, congestion. In other words, does the benefit of a larger wind farm in terms of increased average energy production exceed the lost opportunity cost of brief periods of curtailment? This is likely to occur when wind power production is high, local load is low, and consequently the local voltage is unacceptably high (Dinic *et al.*, 2006). At such times the energy price will undoubtedly be relatively low. For most utilities there are unresolved issues regarding who should suffer the financial costs of lost wind power production. Does the potential need for curtailment suggest that wind generation should be centrally dispatched or profiled? There is, however, a consensus that spilling of wind should only occur when no other alternative strategies exist. Minimum system demand, when generating units are already operating close to their operational limits, is often considered as a period when high wind production may lead to curtailment. However, the value of generation at such times is likely to be low. Market incentives can confuse the matter further because the wind farm operators may receive payment for *green* certificates, in addition to energy (see Chapter 7: Wind Power and Electricity Markets). In countries where an ancillary services market exists it may be possible to profit from the imposed curtailment by supplying spinning reserve and/or load-following duties.

5.3.7 Wind turbine generator inertial response

Following the loss of a major system infeed, the initial ROCOF will depend on the magnitude of the lost generation, and the stored energy of the system, as expressed earlier in Equation 5.1:

$$\frac{\mathrm{d}f}{\mathrm{d}t} = \frac{-\Delta P \times f_0}{2 \times E_{\text{system}}} \text{ Hz/s}$$

For most power systems, about two thirds of the stored energy will be provided by generation, in the form of synchronous machines driven by multi-stage turbines, with the remainder coming from the load. As wind penetration levels increase, these synchronous machines will be displaced gradually with a mixture of wind turbine generators: fixed-speed (induction generators); partial-variable-speed (DFIGs); and full-variable-speed (direct-drive synchronous generators). For a given turbine rating, the inertia, and hence the stored energy, will be of similar magnitude for each design. However, the ease with which this stored energy can be *extracted* by the power system varies greatly depending on the technology.

Until the late 1990s, wind turbines were predominantly based on fixed-speed squirrel-cage induction generators. Such machines operate above synchronous speed, with a small speed variation of around 2 per cent corresponding to an increase from no-load to full-load output (Fig. 3.14). Should the frequency fall by say 1 per cent (0.5 Hz) following a loss of generation event, the wind turbine generator could possibly double its electrical output, as illustrated by the frequency-shifted torque-speed characteristic of Figure 5.32. If the wind speed, and hence the mechanical input power, remain more or less unchanged over a short time period, then the induction machine will decelerate, releasing stored energy to the system. The magnitude of this response will depend on the combined inertia of the induction generator and the wind turbine rotor. Theoretical calculations and practical studies have suggested typical inertial constant, H_{gen}, values of 3–5 seconds, comparable with conventional generation (Holdsworth *et al.*, 2004; Littler *et al.*, 2005). It could, therefore, be expected that large-scale growth in fixed-speed designs will not unduly impact on the system inertia, and, hence, the ROCOF during a system transient. Some wind turbine manufacturers have introduced modifications to the basic induction machine design to increase the operating speed range, such as: enabling the number of poles of the stator winding

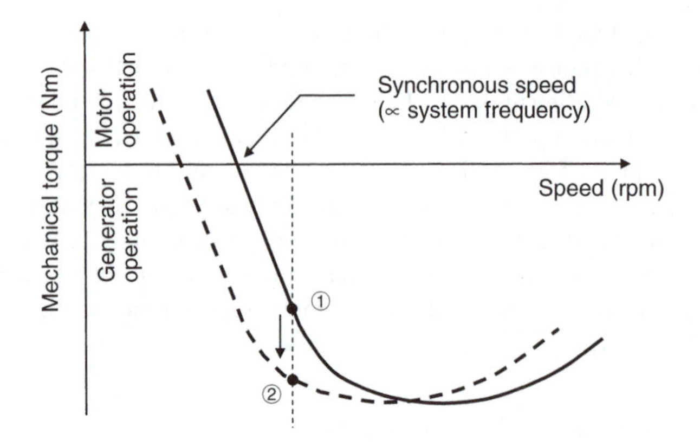

Figure 5.32 Induction machine torque-speed characteristic

to be changed (adjusts the no-load synchronous machine speed); inserting controllable external rotor resistance (allows the torque-speed characteristic to be shaped). However, such machines can still be classified as fixed-speed, and hence they will provide an inertial response.

In contrast, direct-drive, variable-speed wind turbine generators, based on synchronous machines, offer several advantages over fixed-speed machines. Using an AC–DC–AC converter the synchronous machine and the wind turbine rotor are *mechanically* decoupled from the power system. Thus the rotor rotational speed can be varied in accordance with wind speed, and independently of system frequency. As discussed previously in Section 3.6 this increases energy capture, particularly at lower wind speeds, as aerodynamic efficiency can be maintained over most of the operating range. The further ability to cope with gusts by adjusting rotor speed implies a reduction in tower stresses, and hence a reduction in capital cost, as well as voltage flicker. The direct-drive arrangement also allows the gearbox between the wind turbine rotor and generator to be eliminated, along with the associated energy losses and maintenance costs. One disadvantage of this arrangement, however, is that an inertial response will not be provided, since the turbine rotor speed is now independent of the system frequency. However, in principle, this link can be restored by suitably modifying the torque setpoint of the power converter control system, following Equation 5.2 as follows

$$\Delta\tau = \frac{\Delta P_{\text{gen}}}{\omega} = \frac{-2H_{\text{gen}} S_{\text{max}}}{\omega f_0} \times \frac{\mathrm{d}f}{\mathrm{d}t} \text{ Nm}$$

where H_{gen} is the inertial constant of the wind farm. Figure 5.33 illustrates a modified version of the torque-control loop for a variable-speed turbine from Chapter 3 (Fig. 3.22(a)), with the torque setpoint, τ, modified by an input dependent on the ROCOF. Hence, a fall in the system frequency will cause an increase in the torque setpoint, leading to a transient increase in the electrical output of the machine – an inertial response. Since this mechanism is provided *electronically* rather than mechanically,

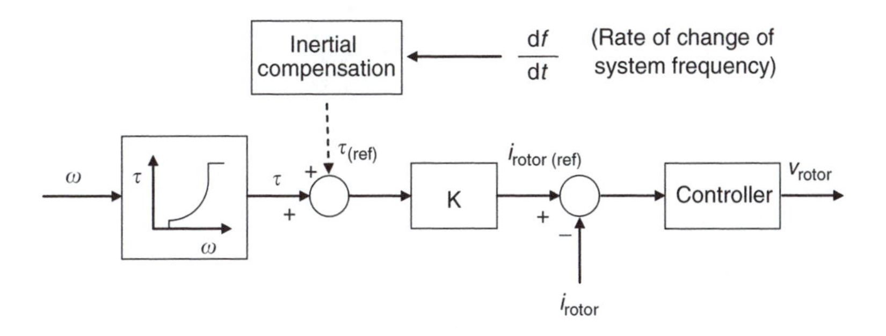

Figure 5.33 Modified DFIG torque-control loop

the user value selected for H_{gen} could actually exceed the physically defined limit. During the initial stages of an event, as much energy as required could be extracted from the rotor inertia, subject to limits on the capacity of the drive train and the power converter limits. Care would also need to be taken that the rotor speed did not decrease so much that the aerodynamic efficiency was critically affected. Following the system transient, H_{gen} and hence the magnitude of $\Delta\tau$ could be reduced, allowing the rotor speed to recover gradually. Indeed, by introducing a compensation element the inertial response can be shaped as required, maximising the system benefit, while minimising the impact on the wind turbine itself (Holdsworth *et al.*, 2004).

The current dominant technology is based on DFIGs, which provide limited variable-speed capability, and thus offer some of the advantages of full-variable-speed designs. As discussed in Section 3.6.1, the stator windings of the induction generator are connected directly to the power system, while the rotor windings are fed via an AC–DC–AC converter to provide a variable-frequency supply. The generator (and turbine rotor) is thus partially decoupled from the power system, with the result that the *natural* provision of an inertial response will depend on a number of factors – these include current wind speed, wind turbine aerodynamic mapping, stall versus pitch power regulation and the torque-control system speed of response (SEI, 2004). Of these factors, the torque-control loop is dominant and so a DFIG machine will not naturally provide an inertial response. However, it is clear that in a manner similar to full-variable-speed machines, it can be introduced artificially by modifying the torque set point as outlined earlier.

At present, however, the control systems of DFIGs have not been modified in this way. It may, therefore, be expected that with increased wind penetration, mainly in the form of such variable-speed machines, conventional generation will be displaced, leading to a reduction in the system inertia and stored energy. For smaller power systems, where the system inertia is comparatively low and individual infeeds may supply a significant fraction of the load demand, any reduction in system inertia will lead to a more dynamic power system, with higher rates of change of frequency and greater frequency dips (Lalor *et al.*, 2004). This is particularly true during periods of low demand, when fewer conventional generators are online (Fig. 5.34). Using a simulation of the Ireland power system, during the summer minimum load, a 350 MW

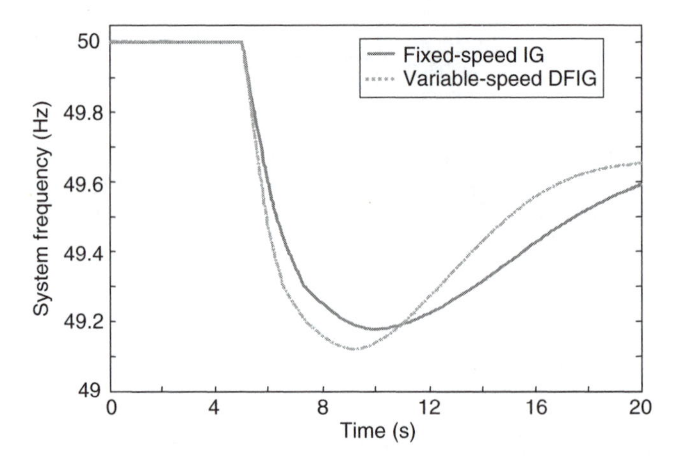

Figure 5.34 DFIG versus fixed-speed turbine inertial response

loss of generation occurs after 5 seconds, causing the frequency to fall. Since the DFIG turbines do not provide an inertial response the frequency falls more quickly and to a lower nadir, as compared with wind farms comprised entirely of fixed-speed turbines. The reduced system inertia, however, does allow the system frequency to recover faster.

With the expected expansion of wind farms sited offshore a further concern arises, irrespective of the wind turbine technology. Existing offshore wind farms have been located near to shore in shallow depths, with high-voltage alternating current (HVAC) connection to the mainland transmission network. The largest of these are Horns Rev (160 MW), located in the North Sea, and Nysted (166 MW) in the Baltic Sea, both off Denmark. Future offshore wind farms, however, are likely to be much larger (250–1,000 MW) and to be located further offshore in deeper waters. This presents a number of difficulties. For alternating current (AC) transmission, losses increase significantly with distance. This can be partially countered by increasing the operating voltage, but at the cost of larger and more expensive transformers, cabling and switchgear. Increased cable lengths (and hence capacitance) will also require greater use of reactive power compensation (e.g. static VAr compensators), placed possibly at both ends of the cable. Beyond a distance of $\approx$100 km, HVDC transmission becomes economically viable – a horizon that is likely to move inshore with advances in technology. DC transmission can reduce cabling requirements and power losses, while also offering flexibility of both real and reactive power control (Kirby *et al.*, 2002). One disadvantage of this approach, however, is that irrespective of the wind turbine technology, the DC connection decouples the stored energy of the turbine rotor from the electrical grid (i.e. an inertial response is not provided).

5.3.8 Distributed generation protection

A significant fraction of existing wind farm installations have been connected at relatively low voltages to the distribution network. Such *embedded* generation is

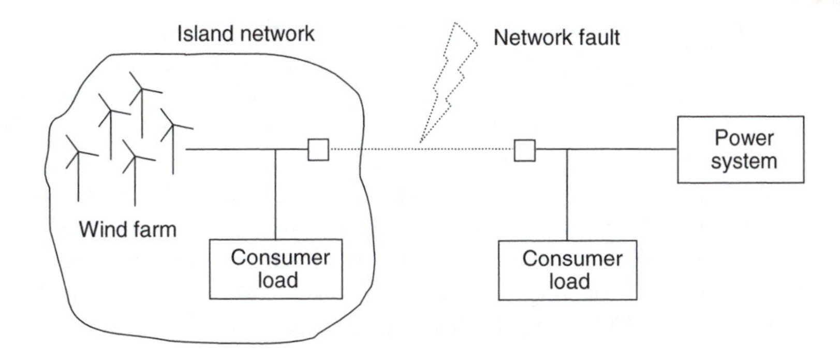

Figure 5.35 Distributed generation protection

assumed to be of small individual rating (<10 MW), scattered across a large geographical area, probably under independent control, and unlikely to be continually operational. In addition to *unit* protection, against turbine over-speed, terminal over-voltage, etc., the wind farm will be fitted with *network* protection, such that the wind farm will be isolated from the power system under certain conditions (see Section 4.9). For example, in the event of a network fault and the operation of line protection, an embedded generator may become *islanded* from the main power system (Fig. 5.35). Although the wind farm could possibly supply some local load, the voltage and frequency levels would be uncertain, potentially exceeding legal and operational limits and leading to equipment damage. The generator may also be unearthed, compromising normal protection devices, while synchronisation procedures would be required later to integrate the islanded part of the network. It is thus advisable that the wind farm disconnects itself in these circumstances. For example, Engineering Recommendations G59/1 and G75 (Recommendations for the connection of embedded generating plant to the public electricity suppliers' distribution systems) in the United Kingdom require that embedded generators are fitted with loss-of-mains (LOM) detection.

The LOM condition is normally detected using either ROCOF or vector shift protection relays. ROCOF protection relies on the assumption that, when an islanding condition occurs, the local generation will not balance the trapped load. Consequently, the system frequency will change at a rate depending on the power imbalance and the stored energy of the island network, E_{island}, as previously expressed in Equation 5.1,

$$\frac{\mathrm{d}f}{\mathrm{d}t} = \frac{-\Delta P_{\text{imbalance}} \times f_0}{2 \times E_{\text{island}}} \text{ Hz/s}$$

When selecting the threshold setting for relay activation, two conflicting factors must be considered. On one hand, since the power imbalance may be small, and hence the ROCOF low, the relay should be made as sensitive as possible. Conversely, if, for example, a large generator is tripped from the system, the system frequency will initially fall rapidly, as outlined in Section 5.2, causing the ROCOF protection to activate, exacerbating further the original event (Littler *et al.*, 2005; SEI, 2004).

Hence, the relay should be made as insensitive as possible. A compromise setting is, therefore, required to meet the conflicting requirements, while ensuring that consumer safety is paramount. The recommended setting in Great Britain is 0.125 Hz/s, while the equivalent figure for Ireland is 0.5 Hz/s. The significant difference arises because Great Britain, being a much larger system, has correspondingly greater stored energy, such that the frequency will change less rapidly when an external disturbance occurs.

However, even the higher threshold figure for Ireland may not always avoid nuisance tripping. Consider a summer night (low demand) period with a generation capacity of 3,000 MW supplying the system load of 2,500 MW. If the system H value is 4 seconds, and a 350 MW infeed is lost, the initial ROCOF can be calculated as

$$\frac{df}{dt} = \frac{-\Delta P \times f_0}{2 \times E_{system}} = \frac{-350 \times 50}{2 \times 4 \times 3000} = -0.73 \text{ Hz/s}$$

There is thus the danger of widespread nuisance tripping, whereby ROCOF protected wind farms, and other embedded generation, may be tripped unnecessarily. The severity of the initial event may therefore be increased significantly, leading possibly to load shedding. Figure 5.36 illustrates the response of a fixed-speed turbine to a frequency transient on the Ireland system. Although an inertial response (see Section 5.3.7) is provided initially, the high ROCOF (≈0.18 Hz/s), as the result of a generator loss elsewhere in the system, causes the ROCOF relay (0.125 Hz/s threshold) to activate incorrectly.

Vector shift protection offers an alternative to ROCOF protection, and is preferred on more recent wind farm installations. Again relying on a power imbalance causing the island frequency to change, the period of the electrical cycle is monitored. The recommended United Kingdom setting is a vector shift of 6° at 50 Hz, equivalent to a 0.33 ms change in the period. Assuming that the vector shift is sustained (following a frequency transient) this figure equates to a ROCOF of 0.83 Hz/s – just about

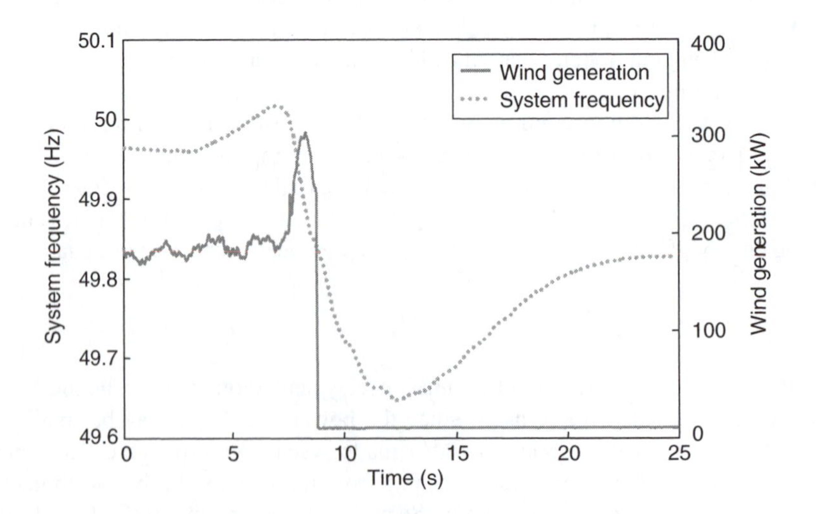

Figure 5.36 Fixed-speed turbine inertial response

conceivable on the Ireland system. However, as discussed earlier in Section variable-speed wind turbines do not naturally provide an inertial response, since the turbine rotor rotational speed is independent of the system frequency. Any reduction in the effective system-stored energy (due to variable-speed turbines displacing conventional generation) will thus further increase the ROCOF, and the likelihood of widespread nuisance tripping.

5.4 Energy storage/flexible load management

For system equilibrium, electrical generation must equal the load demand, as discussed in Section 5.2, so that the ideal system demand profile would be invariant. Under such circumstances, the most efficient generation could be scheduled to operate continuously at the desired level. However, the daily variation in electrical demand and the variable nature of renewable energy sources requires that (thermal) generating units be scheduled to start up/shut down in sympathy with load variation, with most of these units being further required to participate in load-following and cycling behaviour. In Section 5.3, against a background of increasing wind generation, various measures were examined to lessen the load-following burden placed on the remaining synchronous plant: prediction of wind farm output; integrating wind forecasting into unit commitment; geographical dispersion; incorporation of power electronics control (DFIGs); and reduction in wind farm variability by regulating wind turbine output. All of these approaches attempt to overcome the difficulties of predicting wind farm *power* output accurately and reliably, so that an instantaneous balance is achieved between demand and generation. However, it is much more straightforward to predict wind farm *energy* production (e.g. it will be windy *sometime* this afternoon).

Although many utilities possess pumped storage plant, little focus has been placed on the potential roles that management of load behaviour can play in reducing demand variability and/or scheduling of load blocks to *fill* demand troughs or *shave* demand peaks – and partially decouple energy production from energy consumption. Energy storage can perform the same roles, but may also be employed as a generation source, either replacing expensive, low efficiency peaking plant or providing ancillary support services. In theory, with sufficient storage capability and/or load scheduling, the generation capacity would be required to meet the *average* electrical demand only rather than the *peak* demand. Expensive network upgrades can thus be deferred. Through enabling thermal generating units to operate closer to rated capacity, higher thermal efficiencies are obtained, and both system fuel costs and CO_2 emissions are reduced. Further benefits also come from reducing demand variability, and hence the requirement for load cycling of generating units and the need for additional regulating reserves. Consequently, the balancing costs that may be associated with wind variability can be reduced. Expensive standing reserve, in the form of OCGTs, diesel engines, etc. can also be reduced, since both energy storage and load management can provide a similar role.

The degree of benefit obtained from the above measures, however, will depend on the existing plant mix, with the greatest opportunities available for nuclear-based power systems or those consisting of lots of must-run, base-loaded generation. Also, CCGTs and CHP plant may not be best suited to operate part-loaded and to participate in load-following and operating reserve duties (Watson, 1998). If a system consisting of such a generation mix is considered, then, particularly during periods of low demand and high wind generation, there may be insufficient unit flexibility to *load follow* while maintaining adequate reserve levels. As an alternative to curtailing wind generation output at such times, while maintaining the integrity and reliability of the power system, energy storage and load management would enable *excesses* in generation to be stored for later use. The same benefits also apply to small or synchronously isolated power systems, where, in order to ensure system stability, a minimum number of generating units may be required to run at all times. This inevitably leads to part-loaded plant, operating inefficiently at or near minimum stable generation. In both cases, the regulating burden imposed by wind (and demand) variability is reduced. In addition, the creation of a controlled non-critical load enhances the emergency response of the system to the unexpected loss of large conventional generators.

It is important to note that the benefits of energy storage and/or load management should not be considered as a means of *combating the problems* associated with wind generation. This will only have the effect of unduly constraining their operation, for example, a storage system should discharge, rather than charge, during peak demand periods, even though wind generation may be high. Instead, to be economically and operationally viable, both energy storage and load management should be seen as *system* resources utilised for global benefit. Distributed storage, coupled with wind generation, will also lead to lower distribution network loses, and probably lower transmission losses as well. The above argument can be affected by electricity market participation rules (see Chapter 7: Wind Power and Electricity Markets). The nature of any diurnal variation in wind generation, and the correlation with system demand, can influence the role that energy storage will play: greatest benefit is likely to be achieved in parts of the world, where there is a counter-cyclical relationship between wind speed and demand consumption, that is, peaks and troughs in one tend not to coincide with peaks and troughs in the other.

New York state, for example, is a case where the seasonal and daily patterns of wind generation are largely out of phase with the load demand (Piwko *et al.*, 2005). Onshore wind production tends to be high at night, low during the day, ramping down during the morning demand rise and ramping up during the evening fall – accentuating the load-following requirements of conventional plant. Wind production also tends to be lowest during the June – August period when the system load is greatest. Consequently, although the average capacity factor for onshore wind is approximately 30 per cent, the capacity credit is only 10 per cent, that is, the wind tends not to blow during peak demand periods. Interestingly, however, offshore wind generation (sited near Long Island) has a capacity credit approaching 40 per cent, similar to its capacity factor. Here, the wind strength tends to peak several hours later in the day, as compared with the onshore sites, so production is more in phase with the demand pattern. As illustrated earlier in Figures 5.9 and 5.10, depicting a typical electrical demand profile and wind generation profile for Ireland, the same is not true for northwestern Europe. Here, a weak correlation between wind speed and electrical demand exists, suggesting that the benefits of energy storage are less compelling.

5.4.1 Conventional energy storage

A wide variety of energy storage technologies are commercially available and include pumped storage, rechargeable batteries, flow batteries, and compressed air. Potential benefits, of relevance here, include capacity reduction, frequency support, standing reserve provision and blackstart capability. Depending on technical requirements and geographical settings, a particular utility may avail of one or more of these technologies. Research effort has also focussed on ultracapacitors, high-speed flywheels and superconducting magnetic energy storage (SMES). While these are highly responsive, their energy storage capabilities are limited, making such approaches more suitable for power quality applications and for improving system reliability. Of course, even here some niche applications of interest do exist: Palmdale water district, California employs a 450 kW supercapacitor to regulate the output of a 950 kW wind turbine attached to the treatment plant microgrid (Gyuk *et al.*, 2005). This arrangement helps to reduce network congestion in the area, while providing reliable supply to critical loads in the microgrid.

5.4.1.1 Pumped storage

The most widely established large-scale form of energy storage is hydroelectric pumped storage (see Section 5.2). Typically, such plant operates on a diurnal basis – charging at night during periods of low demand (and low-priced energy) and discharging during times of high or peak demand. A pumped storage plant may have the capacity for 4–8 hours of peak generation with 1–2 hours of reserve, although in some cases the discharge time can extend to a few days. Worldwide capacity is almost 100 GW, with facilities ranging up to 2,000 MW. The high construction costs, long development times and environmental considerations (most feasible locations are already being exploited) suggest that future growth in this area will be limited. Traditionally, pumped storage is utilised for energy management and the provision of standing reserve, but more recent installations possess the ability to provide frequency support and operate at partial capacity (Donalek, 2003).

Hydroelectric plants typically have fast ramp-up and ramp-down rates, providing strong regulating capabilities, and their marginal generation cost is close to zero. In many countries, a natural synergy exists between hydroelectric generation/pumped storage and wind power. Clearly, if hydro generation is being *replaced* by wind energy then emission levels will not be directly affected, but the hydro energy can be transformed into potential energy stored for later use. As an example, during periods of high wind power production, network congestion can occur in both Sweden and Norway. The existing hydroelectric plant can reduce their output, using the reservoirs as storage, to avoid wind energy curtailment (Matevosyan and Söder, 2003; Tande and Vogstad, 1999). Similarly, in Portugal, most of the wind farms are located in the north of the country, close to existing hydroelectric plant (Peças Lopes, 2005). However, the major load centres of Lisbon and Porto lie to the south, potentially leading to future congestion problems in the transmission network. A critical concern for the interconnected Iberian system is that a network disturbance may cause a sudden loss of wind generation, leading to an influx of power from neighbouring regions and an overload of the limited interconnection capacity between Portugal and Spain, and between Spain and France. The large differential in payment for wind energy between valley hours and off-valley hours in Portugal does suggest, however, that a pumped storage arrangement could become economically viable (Castronuovo and Peças Lopes, 2004). On a much smaller scale, pumped storage has also been proposed in the Canary Islands to mitigate wind variability using existing water reservoirs (Bueno and Carta, 2006).

5.4.1.2 Secondary batteries

Rechargeable lead-acid and nickel-cadmium batteries have been used widely by utilities for small-scale backup, load levelling, etc. The largest (nickel-cadmium) battery installation is a 45 MW, 10 MWh installation in Fairbanks, Alaska built in 2003, and designed to provide a guaranteed 27 MW for at least 15 minutes following local power outages. For similar reasons, the largest (20 MW, 14 MWh) lead-acid system was installed by the Puerto Rico Electric Power Authority in 1994, and later re-powered in 2004. However, given the fairly toxic nature of the materials involved, low efficiency (70–80 per cent) and the limited life and energy density, secondary batteries based on other designs are being sought for utility-scale applications. The sodium-sulphur battery consists of molten sulphur at the positive electrode and molten sodium at the negative electrode separated by a solid beta-alumina ceramic electrolyte – the electrolyte permits only sodium ions to traverse, combining with the sulphur to form sodium polysulphide. Operating at approximately 300 °C, they have a high energy density, approaching 650 MJ/m^3, and a cycle efficiency of 85–90 per cent including heat losses, making them suitable for large-scale storage applications. Over one hundred projects have been installed worldwide, with over 30 of these in Japan. The latter include 2 $\times$ 8 MW, 60 MWh installations, used for daily peak shaving, with a nominal discharge time of 7.5 hours at rated power (Nourai *et al.*, 2005).

Alternatively, the sodium-nickel-chloride battery employs nickel/sodium chloride at the positive electrode and liquid sodium acting as the negative electrode, with a liquid $NaAlCl_4$ electrolyte contained within beta alumina. In comparison with sodium-sulphur, safety characteristics are improved, along with an ability to withstand limited overcharge/discharge. However, both the energy density (550 MJ/m^3) and power density (300 kW/m^3) are reduced. Although currently aimed at the automotive market, proposals within the 100 kWh – 10 MWh range have been made for load levelling applications. Similarly, lithium ion and lithium polymer batteries are expected to be expandable from their current prevalence in the portable electronics market to larger-scale utility needs. The cathode can be formed from a range of lithiated metal oxides, for example $LiCoO_2$ and $LiMO_2$, while the anode is graphite carbon, and the electrolyte is formed from lithium salts, such as $LiPF_6$, dissolved in organic carbonates (Schaber *et al.*, 2004). The operating temperature is 60 °C, with an energy density of 720 MJ/m^3 and an efficiency exceeding 85 per cent. The main obstacle to further development is the high cost of special packaging and internal overcharge protection circuits.

5.4.1.3 Flow batteries

Flow batteries store and release electrical energy through a reversible electrochemical reaction between two liquid electrolytes. The liquids are separated by an ion-exchange membrane, allowing the electrolytes to flow into and out from the cell through separate manifolds and to be transformed electrochemically within the cell. In standby mode, the batteries have a response time of the order of milliseconds to seconds, making them suitable for frequency and voltage support. Battery capacity depends on the volume of solution, providing economies of scale with larger installations. This contrasts with secondary batteries where both energy and power density are affected by both the size and shape of the electrodes – larger batteries are, therefore, not directly scalable.

There are three types of flow battery that are approaching commercialisation: vanadium redox, polysulphide bromide and zinc bromide. The vanadium redox design utilises vanadium compounds for both electrolytes, which eliminates the possibility of cross contamination and simplifies recycling. The polymer membrane between the two electrolyte tanks is permeable to hydrogen ions, enabling ion exchange during the charge/discharge cycles. Cycle efficiencies over 80 per cent have been achieved, and long cycle lifes are expected, although the energy density is comparatively low. As yet only demonstration systems have been tested – a 1 MW vanadium bromide system has been investigated in Castle Valley, Ohio, United States for voltage/reactive power support (van der Linden, 2004).

Alternatively, the polysulphide bromide battery utilises sodium bromide and sodium polysulphide electrolytes separated by a polymer membrane that only permits sodium ions to penetrate. A storage time exceeding 12 hours has been achieved, with a cycle efficiency of 75 per cent, enabling daily charge and discharge cycles. A demonstration project at Little Barford, England was proposed (but never completed) to provide 120 MWh of energy at 10 MW rating to Southern Electric's 33 kV distribution network. The Regenesys company which originally developed the technology was taken over in 2004. Finally, the zinc bromide configuration has a cycle efficiency of about 75 per cent, and many small-scale units have been built and tested. Here, the two electrolyte storage reservoirs of a zinc solution and a bromine compound are separated by a microporous polyolefin membrane. Multi-kWh designs are available for assembly complete with necessary plumbing and power electronics, while larger installations (2 MWh) have been used for substation peak management.

5.4.1.4 Compressed air storage

In an OCGT or CCGT plant, incoming air is compressed by the gas turbine compressor before being ignited with the incoming fuel supply. The exhaust gases are then expanded within the turbine, driving both an electrical generator and the compressor. If suitable large-scale storage is available, such as an underground mine, salt cavern, aquifer, etc. the compressor alone can be operated at off-peak times to create a ready supply of compressed air. Alternatively, an underground storage complex can be created using a network of large diameter pipes. Later, the compressed air can be *released* as part of the generation cycle, providing a cycle efficiency of approximately 75 per cent (Kondoh *et al.*, 2000). Commercial installations are still few, but include a 290 MW unit built in Hundorf, Germany in 1978 and a 110 MW unit in McIntosh, Alabama, United States, built in 1991. The Hundorf installation was originally intended as fast-acting reserve for a nearby nuclear power station, but has since been modified to provide a grid-support role (van der Linden, 2004). In contrast, the much greater cavern capacity of the McIntosh installation enables it to generate continuously for 24+ hours.

More recently, in 2004, a 2700 MW plant was proposed for Norton, Ohio, United States consisting of 9 × 300 MW generating units and an existing limestone mine 700 m beneath the surface with a cavern volume of 120×10^6 m^3, providing a storage capacity of 43 GWh (van der Linden, 2004). Independent of this, an underground aquifer has been selected near Fort Dodge, Ohio, United States to store compressed air at 36 bar *driven* by a 100 MW wind farm and is expected to be operational by 2010. A separate section of the aquifer provides storage of natural gas, enabling purchase when prices are low, for supply to 200 MW of generating capacity.

All of these plant operate on natural gas, but conceptually, at least, more environmentally friendly schemes can be envisaged using gasified biofuels, forestry residue, etc. as the fuel supply. With the future growth of *clean coal* generation plant, based on integrated gasification combined cycle (IGCC) technology, similar off-peak storage can be envisaged. IGCC plant are likely to be relatively inflexible, so energy storage will also enable such plant to maintain operation during off-peak periods.

5.4.2 Demand-side management

Often neglected by utilities, management of load behaviour can have a major impact on the operation and planning of power systems. Measures can be categorised as either reducing energy consumption or rescheduling to a later time. The former includes techniques such as energy-efficient lighting and household appliances, building energy management systems, and architectural design to reduce heating and ventilation requirements. Additionally, for large industrial and commercial customers, maximum demand charges have encouraged the growth of on-site, embedded generation. In the United States, consumers can also sell *saved* electricity to the power system during periods of peak demand and/or shortages (Sweet, 2006). Consequently, the peak system demand is reduced, delaying, or even avoiding, the construction of new generation plant and expansion of the transmission network.

Of greater interest here is the ability to time shift the electrical demand from periods of high demand, when generation capacity is stretched, to periods of low demand, when spare capacity is available. Utilities attempt to encourage this through *time-of-day* tariffs, with rates often considerably higher during peak periods. Interruptible load tariffs may also be offered to large consumers, whereby a price discount is given in exchange for a commitment to reduce demand when requested by the utility. Triggered by under-frequency sensitive relays, fast standing reserve can thus be provided during emergency conditions. Such schemes reduce demand variability and are thus of great benefit to the utility. In New Zealand, for example, interruptible load has become the primary source of operating reserve. However, although there is likely to be a diurnal variation in wind output, neither of the above measures directly assists wind integration.

The nature of low grade water and space heating, air-conditioning equipment, heating of swimming pools or even refrigeration systems, is that they have inherent storage capabilities, that is, electrical power can be disconnected without any initial, obvious effects. Similar *storage* potential also exists for pneumatic compressed air supply, production line inventory, municipal water pumping systems, water desalination and purification, aluminium smelting, etc. Typically, such loads require no advance warning of curtailment, and provide an instantaneous and full response. For some processes time may be required for valves to operate and/or shut down procedures to be completed. Consequently, the heating/refrigeration load can be *charged* during periods of low demand and switched out during peaks. Alternatively, the system can be activated during periods of high wind generation and switched out during lulls, reducing the impact of high wind energy penetration. However, a common characteristic of load storage is that it is limited in duration, for example, a consumer may accept (or not even notice) a curtailment in hot water/heating load for 30–60 minutes, but a sustained outage period of discomfort could not be accepted (particularly during cold, winter evenings when electrical demand is at its peak). On the positive side, there is often a useful correlation between load behaviour and weather patterns. So, in northern Europe for example, it is generally windier in the winter relative to the summer, and windier during the day than at night (see Fig. 5.9). Windy, winter evenings will also introduce a *wind chill* factor that increases the space-heating requirement further.

Through balancing the output of an individual wind farm against a local (storage) load, network impact can be lessened, and it may even be possible to increase the capacity of the wind farm. For example, at the Mawson base, Antarctica the large, controllable heating load enables 100 per cent wind penetration for up to 75 per cent of the year, with fuel cells providing longer-term storage (AGO, 2003). However, the true benefits of such arrangements only accrue when extending and coordinating the scheme across larger network areas. A study for the Northern Electric regional electricity company (REC) in the United Kingdom indicated that active control of consumer load could enable an additional 1,600 MW of onshore wind farms to be accepted on to the network (Milborrow, 2004). Under system emergency conditions, the ability to switch off load quickly also offers a valuable source of operating reserve. Within the Long Island (New York) Power Authority (LIPA) it has been proposed that responsive load act as fast/emergency reserve (Kirby, 2003). Residential and small commercial air-conditioning load was mainly examined. LIPA has over 20,000 residential consumers and 300 small commercial consumers, who have centrally controlled air-conditioning thermostats for demand peak reduction to alleviate network constraints during the summer. Demand reduction of 25 MW is available, while a conservative estimate of 75 MW of 10-minute *emergency* reserve should be achievable at low cost during periods of high demand – reserve provision is highly correlated with the system demand. In the LIPA scheme demand reduction is achieved by raising the temperature setpoint, while reserve would be achieved by completely shutting off the load. Hence, the same load can provide both demand reduction and reserve capability. Similarly, within the PJM (Pennsylvania, New Jersey and Maryland) control area a curtailment service provider (CSP) provides approximately 50 MW of load reduction in winter (35 MW in summer) through control of residential electric water heaters, water pumps and thermal storage space heaters (Heffner *et al.*, 2006). Communications and dispatch of 45,000 load control switches is achieved using a telephone/radio communications system, managing the daily peak demand.

Forecasting of load behaviour can generally be achieved with high accuracy, as discussed in Section 5.2. Forecasting of the availability of responsive loads should therefore be equally achievable. It is, perhaps, even easier, since there is likely to be less diversity (relative to the total load) in the size and rating of the *storage* devices and they are likely to be driven by the same time-of-day, time-of-year and weather patterns. The difficulty often faced with load management schemes, however, is the ability to control and coordinate switching operations centrally. Traditionally, communications and control technology made it much easier to monitor the operation of a few large resources (generators) rather than the multitude of dispersed small resources (loads). Radio teleswitching and mains signalling ripple control have been applied for many years for water heating, so that, for example, a well-lagged domestic hot-water tank of reasonable size can deliver adequate hot water over 24 hours derived from 7 hours of supply during the off-peak night time period (McCartney, 1993). However, communication is unidirectional and commands are normally only sent every 24 hours. Telecommunications technology has advanced to the point where real-time bi-directional control of air-conditioning loads is now possible (Kirby, 2003). Using a pager system, thermostats can receive curtailment orders while returning acknowledgement/override signals and system status reports. The thermostats may be accessed collectively, individually or even as a group, so that local network constraints can be addressed directly. Approaches using text messaging and mains signalling are also possible. More conveniently, perhaps, loads such as air conditioners, refrigerators, etc. could be designed to be frequency sensitive, increasing the load when the frequency is high and decreasing the load when the frequency is low – just like the governor control on a steam-turbine-generator (Vince, 2005). Many modern offices and buildings incorporate energy management control systems that provide the capability of controlling electrical equipment associated with heating systems, air conditioning and lighting. Off-site monitoring is generally provided, but few utilities currently exploit such features in their own regional control centres.

5.4.3 Hydrogen energy storage

Hydrogen has been proposed as the energy store (carrier) for the future, and the basis for a new transport economy. The reasons for this are simple: hydrogen is the lightest chemical element, thus offering the best energy/mass ratio of any fuel, and in a fuel cell can generate electricity efficiently and cleanly. Indeed, the waste product (water) can be electrolysed to make more fuel (hydrogen). Hydrogen can be transported conveniently over long distances using pipelines or tankers, so that generation and utilisation take place in distinct locations, while a variety of storage forms are possible (gaseous, liquid, metal hydriding, etc.). For transport needs, fuel cells in vehicles combine multi-fuel capability, high efficiency with zero (or low) exhaust emissions and low noise. Portable applications including mobile phones and laptop computers can also employ such compact storage. Fuel cells also encourage the trend towards decentralised electrical generation and/or the growth of CHP schemes, through exploiting the waste heat.

Iceland has set itself the target of being the first hydrogen economy in the world, totally eliminating the need for fossil fuels within the next generation (2030–40). The country meets virtually all its electricity and heating requirements from renewable (hydroelectric and geothermal) sources, with excess energy used to generate hydrogen through electrolysis (Árnason and Sigfússon, 2000). Similarly, the United States Department of Energy is aiming for at least 10 per cent of annual energy consumption to be provided by hydrogen-powered fuel cells by 2030. A study by the Danish Department of Energy has forecast that by 2030 the growth in wind power would allow 60 per cent of transport energy needs to be met by hydrogen (Sorensen *et al.*, 2004). A further 20 per cent of vehicles could be expected to run on methanol, obtained by biomass gasification, again derived from wind power.

5.4.3.1 Hydrogen production

Hydrogen is abundant and distributed throughout the world without reference to political boundaries, but is *locked up* in chemical compounds such as water, methane, ethane and other hydrocarbons. The hydrogen can be released through steam reformation of fossil fuels, metabolic synthesis by unicellular cyanobacteria from water, thermal decomposition (thermolysis) of water at high temperature ($>2,000\,°C$), and photosynthetic conversion of water and carbon dioxide by single-cell organisms such as algae. At present, most of the world's hydrogen is produced through reformation, but from an environmental point of view there is little logic in such an approach – fossil fuels are involved and CO_2 emissions are still released. Of more relevance here, electrolysis of water enables the hydrogen atoms to be split from the oxygen atoms. Much of the electricity required for electrolysis currently comes from the burning of fossil fuel sources, with an associated release of various pollutants into the atmosphere. However, electrical generation from wind, tidal stream, biomass, photovoltaic cells, etc. can equally be used to provide hydrogen, so that over the complete cycle no emissions are released. Conversion efficiencies of 77–82 per cent are representative, although techniques in development, such as advanced alkaline electrolysis, may advance this figure towards 90 per cent (Sherif *et al.*, 2005). Typically, electrolysis plants operate well over a range of load factors, making them suitable for wind *balancing*, although electrode lifetime and cycle efficiency is enhanced by operating at constant output.

5.4.3.2 Hydrogen storage and containment

Having created the hydrogen (from electrolysis) the issue of storage or containment immediately needs to be addressed. In comparison with conventional fossil fuels, hydrogen has a high specific energy (energy/mass), but a low energy density (energy/volume). The most obvious way of improving the density is by compression. Across the world, gases such as helium, town gas and hydrogen have been stored in rock strata, abandoned salt mines and gas caverns. For example, Gaz de France stores hydrogen-rich refinery byproducts in an aquifer near Beynes, France, while the Kiel public utility in Germany stores town gas (60–65 per cent hydrogen) in an underground cavern. The storage potential of these repositories is huge; however, the availability and accessibility of such natural stores is likely to be limited. Alternatively, suitably sized containers can be used, using standard storage pressures of 350 and 700 bar. The storage efficiency achievable is approximately 90 per cent, comparable with underground storage (Schaber *et al.*, 2004). Since electrolysers can operate at pressures up to 200 bar, additional compression equipment may thus be avoided or reduced in cost. Energy densities achievable are, respectively, 3,000 and 5,000 MJ/m^3, which greatly improves upon comparable figures for rechargeable batteries (600–800 MJ/m^3) and flow batteries (100–150 MJ/m^3). Other techniques under development include cryogenic storage as a liquid, adsorption in nano-structured materials, storage between thin layers of closely spaced ($\approx 6 \times 10^{-10}$ m) graphite or graphene sheets, covalent organic frameworks (rigid plastics) and hydrogen bonding with various atoms or molecules, including metal hydrides (Crabtree *et al.*, 2004). These solutions improve energy density further but are likely to be expensive and/or inefficient. For utility needs energy storage density is only important relative to the size of the energy production facility itself, so the more economic compression methods are likely to be pursued. For small-scale, portable applications, including vehicles, where energy density is paramount, the alternative methods may be considered.

As a consideration for the future, hydrogen pipeline infrastructures are likely to be developed around the world. Excess hydrogen (i.e. energy) could be stored by temporarily increasing the gas pressure. One proposal from the United States is for a hydrogen pipeline *connecting* wind farms (electrolysing water to form hydrogen) in South Dakota to Chicago, 1,600 km away (Keith and Leighty, 2002). If the gas pressure were varied between 35 and 70 bar, 120 GWh of hydrogen storage could be obtained, equivalent to a constant 5 GW load for a 24-hour period. As well as providing a storage capability, the scheme has the further benefit of reducing wind power variability, since the wind energy is not directly used for electrical generation. For distances greater than 1,000 km, *energy* transportation by hydrogen carrier should be more economical than high-voltage electrical transmission (Sherif *et al.*, 2005).

5.4.3.3 Fuel cell generation

The stored hydrogen can be converted back to electricity using an OCGT arrangement. However, electrical efficiency tends to be low (30–35 per cent), even ignoring transportation losses and those associated with converting the electricity to hydrogen in the first place. Fuel cells offer an alternative approach, and essentially consist of an electrolyte (liquid or solid) membrane sandwiched between two electrodes. Fuel cells are classified according to the nature of the electrolyte used and the operating temperature, with each type requiring particular materials and fuels (Sherif *et al.*, 2005), as

Table 5.2 Categorisation of fuel cell types

Fuel cell type	Electrolyte	Efficiency (%)	Operating temperature (°C)	Fuel
Alkaline	Potassium hydroxide (conc.)	50–60	50–90	H_2
Direct alcohol	Polymer membrane/ liquid alkaline	35–40	60–120	Methanol, ethanol
Molten carbonate	Alkali (Li, Na, K) carbonates	45–55	600–750	H_2, CH_4, biogas, coal gas, etc.
Phosphoric acid	Phosphoric acid (conc.)	35–45	150–220	H_2, CH_4
PEM	Polymer membrane	35–45	60–90	H_2
Solid oxide	Metal oxide (Y_2O_3/ZrO_2)	45–55	700–1,000	H_2, CH_4, biogas, coal gas, etc.

summarised in Table 5.2. The electrochemical efficiency tends to increase with fuel cell temperature. It is often the nature of the membrane that dictates the operating temperature, and expensive catalysts, such as platinum, may be required to speed up the rate of electrochemical reactions. Proton exchange membrane (PEM) fuel cells predominantly employ perfluorosulphonic acid polymer as the electrolyte, which, since the membrane must incorporate mobile water molecules, restricts operation to less than 100 °C. Conversely, solid oxide (SO) fuel cells consist of perovskite-based membranes that only become sufficiently conductive above 800 °C – this requires expensive construction materials and limits the operating environments in which they can be used (Crabtree *et al.*, 2004).

In addition to the obvious raw fuel of hydrogen, hydrocarbons, alcohols, coal, etc. can also be seen as exploitable sources of hydrogen. More environmentally friendly fuels, such as biogas and biomass, may also be employed. For example, the New York port authority employs anaerobic digester gas as the fuel, derived from its wastewater treatment plants (Kishinevsky and Zelingher, 2003). For most fuel cells, such fuels must be transformed into hydrogen using a reformer or coal gasifier (possibly including CO_2 sequestration). However, high temperature fuel cells can generally use a fossil fuel (natural gas, coal gas, etc.) directly. Polluting emissions are produced, but since hydrogen is released chemically (and not by combustion) the quantities are low. The hydrogen is passed over one electrode (anode), where the hydrogen molecules separate into electrons and protons. Only the protons can pass through the electrolyte membrane to the cathode where they combine with oxygen to form water. The oxygen supply may be derived from air, or as a stored byproduct from the water electrolysis (forming hydrogen). The waste heat generated may also be exploited in a CHP scheme, boosting the conversion efficiency.

By covering one of the electrodes with a hydriding substance the fuel cell and hydrogen storage functions can be combined into a single device (Sherif *et al.*, 2005).

So, during water electrolysis, hydrogen is absorbed by titanium nickel alloy, for example, coating one of the electrodes. Later, the device can reverse its operating mode to become a fuel cell by *releasing* the stored hydrogen in the metal hydride. It is probable, however, that hydrogen storage will remain largely distinct from fuel cell generation, in order that the stored energy (capacity) of the former can be independent of the power rating of the latter.

For large-scale utility storage applications, the choice of technology will depend on the ability to use pure hydrogen (electrolysed from water) as the fuel, the electrical efficiency of conversion, and the load-following capability of the fuel cell, thus providing a degree of regulation from fluctuating wind or other renewable sources. Of the various options available, SO and PEM seem most likely to succeed (Steinberger-Wilckens, 2005). SO fuel cells offer high electrical efficiency, although they cannot be shut down easily and are perhaps more suited for *base* load rather than *flexible* load operation. In contrast, PEM fuel cells operate at a comparatively low temperature, which enables both fast starting, flexibility of operation and increased durability. Use of a solid polymer as the electrolyte improves safety and eases transportation concerns. PEM technology is, therefore, likely to be highly favoured for transportation applications, leading to a substantial reduction in cost as fuel cell vehicles gradually proliferate.

5.4.3.4 Power system integration

At this early stage, the final structure of any *hydrogen economy* is unclear and it remains uncertain how it should be best integrated with the existing electrical infrastructure. One of the limiting factors in future development is fuel cell efficiency – a likely target figure is 60 per cent. If this is combined with an electrolyser efficiency of 90 per cent and a compression storage efficiency of 90 per cent, then an overall cycle efficiency of about 50 per cent is achievable. This compares unfavourably with cycle efficiency figures for pumped storage (60–80 per cent), secondary batteries (75–85 per cent), flow batteries (75–80 per cent) and compressed air storage (75 per cent). The one advantage offered by hydrogen not available to the other technologies is the ease with which energy generation and storage can be physically separated, assuming that a pipeline/transport infrastructure is in place. The low efficiency ($\approx$20 per cent) of today's internal combustion engines and their noxious emissions strongly suggest, however, that fuel cell vehicles have a bright future.

A number of options for system integration are available: hydrogen could be stored at a utility level and then distributed *electrically* along transmission lines during peak demand periods; hydrogen could be pumped by pipeline to widespread locations and then transmitted electrically along distribution lines; or hydrogen could be delivered direct to the end consumer where a fuel cell would provide electricity and (waste) space heating. Alternatively, the electrolysed hydrogen could be used to drive a fleet of hydrogen-powered cars, buses, and other vehicles. In all likelihood, a combination of the above scenarios will evolve, with government policy and topographical issues as well as technical and economic arguments influencing the final balance (Anderson and Leach, 2004).

Wind generation can fit into this proposed hierarchy in a number of ways: low-scale, on-site hydrogen production would reduce wind generation variability, protect against regional network overloading and/or provide a short-term backup supply during becalmed periods. ScottishPower, for example, is aiming to exploit excess wind output (which cannot be accepted on to local grids in Scotland) for hydrogen production, from 200 MW of installed wind capacity, beginning in 2007. Alternatively, wind farms may be built with the primary objective of providing hydrogen rather than utility connected electricity. Any such exploitation of hydrogen for transport or other applications (rather than fuel cell generation) will lessen the power system, if not commercial, benefit. With the expected growth of large offshore wind farms, a hydrogen pipeline offers an alternative method of transporting *wind energy* ashore. The necessary equipment costs and transportation losses are comparable with laying a high voltage cable. However, given the low cycle efficiency of converting electricity to hydrogen and back again, this can only be justified if there is a benefit to the system or a market exists for the hydrogen supply.

Chapter 6

Wind power forecasting

6.1 Introduction

The increasing penetration level of wind power into power systems has resulted in the requirement for accurate, reliable, online wind power forecasting systems. The typical transmission system operator (TSO) requirements for a forecasting system have been identified by Schwartz (2000) as follows:

- The forecasts should be of wind power output (in MW), rather than wind speed, with look-ahead times extending out to 48 hours.
- The forecasts should be available for individual wind farms, for regional groupings of wind farms and for the total wind power installed in a TSO's area.
- The forecasts should be accurate and supplied with an associated level of confidence – clearly dispatchers would tend to be more conservative when dealing with large forecast uncertainties.
- The forecast should predict changes in wind power reliably.
- There should be a good understanding of the meteorological conditions which would lead to poor quality forecasts.
- Historical data should be used to improve the forecast over time.

Wind power forecasts are used as inputs to the various simulation tools (market operations, unit commitment, economic dispatch and dynamic security assessment) available to the power system operators to ensure that economic, efficient and secure operation of the power system is maintained.

Over the last decade or so there has been considerable activity and progress in the development of wind power forecasting. However, there is still much scope for improvement. Two broad strands can be identified in the systems that have been developed so far: those using predominantly physical modelling techniques; and those using predominantly statistical modelling techniques. In the physical modelling approach the physical atmospheric processes involved are represented, in as far as this is possible. The statistical modelling approach is based on the time series of wind farm power measurements which are typically available on-line. If purely statistical modelling techniques are used, good forecast results can be achieved for the short-term look-ahead times (from 0 to 6 hours) only. Beyond this, and especially in the 12- to 48-hour range of look-ahead times, it is essential to use an input from a numerical weather prediction (NWP) model if successful results are to be achieved. As the development of wind power forecasting techniques has progressed over the last decade there has been significant cross fertilisation between the physical and statistical approaches, with the result that most modern advanced wind power forecasting systems use a combination of both physical and statistical modelling.

In this chapter, some relevant meteorological background is first outlined, leading to a brief description of the very complex field of NWP. The most basic forecasting model, that is, persistence forecasting, is then explored, and the opportunity is taken to introduce the various error measures that are used to quantify the performance of wind power forecasting systems. The more advanced wind power forecasting systems are then considered. In the discussion of the different systems results from particular applications are provided to illustrate important features.

6.2 Meteorological background

6.2.1 Meteorology, weather and climate

Meteorology is the science of the atmosphere of planets. As applied to the earth, it is concerned with the physical, dynamical and chemical state of the atmosphere and the interactions between the atmosphere and the underlying surface including the biosphere (Meteorological Office, 1991). The methods employed are both qualitative and quantitative, based on observation but also on analysis and the use of complex mathematical models. Understanding and predicting the internal motion of the atmosphere, which is caused by solar radiative heating, lies at the heart of meteorology.

Meteorology includes the study of weather and climate. Weather is the changing atmospheric conditions at a particular location and time as they affect the planet. The elements of the weather are temperature, atmospheric pressure, wind, humidity, cloudiness, rain, sunshine and visibility. Climate is a summary of the weather experienced at a location in the course of the year and over the years. Climate is usually described not only just by average values but also extreme values and frequency of occurrence of the weather elements. As the average condition of the weather elements change from year-to-year, climate can only be defined in terms of some period of time. Climate data are usually expressed for individual calendar months and are calculated over a period long enough (e.g. 30 years) to ensure that representative values for the month are obtained. The climate at a particular location is affected by its latitude, by its proximity to oceans and continental land masses and large-scale atmospheric circulation patterns, its altitude and local geographical features.

Wind power meteorology is a new term used by Petersen *et al.* (1997) to describe the theories and practice of both meteorology and climatology as they apply specifically to wind power. This field has developed in the last few decades and includes three main areas: (1) regional wind resource assessment, (2) micro-siting of wind turbines and (3) wind farms and short-term wind power forecasting.

6.2.2 *Atmospheric structure and scales*

The atmosphere is a thin film of gas clinging to the earth's surface under gravitational attraction. The horizontal and vertical spatial scales in the atmosphere differ widely. In the horizontal the spatial scale is very large, which can readily be appreciated from photographs of the earth from geo-stationary satellites. This horizontal scale is quite large, of the order of the earth's size (the earth's mean radius is 6,370 km). In the vertical the spatial scale is very much smaller. Although the atmosphere does not have any definite upper surface, 90 per cent of its weight is concentrated in the lower 16 km, which corresponds to only about 0.3 per cent of the earth's radius (McIlveen, 1992). The lower 16 km, although small in scale, is very significant in its effect in meteorology, since it is here that most of the cloud activity and the weather is produced. This part of the atmosphere is called the troposphere.

The atmospheric structure is also quite different in the horizontal and the vertical, with higher gradients in the vertical than in the horizontal, giving rise to a markedly stratified appearance. An example of this difference is how temperature decreases with height in the vertical at about 6 °C/km, whereas the strongest horizontal gradients associated with fronts in mid-latitudes would rarely exceed 0.05 °C/km (McIlveen, 1992).

The region of the atmosphere of most interest from a wind power perspective is the planetary boundary layer, which is the layer directly above the earth's surface. It is variable in depth but of the order of 500–2000 m. This region is dominated by turbulent interactions with the surface, with turbulent eddies in the spatial range from 500 to 5 m. The atmosphere above the planetary boundary layer is called the free atmosphere and is much less affected by friction with the earth's surface. The free atmosphere is dominated by large-scale disturbances.

Air motion is a very complicated phenomenon and covers a wide range of scales, both spatial and temporal. Small-scale features are visible in the movements of smoke from a burning candle, medium-scale features in the plumes from fossil fuel burning power stations and the large-scale circulations of the atmosphere are visible from geo-stationary satellites. In the atmosphere there is a range of eight orders of magnitude in spatial scale and an almost equally large range of temporal scales corresponding to the lifetime of the disturbance phenomena. This can be seen in Figure 6.1, which is adapted from McIlveen (1992). The disturbance phenomena range from short-lived and small spatial scale turbulence through convection processes in the formation of clouds, to meso-scale systems such as sea breezes, through to the synoptic scale growth and decay of extra-tropical cyclones which dominate the weather patterns of north-western Europe. Over most of the range of these phenomena the ratio of spatial to time scales is approximately 1 m/s, and is shown as the straight dashed line in Figure 6.1. It is a measure of the intensity of the activity of the atmosphere.

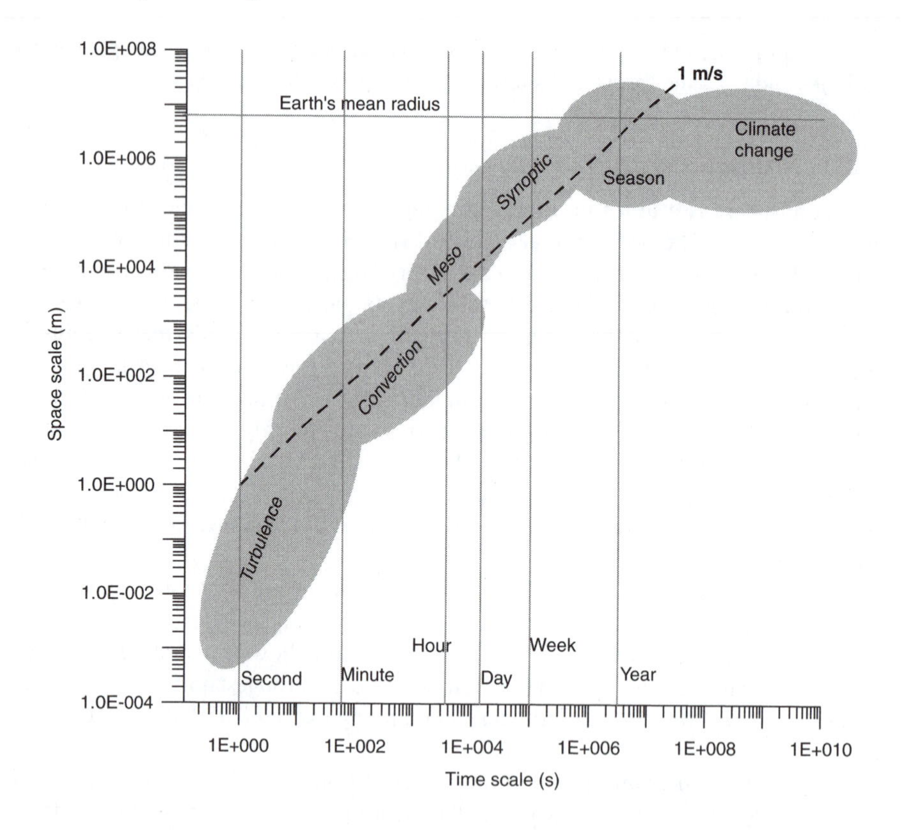

Figure 6.1 Spatial and temporal scales of disturbance phenomena in the atmosphere

6.3 Numerical weather prediction

The state of the atmosphere can be described by seven meteorological variables: pressure, temperature, amount of moisture, air density and wind velocity (two horizontal components and a vertical component). The behaviour of these variables is governed by seven physical equations, three arising from thermodynamic considerations and four arising from hydrodynamic considerations (Atkinson, 1981). The thermodynamic equations are the gas law, also known as the equation of state, the first law of thermodynamics and an equation representing the conservation of moisture. The hydrodynamic equations are the equation of continuity and the three equations of motion corresponding to the components of Newton's second law in three directions. The seven governing equations involve the atmospheric variables and their spatial and time derivatives. In order for the atmospheric model to provide a usable description of the behaviour of the atmosphere, a solution must be found to the equations. In general, no analytical solution is possible and numerical methods must be adopted. The task involved is to calculate how each of the meteorological variables will change as the simulation runs forward in time.

NWP is an objective forecast in which the future state of the atmosphere is determined by the numerical solution of a set of equations describing the evolution of meteorological variables which together define the state of the atmosphere (Meteorological Office, 1991). In the excellent review of the developments in NWP to be found in Kalnay *et al.* (1998) it is pointed out that, while there has been huge progress in the last two decades, with a doubling of forecast skill, there are three major requirements for improved NWP: (1) better atmospheric models, (2) better observational data and (3) better methods for data assimilation. Data assimilation involves the quality check run on the weather observations received via the Global Telecommunication System (GTS), conversion of the accepted data to a format required by the analysis scheme, processing of boundary values at the lateral boundaries of the model domain and statistical procedures employed to make corrections to first guess forecasts (usually forecasts from the previous analysis time), so that the differences between the corrected first guess and the accepted observations at the analysis time are minimised. All of these improvements are computer intensive and will be facilitated by ever increasing availability of computer power.

One of the basic principles underlying the numerical modelling of the atmosphere is that the atmosphere as a continuous fluid can be represented by the values of the seven meteorological variables at a finite set of discrete points. These points are arranged in a three-dimensional grid with fixed spatial coordinates. The spacing of the grid points is also known as the spatial resolution or the grid length. Different spatial resolution is generally used in the vertical compared to the horizontal, due to the stratified structure of the atmosphere in the vertical. Increasing the spatial resolution increases the accuracy of the atmospheric model, but at the cost of additional computational effort. Thus, if the grid size is halved, the computational effort increases by a factor of $2^4 = 16$ as the number of grid points is doubled in each space direction and the number of time steps must also be doubled.

The choice of vertical and horizontal spatial resolutions is clearly very important. In the atmosphere there are motions with spatial scales ranging from many thousands of kilometres down to a millimetre. In order to capture all atmospheric processes, the atmospheric model should ideally be constructed with spatial resolutions down to a millimetre. Clearly this would impose impossibly high burdens on computer resources. On the other hand, a particular spatial resolution will exclude from consideration any physical atmospheric process whose spatial scale is smaller than the spatial resolution. The choice of spatial resolution is therefore a trade-off between an acceptable level of accuracy and available computer power.

For a particular grid resolution there will always be sub-grid scale physical processes that the model cannot resolve. In order to be realistic, the model must take account of the net effect of these processes. This is done by parameterisation of these effects, based on statistical or empirical representations of the sub-grid size physics. These parameterisations will also contribute to the computational effort.

NWP models can be global or limited area models. The limited area model is nested within the global model. The high-resolution limited area model (HIRLAM) is used widely in Europe and is a result of cooperation between the meteorological institutes of Denmark, Finland, Iceland, Ireland, Netherlands, Norway, Spain,

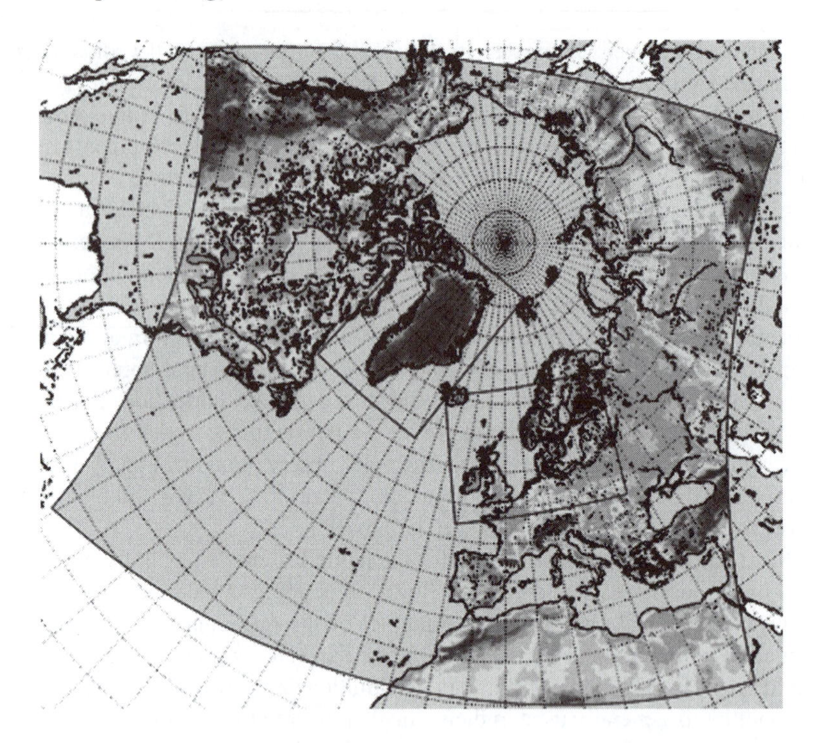

Figure 6.2 DMI-HIRLAM domains

Sweden and France. HIRLAM receives its lateral boundary conditions from the European Centre for Medium Range Weather Forecasting (ECMWF) global atmospheric model every six hours. The Danish Meteorological Institute (DMI) HIRLAM domains are shown in Figure 6.2. The inner domains (around Denmark and around Greenland) are of higher resolution, with 5.5 km grid length, and use lateral boundary conditions from the outer, lower resolution domain that has a 15 km grid length. All domains have 40 layers in the vertical. The time step for the outer domain is 300 seconds while for the inner domains it is 90 seconds. The physical parameterisations are applied every third time step.

6.4 Persistence forecasting

A large number of advanced wind power forecasting systems have been developed over the last two decades, but it is worthwhile to begin by considering the simplest forecasting system, which is persistence forecasting. This also provides an opportunity to introduce the error measures that are used to quantify the performance of different wind power forecasting methods. The persistence model states that the forecast wind power will be the same as the last measured value of wind power.

$$\hat{P}_P(t + k|t) = P(t)$$

$\hat{P}_P(t+k|t)$ is the forecast for time $t+k$ made at time t, $P(t)$ is the measured power at time t and the look-ahead time is k. Although this model is very simple, it is in fact difficult to better for look-ahead times from 0 to 4–6 hours. This is due to the fact that changes in the atmosphere take place rather slowly. However, as the look-ahead time is increased beyond this time the persistence model rapidly breaks down. The persistence model is often used as a reference against which the more advanced and elaborate forecasting systems are compared.

6.4.1 Error measures

There are a number of different measures of the errors of a wind power forecasting system, and it is important to be clear about the differences between them when comparing performance results from different forecasting systems. These errors have been summarised in Madsen *et al.* (2004) and the notation used there has been adopted below.

The forecast error at a particular look-ahead time k is defined as the difference between the measured value and the forecast value at that time,

$$e(t+k|t) = P(t+k) - \hat{P}_P(t+k|t)$$

where $\hat{P}_P(t+k|t)$ is the forecast for time $t+k$ made at time t and $P(t+k)$ is the measured value at time $t+k$. Note that this definition produces the counter-intuitive result that, if wind power is over-predicted, then the error is negative, whereas an under-prediction results in a positive error. In order to produce results which are independent of the wind farm size, the power specified is usually the normalised power, that is, the actual power (MW) divided by the installed capacity of the wind farm (MW).

Forecast errors can be resolved into systematic and random components:

$$e = \mu_e + \nu_e$$

The systematic component μ_e is a constant and the random component ν_e has a zero mean value. The model bias or systematic error is the average error over all of the test period and is calculated for each look-ahead time.

$$\text{BIAS}(k) = \hat{\mu}_e(k) = \overline{e(k)} = \frac{1}{N}\sum_{t=1}^{N} e(t+k|t)$$

Two of the measures most widely used for forecast performance are the mean absolute error (MAE) and the root mean square error (RMSE). The MAE is given by

$$\text{MAE}(k) = \frac{1}{N}\sum_{t=1}^{N} |e(t+k|t)|$$

The mean square error (MSE) is given by

$$\text{MSE}(k) = \frac{\sum_{t=1}^{N}(e(t+k|t))^2}{N}$$

The RMSE is given by

$$\text{RMSE}(k) = \sqrt{\text{MSE}(k)} = \sqrt{\frac{\sum_{t=1}^{N} (e(t+k|t))^2}{N}}$$

It should be noted that both systematic and random errors contribute to both the MAE and the RMSE.

An alternative to the RMSE which is also widely used is an estimate of the standard deviation of the error distribution, that is, the standard deviation of errors (SDE) given by

$$\text{SDE}(k) = \sqrt{\frac{\sum_{t=1}^{N} \left(e(t+k|t) - \overline{e(k)}\right)^2}{N}}$$

In the case of the SDE, only random errors make a contribution to the error measure. For both the RMSE and the SDE large forecast errors make a stronger contribution to the measure than small forecast errors.

To illustrate the way these measures are typically presented, consider the application of the basic persistence model to two cases: (1) a single wind farm and (2) 15 geographically dispersed wind farms. The single wind farm has a high capacity factor and the total capacity of the fifteen geographically dispersed wind farms is about 20 times that of the single wind farm. Using the normalised measured power the MAE, RMSE and SDE for the basic persistence forecast model are calculated for a test period of six months and plotted in Figure 6.3 against look-ahead times from 1 out to 48 hours. As can be seen, and as was mentioned earlier, the persistence forecast deteriorates rapidly as the look-ahead time is increased. The MAE measure is easy to interpret directly. It is clear from Figure 6.3 that the averaged error over the test period for the 6-hour look-ahead time is 16.8 per cent of installed capacity for the single wind farm and 11.2 per cent for the 15 geographically dispersed wind farms. The RMSE and SDE measures cannot be interpreted so directly as they involve squared errors.

It is also clear that the forecast error measures are reduced for all look-ahead times for the 15 geographically dispersed wind farms compared to the single wind farm. This improvement in the error measures expressed in per unit of the single wind farm value is shown in Figure 6.4. It is very significant at shorter look-ahead times and drops off as look-ahead time increases. The improvement is due to the smoothing effects caused by the geographic dispersion of the wind farms.

6.4.2 Reference models

The persistence model is often used as a reference model in wind power forecasting against which other wind power forecasting models can be evaluated.

A refinement of the persistence model is a moving average (MA) predictor model. An example of this is where the last measured power value is replaced by the average

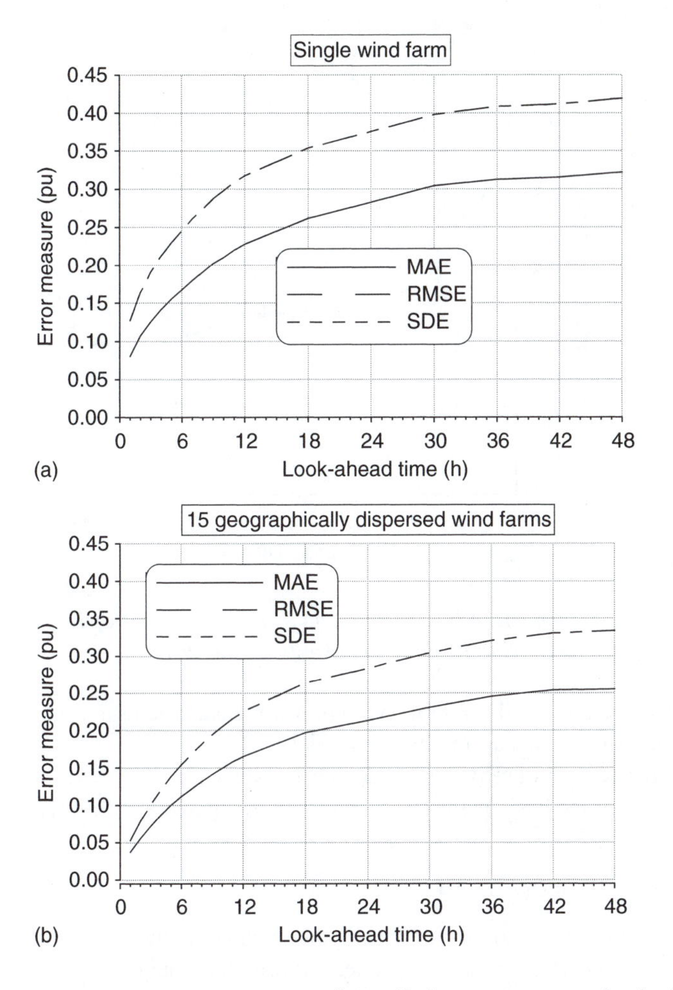

Figure 6.3 Error measures in per unit of installed capacity versus look-ahead time for (a) a single wind farm and (b) 15 geographically dispersed wind farms, calculated over a six-month test period using a basic persistence model. Note that in the above plots the SDE and RMSE are indistinguishable from each other.

of a number of the most recent measured values.

$$\hat{P}_{MA,n}(t+k|t) = \frac{1}{n}\sum_{i=0}^{n-1} P(t-i)$$

Although the performance of the moving average forecast model at short look-ahead times is very poor, it is better than the basic persistence model at longer look-ahead times. As n goes to infinity this model tends to the global average $\overline{P(t)}$,

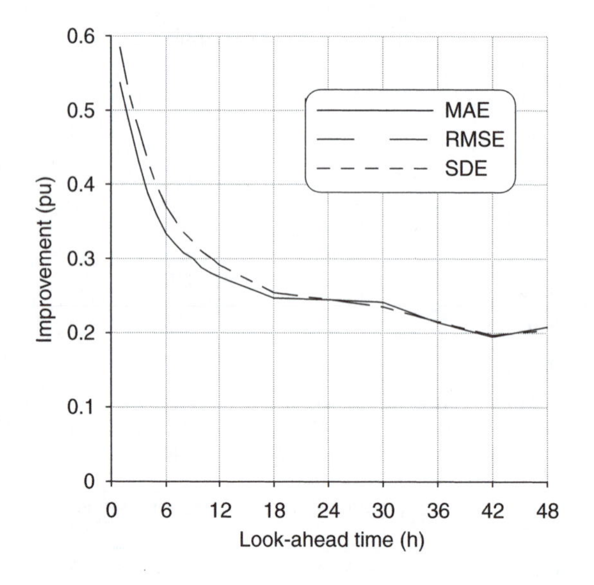

Figure 6.4 Improvement in basic persistence error measures of 15 geographically dispersed wind farms over a single wind farm

which is the average of all available wind power measurements at time t. This global average can also be used as a reference model.

A new reference model has been proposed by Nielsen *et al.* (1998) which combines the advantage of the basic persistence and the moving average forecast models. It is given by

$$\hat{P}_{\mathrm{NR}}(t+k|t) = \alpha_k P(t) + (1-\alpha_k)\overline{P(t)}$$

where subscript NR stands for new reference and α_k is the correlation coefficient between $P(t)$ and $P(t+k)$. This model requires the analysis of a training set of measured data to calculate the required statistical quantities $\overline{P(t)}$ and α_k.

An example of a time series of measured power data in per unit of the installed capacity for a single wind farm and for 15 geographically dispersed wind farms over a period of one year is shown in Figure 6.5. Each time series is divided into two six-monthly sets – a training set and a test set. It is clear that the time series of the 15 wind farms is smoother than the single wind farm and also that the rated output is never reached in the case of the 15 wind farms (see Section 5.3.2).

The training sets are used to calculate the required statistical quantities and the new reference persistence model is then applied to the test data sets. The resulting error measures are plotted against look-ahead time in Figure 6.6. By comparing with the previous result for the basic persistence forecast model in Figure 6.3, which also used the same six-month set of measured test data, it is clear that the performance at short look-ahead times has been maintained, while the poor performance of the basic persistence model at longer look-ahead times has been improved upon considerably. It is also clear that there is a substantial improvement in all error measures over

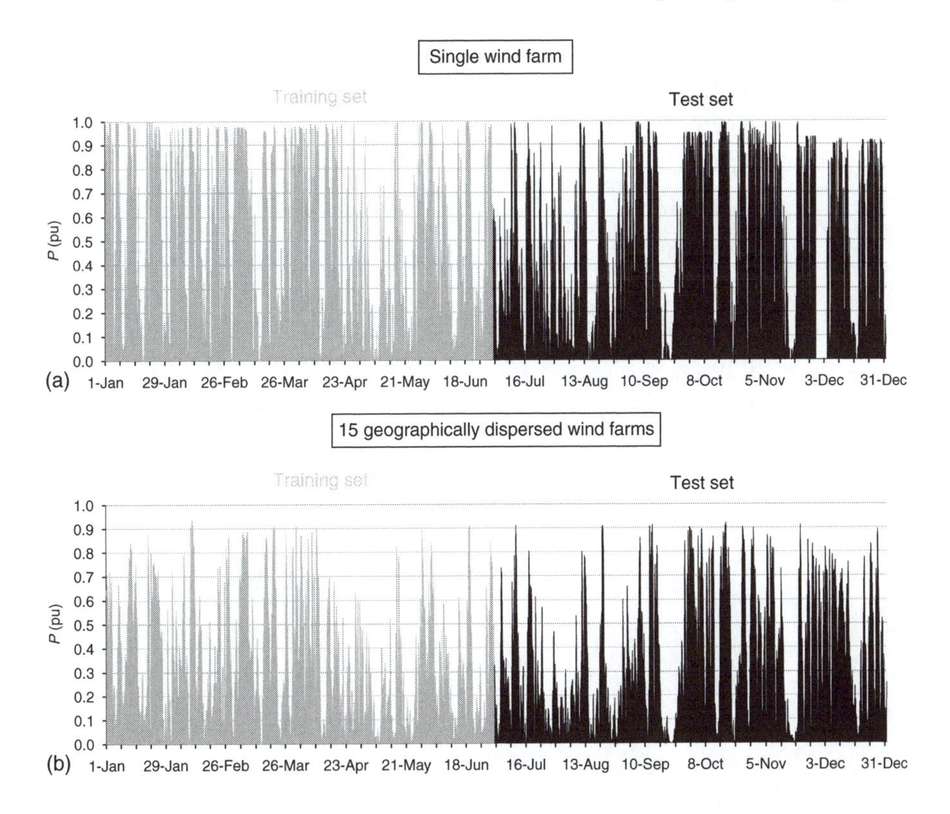

Figure 6.5 Time series of measured power at (a) a single wind farm and at (b) 15 geographically dispersed wind farms divided into training and test sets

all look-ahead times if the results for the 15 wind farms are compared to the single wind farm for the new reference persistence model, as shown in Figure 6.7. The improvement is significant at the shorter look-ahead times, dropping off as look-ahead time increases.

The ease with which more advanced wind power forecasting models can improve on basic persistence at longer look-ahead times can give a false sense of their performance quality. This new reference model is a more challenging benchmark against which to measure the performance of more advanced wind power forecasting models.

Another tool which is widely used for exploratory analysis is the plot of the error distribution. The error distribution for the basic persistence forecast model at look-ahead times of 1 and 6 hours, for the same six months of test data, is shown in Figure 6.8 for the single wind farm and for the 15 wind farms. The bins are 0.05 pu of installed power. In each graph the sharpness and skewness of the distribution can be observed, as can the presence and extent of large forecast errors which are, of course, very significant to power system operators. It is clear that the error distribution is sharper for short look-ahead times and for aggregation of wind farms.

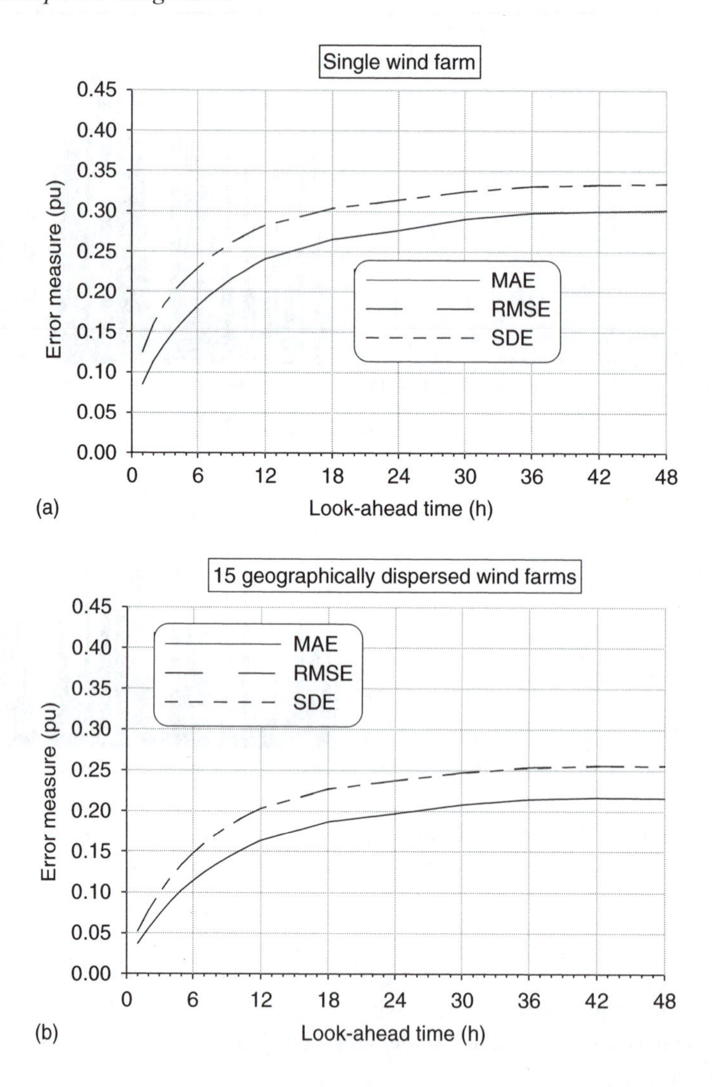

Figure 6.6 Error measures in per unit of installed capacity versus look-ahead time in hours for (a) single wind farm and for (b) 15 wind farms over a six-month test period using the new reference persistence model

The improvement obtained by using an advanced forecast model over a reference forecast model can be quantified using the expression

$$\mathrm{Imp}_{\mathrm{ref,\,EC}}(k) = \frac{\mathrm{EC}_{\mathrm{ref}}(k) - \mathrm{EC(k)}}{\mathrm{EC}_{\mathrm{ref}}(k)}$$

where EC is the evaluation criterion which can be either MAE, RMSE or SDE. Using the test data again for the two cases of (i) a single wind farm and (ii) 15 geographically

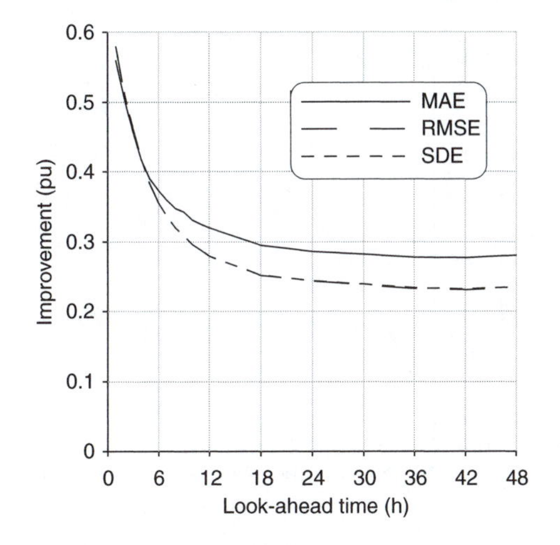

Figure 6.7 Improvement in new reference persistence error measures due to aggregation of wind farm output

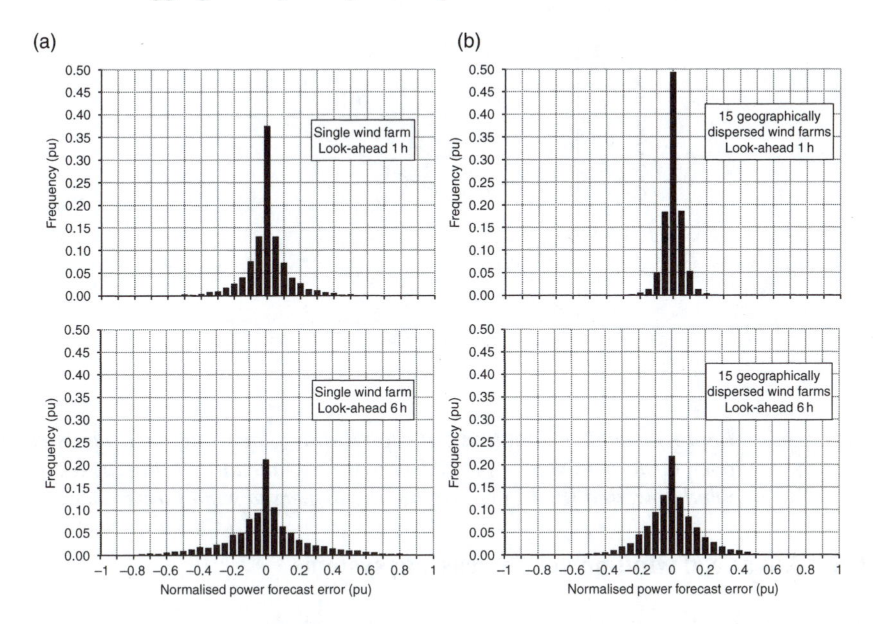

Figure 6.8 Error distribution for the basic persistence model for look-ahead times of 1 and 6 hours for (a) single wind farm and (b) 15 geographically dispersed wind farms

dispersed wind farms, the improvement in the RMSE evaluation criterion of the new reference model over the basic persistence model is plotted in Figure 6.9 against look-ahead time. The improvement in both cases is clearly seen at longer look-ahead times.

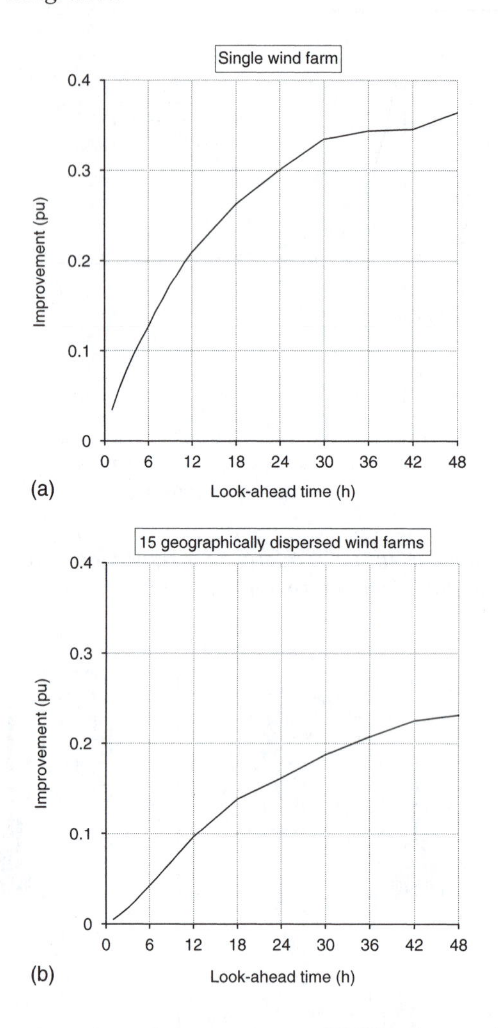

Figure 6.9 Improvement in evaluation criterion RMSE of new reference persistence over basic persistence plotted versus look-ahead time for (a) a single wind farm and (b) 15 geographically dispersed wind farms

6.5 Advanced wind power forecasting systems

The flows of information into and out of a typical wind power forecasting system are shown in the schematic diagram in Figure 6.10. At the heart of the system is the wind power forecasting model which takes inputs from the meteorological service and the wind farms, and provides the wind power forecasts to the main users, that is, the TSO, energy traders and the wind farm owners. In doing so it performs a number of key tasks: (i) downscaling, (ii) conversion of wind forecast to power forecast and (iii) up-scaling. These are described later. The system works online, updating the forecasts at least as often as the NWP forecast is updated. For

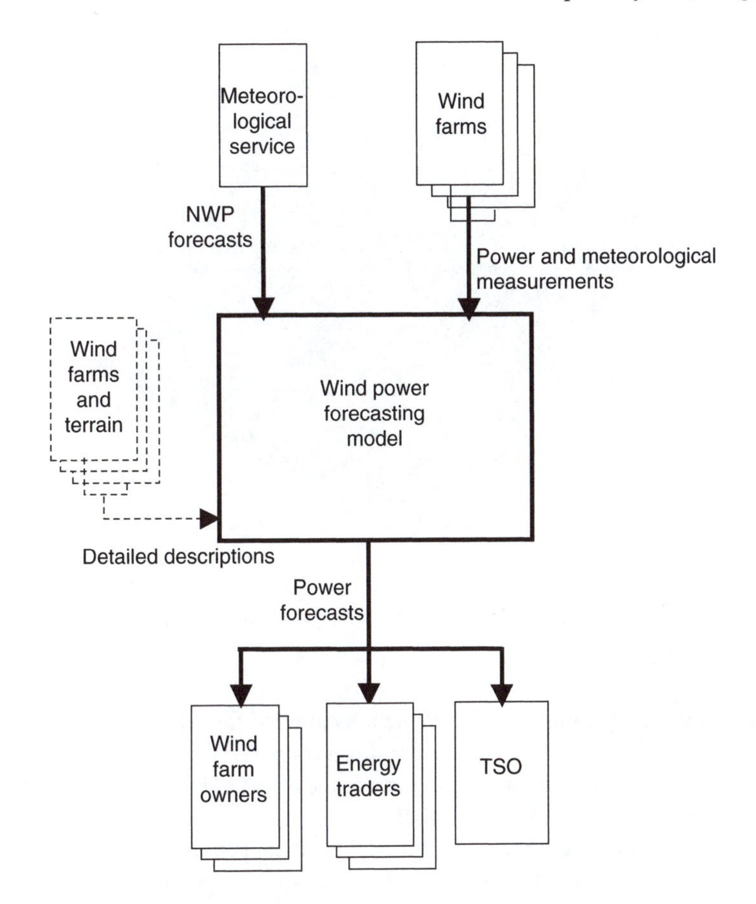

Figure 6.10 Flows of information in a wind power forecasting system

a typical application, covering many wind farms dispersed over a large geographic area, the importance of a robust, efficient, secure and reliable communication system is clear.

The main input is the NWP forecast, usually provided by a national meteorological service. The accuracy of the NWP model is critical in determining the overall accuracy of the wind power forecasting system for look-ahead times greater than 6 hours. The NWP forecast is dynamic in the sense that it might be provided typically every 6 hours, with hourly look-ahead times extending out to 48 hours.

The model output is provided in the form of wind speed and wind direction forecasts for a three-dimensional set of grid points. The grid has rather coarse horizontal resolution (typically in the range 5 to 40 km) and a number (typically 20 to 40) of vertical levels down to the surface level. These vertical levels are more concentrated in the planetary boundary layer. The first task of the forecasting model is downscaling, that is, converting the four closest NWP grid point forecasts to local wind forecasts at the wind farm site. A typical situation is shown in Figure 6.11, with NWP forecasts

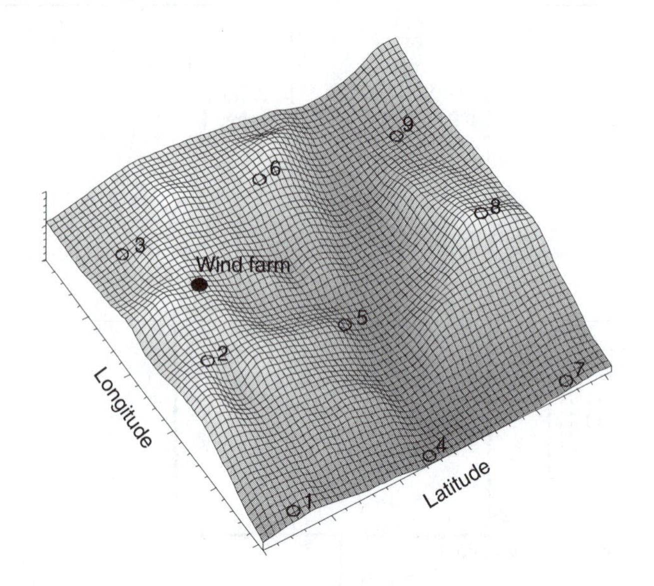

Figure 6.11 Wind farm located in complex terrain with surrounding NWP grid-points

available at the grid points 1, 2, 3,...9 and a local wind forecast required at the wind farm, which is located in reasonably complex terrain. Clearly, the horizontal resolution of the NWP model cannot hope to capture the effect of processes due to the local terrain and local thermal effects on the flow.

One downscaling method involves a mathematical spatial interpolation between the four closest NWP grid points. Another more complex approach, whose use might be justified in the case of very complex terrain, would be to determine the flows over the local terrain using a much higher resolution meso- or micro-scale physical model. The NWP model results are used to initialise and set the boundary conditions for this meso- or micro-scale model. The meso- or micro-scale model is in effect nested within the NWP model. This clearly involves additional modelling complications and meteorological expertise in setting up, operating and interpreting the results. A trade-off needs to be made between the additional cost of the added complexity and the gains in accuracy which might be achieved.

The second input to the wind power forecasting model is a detailed description of the wind farm and its surrounding terrain. This information is static in nature and typically includes the number and location of the wind turbines, the wind turbine power and thrust curves and details of the surrounding orography, roughness and obstacles. The terrain descriptions are required for the downscaling task, whereas the wind farm details are required for the second key task of the wind power forecasting model which is the conversion from a local wind speed and direction forecast to a wind farm power output forecast. One simple method might involve the wind turbine manufacturer's power curve, but an alternative approach is to use a wind farm power curve based on the forecast wind speed and the measured wind farm power output.

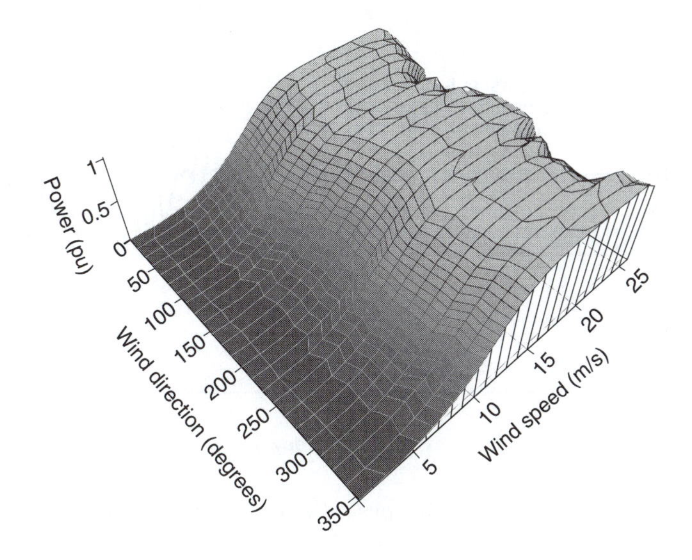

Figure 6.12 Directionally dependent wind farm power curve

The conversion can be carried out via a matrix of wind farm power outputs for a range of wind speed and wind direction bins. An example of such a power curve is shown in Figure 6.12.

The third major input to the wind power forecasting model is the measured power from the wind farms. This can be dynamic if the power measurements from the wind farms are available online or static if they are available off-line only. Measurements of other meteorological parameters from the wind farm might also be included (e.g. wind speed, wind direction, atmospheric pressure and temperature) (see Section 5.3.4). The online measurements are usually available from the SCADA system of the wind farm or TSO. Such measurements are not required for all the wind farms connected to the TSO, but rather for a selected set of representative wind farms. These online measurements then need to be up-scaled to represent the total output of the wind farms in the TSO's area. This up-scaling is the third key task of the wind power forecasting system. In addition, the online measurements need to be pre-processed to account for consistency and for whether wind farms or wind turbines are disconnected due to breakdown, or indeed for failures in the SCADA system itself.

The output of the wind power forecasting model is the wind power forecast, which typically provides hourly predictions for look-ahead times up to 48 hours. The forecasts can be provided for specific wind farms, for a portfolio of wind farms or for all wind farms in a particular region or supply area. In Figure 6.13, four sets of successive wind power forecasts for a group of wind farms (at *t*, *t*-6, *t*-12 and *t*-18 hours) are overlaid, together with the measured power. The forecasts are for look-ahead times in the range 0–48 hours.

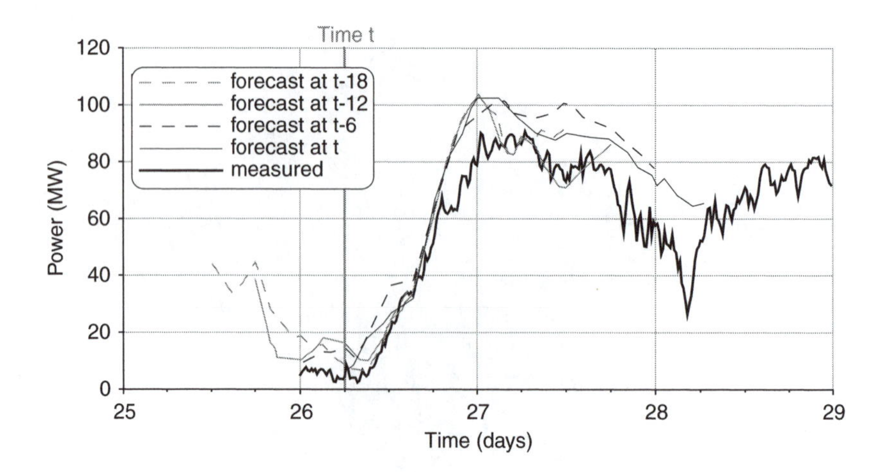

Figure 6.13 Wind power forecasts for look-ahead times of 0–48 hours provided at times t, t-6, t-12 and t-18 hours

6.5.1 Prediktor

Prediktor, one of the earliest wind power forecasting systems, was developed by Landberg at Risø National Laboratory in Denmark (Landberg, 1994; Landberg *et al.*, 1997). Prediktor was applied to 15 wind farms connected to the Electricity Supply Board (ESB) system in Ireland from February 2001 through to June 2002. A description of the application of Prediktor in Ireland can be found in Watson and Landberg (2003). A brief description of Prediktor together with key results from its application in Ireland are presented below to illustrate some important features of typical wind power forecasting systems.

The Prediktor methodology, outlined in Figure 6.14, is predominantly a physical modelling one. The large-scale flow is modelled by an NWP model, in this case a version of HIRLAM run by Met Éireann (the Irish national meteorological service). The version of HIRLAM used had a horizontal resolution of 33 km with 24 vertical levels. HIRLAM was run at 0, 6, 12 and 18 UTC (co-ordinated universal time) each day, taking about 1.5 hours to complete its run after a 2-hour wait for the global meteorological data from the GTS. Results were emailed to Prediktor less than an hour after the run had completed. HIRLAM produced forecasts of hourly wind speed and direction at a selected number of vertical levels for each wind farm for look-ahead times out to 48 hours. These forecasts were spatially interpolated from the four nearest HIRLAM grid points to the wind farm locations.

No online power measurements were available at the time of the trial. The HIRLAM forecasts were thus the only online inputs to Prediktor. Considering Figure 6.14 again, Prediktor firstly transforms the HIRLAM winds to the surface using the geostrophic drag law and the logarithmic profile. The horizontal resolution of HIRLAM is such that it cannot model local effects at the wind farm. Prediktor represents these local effects using correction factors calculated using

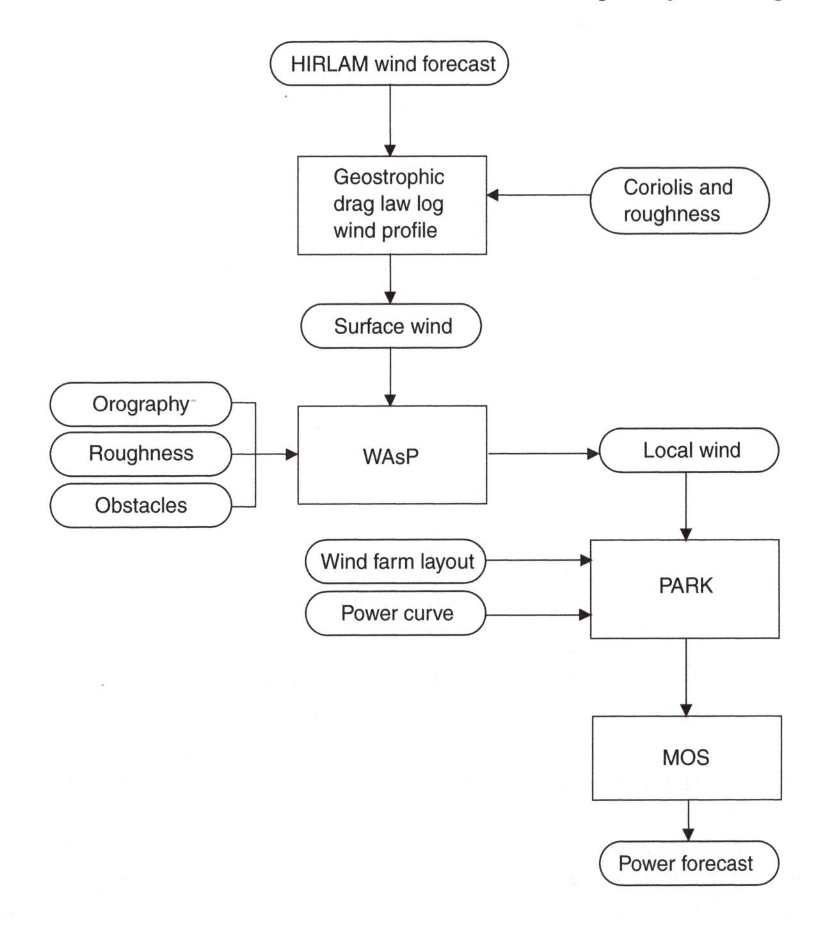

Figure 6.14 Flowchart of Prediktor methodology

Wind Atlas Analysis and Applications Program (WAsP) which was also developed at Risø National Laboratory and is the basis of the wind resource assessment technique of the European Wind Atlas (Troen and Petersen, 1989). The local effects modelled in WAsP are: shelter from obstacles, the effects of roughness and roughness change and the speed-up/ slow-down effects due to the height contours of the surrounding terrain. Prediktor also takes account of wind farm array effects and converts from wind speed to wind power output using correction factors calculated using the PARK model which is an integral part of WAsP.

Prediktor also uses a model output statistics (MOS) module to take into account any effects not modelled by the physical models. The MOS correction factors are calculated off-line from a comparison of predicted and observed power data.

The 15 wind farm locations are shown in Figure 6.15 and have a total capacity of 106.45 MW, representing most of the installed wind power capacity in the country at the time. There is clearly a wide geographic spread over the country. The wind farms were divided into regional groupings labelled 'nw' (north-west), 'w' (west) and 'sw'

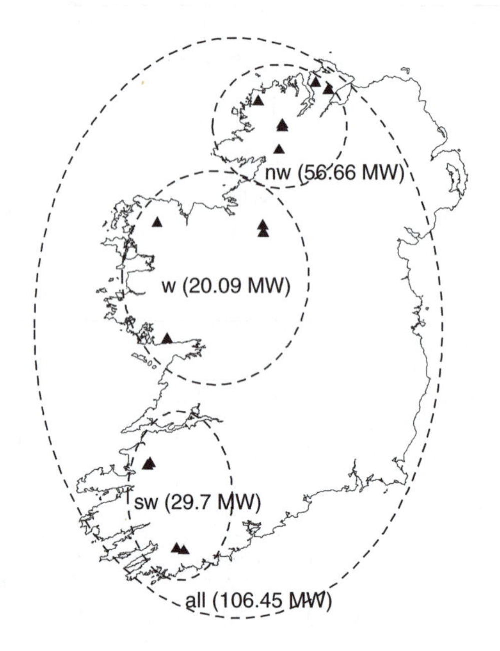

Figure 6.15 Location of wind farms (marked with triangles) and the division into regional groupings (nw, w and sw)

(south-west). The wind farm capacity in each of these groupings is also indicated in Figure 6.15.

The only measured power data available were the metered 15-minute time-resolution energy data from each of the wind farms. This data set did not contain operational data on the availability of wind farms or indeed the availability of individual wind turbines, nor did it contain wind speed or wind direction data from the wind farms. As a consequence, Prediktor could not take the operational status of the wind farms or the wind turbines into account and operated on the basis of an assumed 100 per cent availability.

The results presented below are for the period October through to December 2001, with a training period for the MOS corrections covering February through to September 2001. In Figure 6.16 the SDEs for individual wind farms are plotted against look-ahead times for persistence and Prediktor. Prediktor performed better than persistence for all look-ahead times except those less than 6 hours. As can be seen for Prediktor, the SDEs lie in the range 18 to 33 per cent, whereas the corresponding SDEs for persistence can be up to 52 per cent.

The beneficial effect on forecast error of both persistence and Prediktor models due to the aggregation of wind farm power output from geographically dispersed wind farms is shown in Figure 6.17. SDEs for persistence and Prediktor are plotted against look-ahead time for a single wind farm (the line) in the north-west region, for the 'nw' group of wind farms (the white column) and for all the wind farms (the black column).

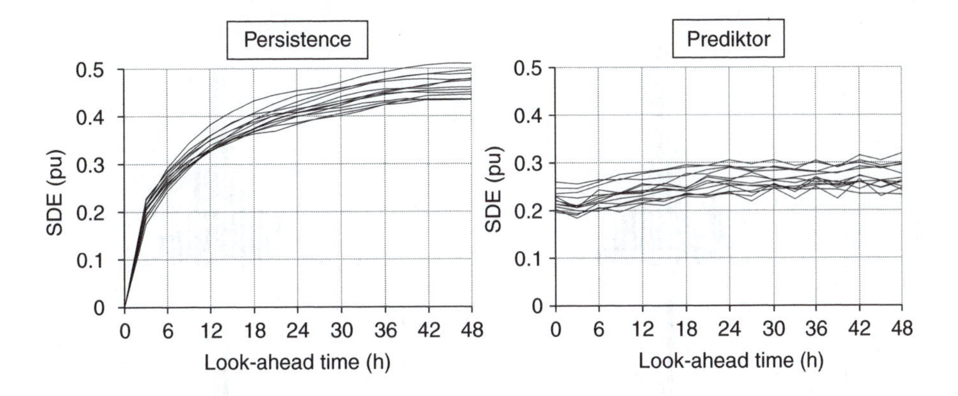

Figure 6.16 Error measure versus look-ahead time for individual wind farms for persistence and for Prediktor

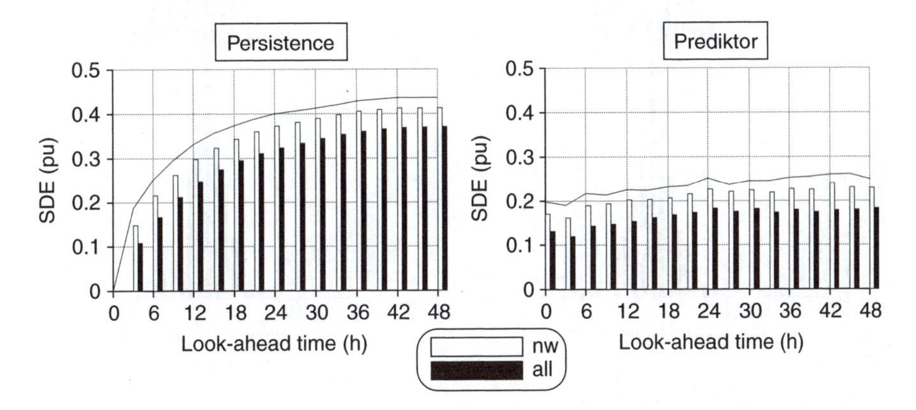

Figure 6.17 Effect of aggregation of wind farm output on forecast error for the persistence and Prediktor models

The effect of aggregation of wind power output is explored further in Figure 6.18, which shows six distributions of errors for the Prediktor model, three on the left for the 6-hour look-ahead time and three on the right for the 12-hour look-ahead time. The top graphs show the distribution for a single wind farm in the north-west region, the middle graphs show the distribution for all wind farms in the north-west region and the bottom graphs show the distributions for all wind farms. It is clear that the more distributed the wind farms, the tighter and more symmetric is the error distribution. The distributions for the 6-hour look-ahead time are seen to be marginally tighter than those for the 12-hour look-ahead time. However, it is also clear that there were low occurrences of large errors, which would be very significant from a power system operator point of view. The maximum forecast errors of individual wind farms, both positive and negative, were found to be very large. Aggregation was found to reduce the maximum errors, but only marginally.

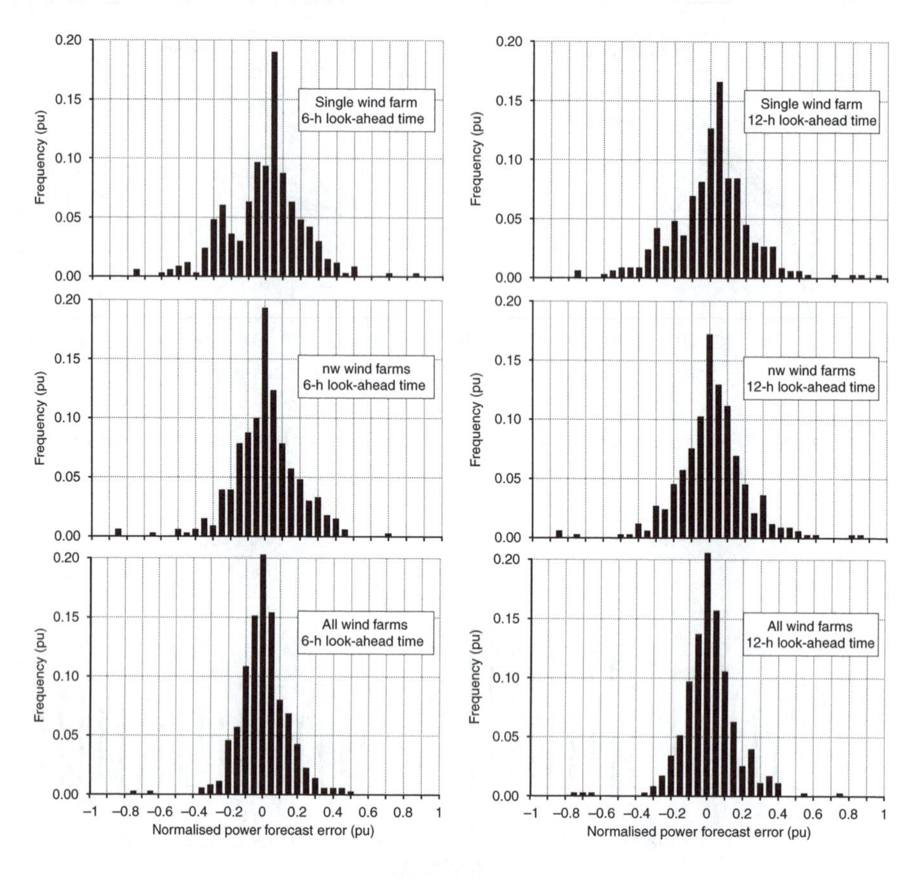

Figure 6.18 Distribution of errors

Another interesting feature of aggregation can be explored by plotting the correlation coefficient between the errors at all pairs of wind farms against distance between the wind farms. This is shown in Figure 6.19 for four different look-ahead times. It is clear that correlation between forecast errors drops off markedly with distance, so that at a distance of 100 km or more the correlation is very weak. Strong correlation between the forecast errors at wind farms would be detrimental from a power system operator point of view. The rapid drop off with distance confirms the benefits in having the wind power generation resource as geographically dispersed as possible (see Section 5.3.2).

6.5.2 Statistical models

In the statistical modelling approach, the key tasks of downscaling and the conversion from wind forecast to power forecast described earlier for the physical approach are replaced by a direct conversion from the input data, that is, the NWP forecast and the online wind power measurement, to a wind power forecast.

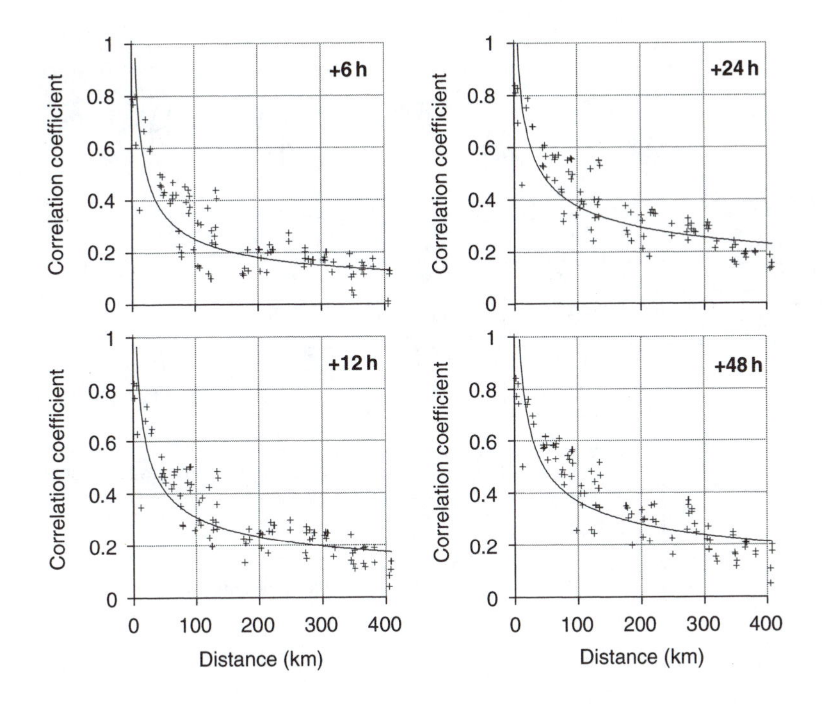

Figure 6.19 Correlation between forecast errors at pairs of wind farms plotted against distance between them

A simple statistical model can be based on the following form (Kariniotakis *et al.*, 2006b):

$$\hat{P}(t+k|t) = f\big(P(t),\ \hat{u}(t+k|t_{\text{NWP}}),\ \hat{\theta}(t+k|t_{\text{NWP}}),\ \hat{x}(t+k|t_{\text{NMP}})\big)$$

$\hat{P}(t+k|t)$ is the power forecast for time $t+k$ made at time t, $P(t)$ is the power measurement at time t, $\hat{u}(t+k|t_{\text{NWP}})$ is the NWP wind speed forecast for time $t+k$ made at time t_{NWP} (the time of the last NWP run), $\hat{\theta}(t+k|t_{\text{NWP}})$ is the NWP wind direction forecast for time $t+k$ made at time t_{NWP} and $\hat{x}(t+k|t_{\text{NWP}})$ is another meteorological variable forecast by the NWP model for time $t+k$ made at time t_{NWP}. The function f can be linear, e.g. ARMA (autoregressive moving average) or ARX (autoregressive with exogenous variables), or it could be non-linear, e.g. NARX (non-linear autoregressive with exogenous variables), NN (neural network) or F-NN (fuzzy neural network). These statistical models require extensive use of training sets of forecast and measured data to capture the spatial and temporal dependencies in the time series data, by identifying the model parameters and functions needed to reproduce the relationships between the input explanatory variables and the forecast wind power.

The wind power prediction tool (WPPT) developed by Informatics and Mathematical Modelling (IMM) at the Danish Technical University (DTU) is an example of a wind power forecasting system which uses the statistical approach to the downscaling

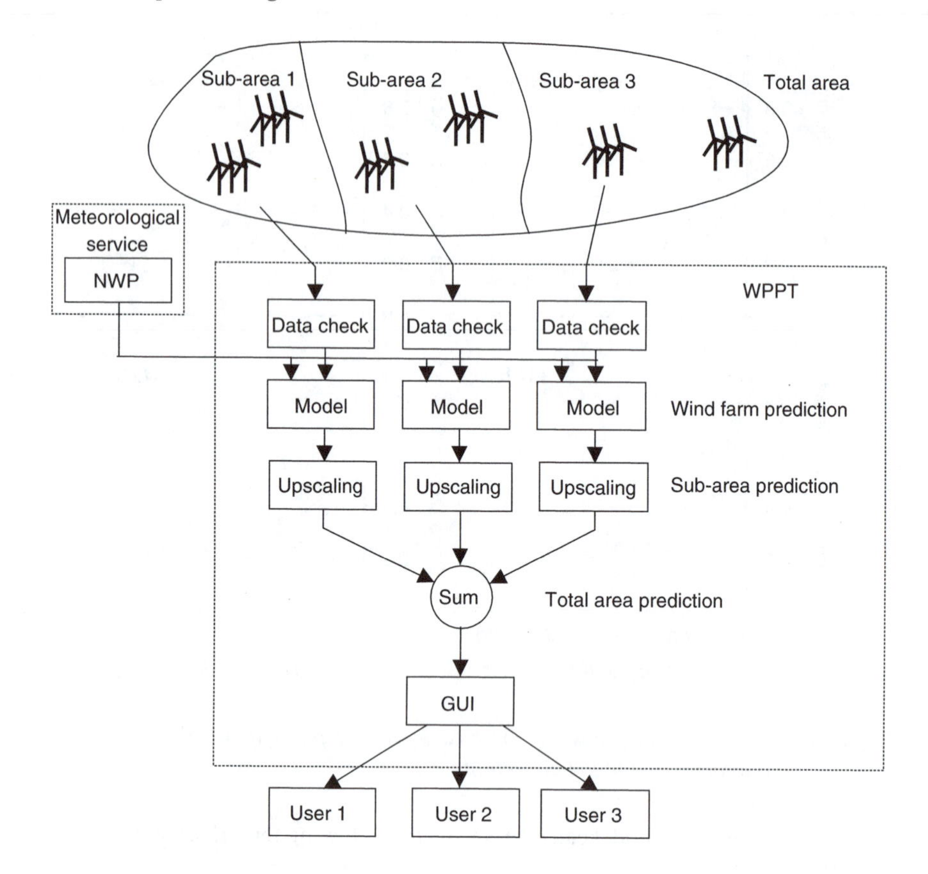

Figure 6.20 Schematic of WPPT wind power forecasting model

and wind to power forecast tasks. A schematic of the WPPT system is shown in Figure 6.20. WPPT has been operational in western Denmark since 1994 and uses statistical non-parametric adaptive models for prediction of power from selected wind farms. Data checking is applied to the online input power measurements from the selected wind farms. It also uses statistical up-scaling to the full installed capacity in the region or sub-region.

A typical result taken from Giebel *et al.* (2003) for the application of WPPT to the western part of Denmark is shown in Figure 6.21, showing a comparison between the SDEs for WPPT and persistence plotted against look-ahead time for the period June 2002 to May 2003. As can be seen, WPPT performs much better than persistence, with the exception of the first few hours. The WPPT SDE rises rather slowly with increasing look-ahead time in the range 5–10 per cent (of installed capacity).

The physical system is non-stationary, so the forecasting method used should be able to adapt to changes in the physical system. Changes occur in the NWP models, the number of wind farms or wind turbines within a wind farm that are not operating due to

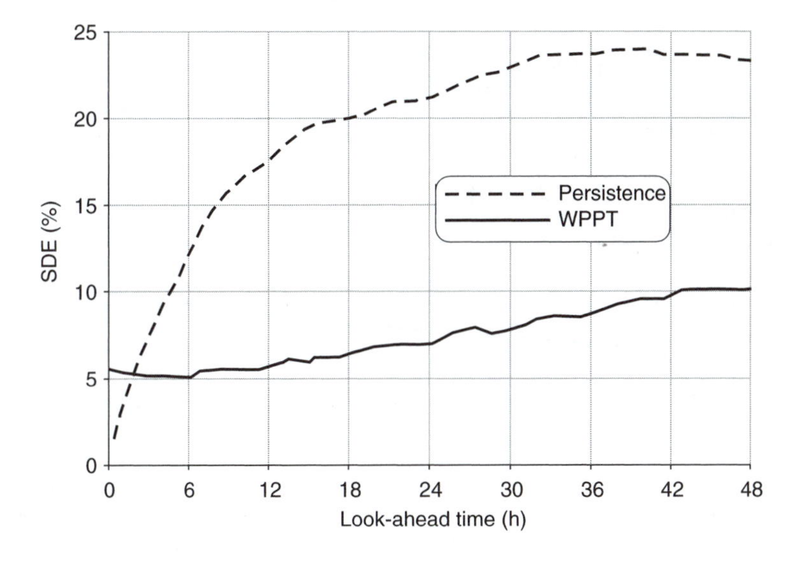

Figure 6.21 Comparison of persistence and WPPT for regional wind power forecasting for western Denmark

forced or scheduled outages, extensions to wind farms, performance due to build-up of dirt on turbine blades, seasonal effects, terrain roughness, and sea roughness for offshore wind farms.

A wind power forecasting system which incorporates this adaptive property, in that the model can fine-tune its parameters during online operation, is AWPPS (ARMINES wind power prediction system), which was developed at Écoles des Mines in France. AWPPS integrates (i) short-term forecasts based on statistical time series approaches for look-ahead times up to 10 hours, (ii) longer-term forecasts based on F-NNs with inputs from SCADA and NWP for look-ahead times out to 72 hours and (iii) combined forecasts produced from an intelligent weighting of short- and long-term forecasts to optimise the performance over all look-ahead times. The results shown in Figures 6.22 and 6.23 are taken from Pinson *et al.* (2004) and are for a single offshore wind farm of 5 MW capacity and are based on a 13-month set of data which was divided into test and training subsets. Figure 6.22 shows the comparison with persistence using two different normalised error measures – the MAE and the RMSE. The typical pattern is again to be seen, with AWPPS outperforming persistence, as would be expected. The results for a single offshore wind farm are clearly not as good as for a whole region. In Figure 6.23 the improvement obtained over persistence is plotted against look-ahead time.

The largest source of error for a wind power forecasting system is in the NWP. It is important to indicate to users the uncertainty that is attached to a particular forecast. Figure 6.24, again taken from Pinson *et al.* (2004), shows a sample plot of forecast power with upper and lower 85 per cent confidence intervals and the measured power plotted against look-ahead time. Tools for the online estimation of

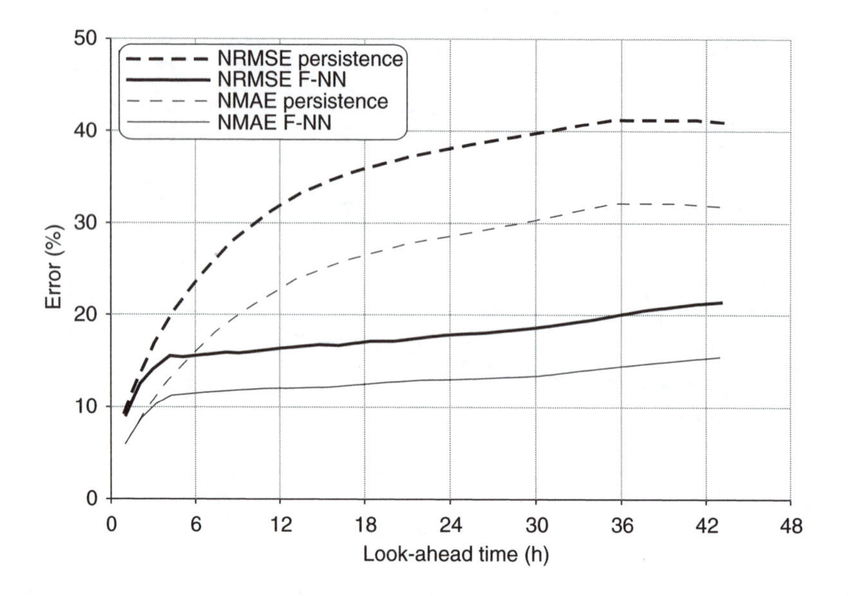

Figure 6.22 Comparison of persistence and Fuzzy-NN models for an offshore wind farm

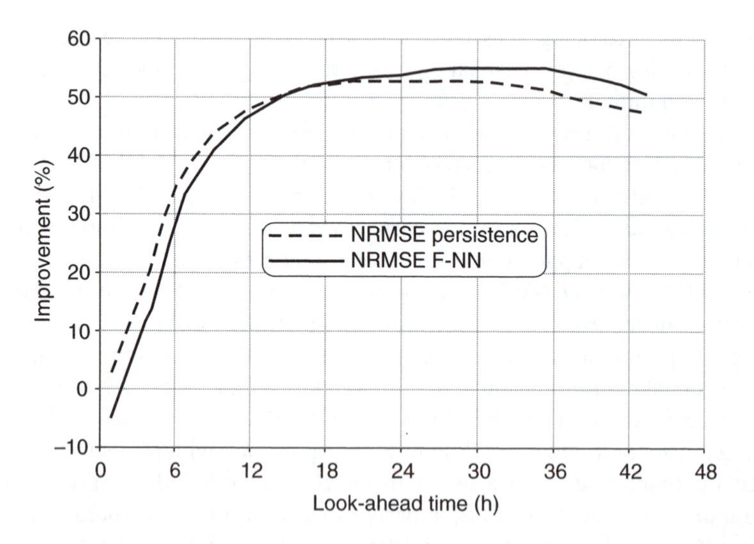

Figure 6.23 Improvement of Fuzzy-NN model over persistence for an offshore wind farm

the forecast uncertainties will no doubt play an important role in trading of wind power in electricity markets, since they can prevent or reduce penalties in situations of poor forecast accuracy (Nielsen *et al.*, 2006). Another example of a forecast with multiple confidence intervals is shown in Figure 6.25.

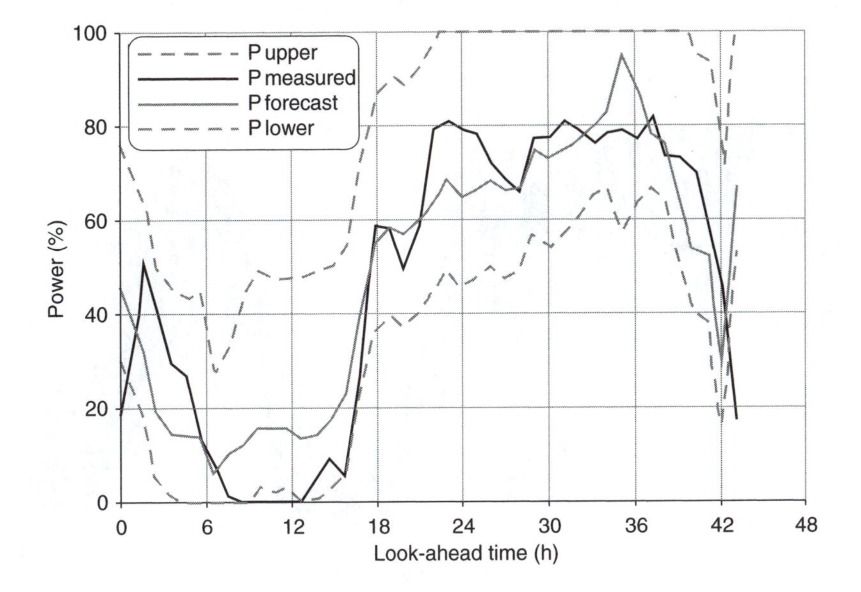

Figure 6.24 Forecast for offshore wind farm with 85 per cent confidence intervals

6.5.3 Ensemble forecasting

Ensemble forecasting is well developed in medium range NWP where it has been shown that the ensemble average forecast is more accurate than individual forecasts after the first few days, but more importantly forecasters are provided with an estimation of the reliability of the forecast, which because of changes in atmospheric predictability varies from day to day and from region to region (Kalnay *et al.*, 1998). Ensemble forecasting involves producing an ensemble of forecasts instead of just an individual forecast. The members of the ensemble can arise from different variants of the same NWP model, e.g. different physical parameterisation of the sub-grid physical processes, or different initial conditions, or different data assimilation techniques; or they can arise from completely different NWP models. The technique of ensemble forecasting has recently been applied to wind power forecasting. The basic assumption is that when the different ensemble members differ widely, there is a large uncertainty in the forecast, while, when there is closer agreement between the ensemble member forecasts, then the uncertainty is lower (Giebel, 2005). A multi-scheme ensemble prediction scheme (MSEPS), which was originally developed at University College Cork, has been implemented at Energinet (the TSO in western Denmark) in co-operation with the research company WEPROG. It consists of 75 ensemble members produced by perturbations of the initial conditions and variations in the parameterisation of selected physical processes. The results for the first year's trial are quite promising, with at least 20 per cent better forecasts of wind power compared to a single forecast averaged over the year. In addition, periods with high wind power output (defined as >70 per cent rated) and low uncertainty were predicted

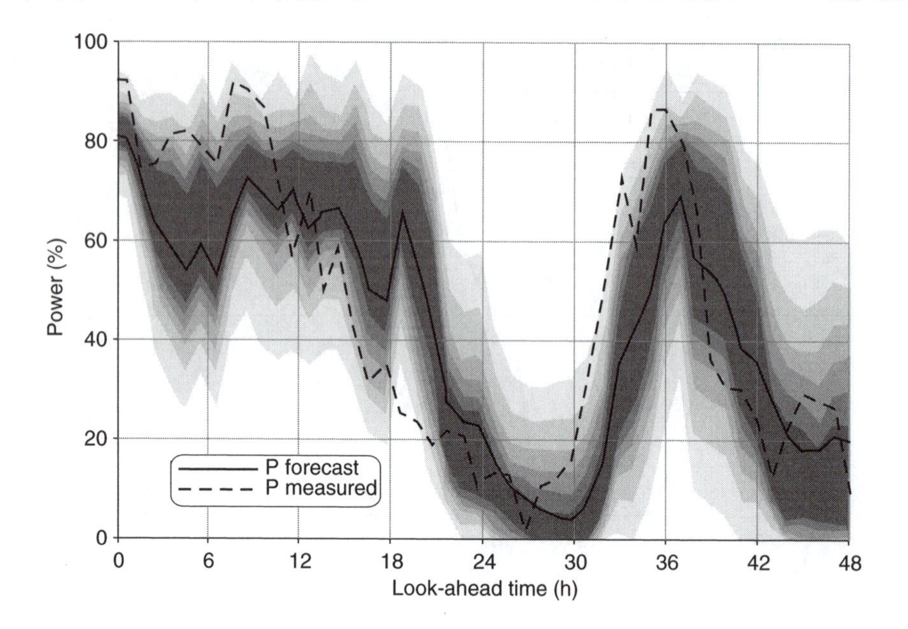

Figure 6.25 Example of forecast with uncertainty information

accurately up to two days ahead, while periods with low wind power and low uncertainty were also predicted accurately (Akhmatov *et al.*, 2005). The MSEPS has also been applied in Ireland for a four-month test period to both a single wind farm of 15 MW capacity (Lang *et al.*, 2006a) and the aggregated output from the wind farms connected to the Irish TSO totalling 500 MW (Lang *et al.*, 2006b), with promising results. In the case of the single wind farm the normalised MAE was 11.4 per cent and the SDE 15.7 per cent, while for the total wind power connected to the TSO the normalised MAE was 6.2 per cent and the SDE was 8.3 per cent. All statistics refer to 24–48 hour forecasts for each day's 00:00 UTC model run.

6.6 Conclusions

While considerable progress has been made in wind power forecasting in the last decade or so, in terms of better understanding of the processes involved and higher accuracy of the forecasts, there is still plenty of scope for improvement. Clearly forecasts for the wind power production of whole regions or supply areas are more accurate than forecasts for particular wind farms due to the smoothing effect of geographic dispersion of wind power capacity. However, with the trend towards large wind farms, and particularly large offshore wind farms, there is a growing requirement for accurate forecasts for individual wind farms. There are considerable differences in the accuracy that can be achieved for wind power forecasts for wind farms located in very complex terrain compared to open flat terrain. Current research on mesoscale modelling aims to address this problem. The development of offshore wind power also presents difficult challenges for wind power forecasting. The uncertainty attached to a forecast is a key concern, particularly when looking to the participation of wind power in electricity markets, as discussed further in Chapter 7 (Wind Power and Electricity Markets). The area of ensemble forecasting shows potential for further progress. The other area of interest will be the integration of wind power forecasting tools into the energy management systems of TSOs.

Chapter 7

Wind power and electricity markets

7.1 Introduction

The electricity industry is over 100 years old. During this time it has grown dramatically and has developed organisational structures and methods of doing business. The industry has particular characteristics that have had a large impact on these structures. These characteristics include the real-time nature of electricity, that is, it is typically generated and consumed at the same instant in time. Storage of electricity is possible, and is performed on a limited scale, but to date large-scale storage has proved to be uneconomic (Kondoh *et al.*, 2000). Another important characteristic of the industry is that some elements are natural monopolies, for example, the transmission and distribution infrastructure – *the wires*. Until recently, economies of scale dictated that the generation assets in the electricity industry would typically be large-scale, capital-intensive, single-site developments. The industry evolved by adopting structures that could be best described as vertically integrated monopolies. The businesses owned and operated entire power systems from generation, transmission and distribution down to domestic customer supply. They were regulated on a cost basis to ensure that they would not abuse their dominant position. In many cases national or local governments owned these business entities, and in others they were investor-owned utilities. Such structures dominated the industry until very recently (Baldick *et al.*, 2005).

There has been concern among many politicians, policy makers, engineers and economists that these types of monopolistic structures were inefficient. As a consequence of cost-based regulation there is a tendency for monopolies to over-staff and over-invest. In recent years, and in many regions of the world, there has been a trend away from cost-based regulation towards competitive market structures (Baldick *et al.*, 2005). The move towards competition is sometimes described as *deregulation*. However, this is a misnomer, as it has led to the creation of regulators where they did not exist previously, and the term *re-regulation* has also been used. A more useful and meaningful term is *restructuring*, which will be used here. The restructuring is largely characterised by the replacement of monopolistic, vertically integrated utilities (VIUs) with a number of smaller entities, and the introduction of competitive electricity markets. The smaller entities can be categorised as generation, transmission, distribution and supply.

In parallel with the radical changes that are occurring in the structures of the electricity industry, there is a dramatic increase in the amount of wind power being connected to many electricity grids (Keane and O'Malley, 2005). Wind power development is being driven by the desire to reduce harmful greenhouse gas emissions and by substantial cost increases in fossil fuels and security of supply issues which make wind power more attractive and more competitive. The convergence of electricity industry restructuring with an increase in wind generation capacity poses a significant challenge to the industry. While the technical issues related to wind power described elsewhere in this book are important, it is the economic and market issues that will ultimately decide the amount of wind power that is connected to electricity grids. There is an opportunity to develop robust solutions to secure the long-term sustainability of the electricity industry, with wind power making a significant contribution. Failure to rise to these challenges will threaten the reliability and efficiency of the electricity industry that underpins modern industrial society.

In contrast to Chapter 5, where it was assumed that wind power was operated in a vertically integrated environment, here it is assumed that wind power is one player in a competitive market place. The players and the market place will now be described briefly.

The monopoly characteristic of the transmission and distribution systems requires independent transmission system operators (TSOs) and distribution system operators (DSOs) who will operate the transmission and distribution systems. The transmission and the distribution network assets may be owned by the system operators or by another entity – the network owner.

As natural monopolies, transmission and distribution entities are regulated under the new structures.

Generation and supply entities are suitable for competition in certain circumstances, and this has led to the development of competitive electricity markets. Electricity markets come in all shapes and sizes. There is no standard design for an electricity market, although attempts have been made to encourage some level of standardisation (FERC, 2003). They can range in size from less than 1 GW with a few participants on small islands to arrangements hundreds of times larger with many participants on large interconnected systems. Physical electrical interconnection, direct (DC) or alternating current (AC) (Section 5.3), is a limiting factor on the size of the market. Electricity markets are a product of their historical and political backgrounds, but do share a number of fundamental characteristics which will be described below.

There is a wide range of generation technologies, ranging from nuclear and fossil fuel to renewable, including wind power. Further details of these other technologies can be found elsewhere (Cregan and Flynn, 2003). The restructuring has been facilitated by technological changes which have made smaller generating plant more cost effective (e.g. gas plant in particular) (Watson, 1997). The competitive market structures have also spawned a degree of innovation in very small distributed plants such as combined heat and power (CHP) and wind.

Generators typically sell electricity, directly or indirectly, in bulk quantities to suppliers who then sell it on to individual consumers. Alliances between generators and suppliers will be based on contracts and/or on mergers into single entities , that is, a level of vertical integration. Consumers, sometimes referred to as the load, can range in size from individual households to large industries with multiple sites consuming significant quantities of electricity. Wind generation has features of both generation and load. It is largely uncontrollable (although it can be curtailed) which is similar to load, and it is uniquely unpredictable over long-time frames (hours/days).

Some form of regulation is a requirement in all electricity markets. The level of regulation is determined by the particular economic and political circumstances of each region. At a very minimum, the monopoly parts of the business, that is, transmission and distribution, need to be regulated. Electricity markets are also subject to regulatory oversight. This is particularly important in markets where competition is in its infancy where, for example, an incumbent VIU still maintains a dominant position. Regulatory decisions can and do have very large financial consequences for the participants and poor regulation can lead to risks for participants. With large amounts of wind power, with its unique characteristics, being connected to electricity systems, there is a danger that regulators will fail to grasp the issues and make decisions that will stifle the development of the industry as a whole and wind power in particular. If the regulatory regime is too favourable to wind power, then it will thrive at the expense of the other parts of the industry, which will be damaging. If the regulator is too severe with wind power, then investment in the sector will suffer, and society will fail to benefit from the emissions reduction and security of supply potential of the technology. Regulators need to encourage innovation and to help the industry to seek out the most effective means of wind power integration.

7.2 The electrical energy market

The principal product being sold in an electricity market is energy. There are a number of possible mechanisms for selling the energy. The choices may be limited by the particular market structure and local circumstances. There may be a pool market where electrical energy is bought and sold centrally (Fig. 7.1), where participation can be mandatory, sometimes known as a gross pool. Buyers and sellers will bid in and the market will clear, setting prices and quantities. The prices and quantities will be applied in market settlement (Baldick *et al.*, 2005). The bidding process will occur in advance of the actual generation and consumption of electricity, and will end some time in advance of real time, known as gate closure. For example, in Ireland half-hourly bids for the day ahead will be accepted up to 12 hours in advance (CER, 2005).

The mechanisms used for clearing can vary from markets based on centralised unit commitment (see Section 5.2), where a lot of detailed bid information on costs and technical limitations is required, to markets of the self-commitment type where price–quantity pairs are all that are required, with the generators making their own commitment decisions. Broadly speaking the market prices will be set on a marginal

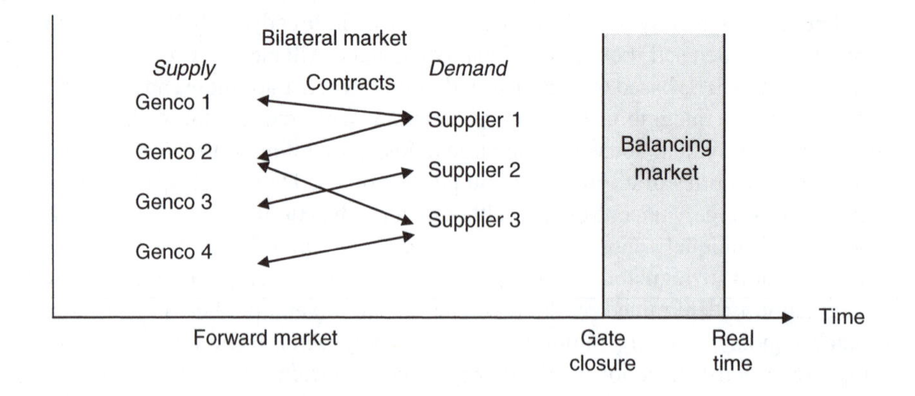

Figure 7.1 Representation of a gross pool electricity market

cost basis, that is, the cost of an additional MWh over a specified time period (typically 30 minutes). The price will be paid to all generators and will be paid by all suppliers for their respective volumes of energy. Variations on this principle will apply for central unit commitment markets where the price will be based on the marginal cost as defined over a longer time frame (for example 24 hours), where start-up and no-load costs are accounted for (see Section 5.2). Variations will also occur in markets that attempt to account explicitly for technical constraints: for example, when transmission congestion occurs the marginal costs of 1 MWh at specified locations will differ and locational prices will apply (Hamoud and Bradley, 2004).

In a centralised unit commitment market, the cost information – coupled with the technical limitations – should allow the market operator to commit and dispatch the units in an efficient manner. In the self-commitment market the simple bids need to be chosen to reflect the underlying costs and are designed so that the resulting dispatched quantities are technically feasible to deliver. Whether or not these pricing mechanisms are adequate to cover all the costs of the generators is a matter of debate (Baldick *et al.*, 2005). Pricing can also occur in advance of delivery (ex ante) or after delivery (ex post). The Single Electricity Market (SEM) that is being proposed for the island of Ireland is a gross pool market where prices are set ex post using a centralised unit commitment approach (CER, 2005).

As wind turbines have negligible operating costs, they should bid zero into a pool market. Alternatively wind can be a *price taker*, that is, it offers its volume with no cost information and is willing to take the price that is set by other bids. Bidding zero and being a price taker are not always equivalent: for example, in certain cases negative prices can occur (Pritchard, 2002). If wind generation decides to bid above zero into the market then it runs the risk of not being physically dispatched and not receiving any revenue from the market settlement.

Large-scale deployment of wind power will affect the price of electricity. When the wind is blowing (and generating) and, provided there are no negative prices or technical constraints that would curtail production, it will be accepted by the market, assuming it is bidding zero. This will displace other generation and the marginal unit

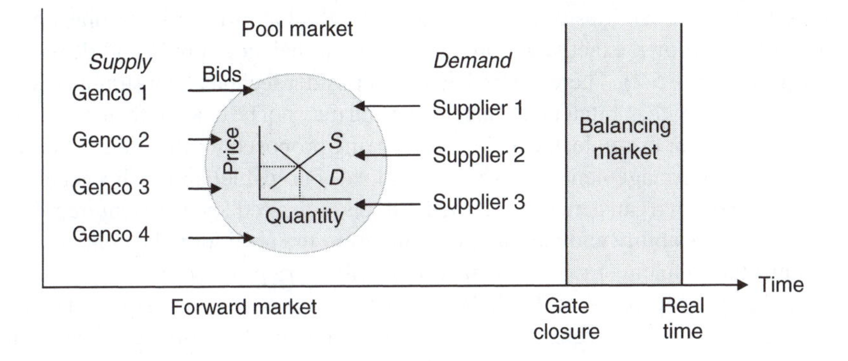

Figure 7.2 Representation of a bilateral contracts market

will be lower down the merit order, and hence the marginal price will be lowered. When the wind is not blowing, then the marginal unit that sets the price will be higher up the merit order and hence the price will rise. Thus wind power will introduce a degree of price volatility in the energy market. It will also mean that the expected revenue for wind from the electricity market will be somewhat lower than the expected price would indicate, as there is a negative correlation between volume and price. Other correlation effects will also affect the revenue. Typically, it is windier during the day than at night (Fig. 5.9), and typically electricity demand is higher during the day than at night (Fig. 5.10), and hence the price is higher. The positive correlation will tend to increase the expected revenue from wind power. Seasonal patterns will also be influential. In some geographical areas, Ireland for example, it is windier during the winter months (Fig. 5.9), when electricity demand is also higher. Such a positive correlation will tend to increase wind revenue. In contrast, there are other geographical areas where the load is higher during the summer, and the opposite may occur.

Another option is to sell the energy bilaterally to another party, typically a supplier. Such bilateral arrangements (Fig. 7.2) can be agreed years in advance. Bilateral trading will continue until gate closure, when all bilateral trades are reported to the system operator. The British Electricity Trading and Transmission Arrangements (BETTA) is a market of the bilateral type (OFGEM, 2004). For bilateral deals, however, there will be a need to match the generator output with the suppliers' needs. These will be related to an aggregation of their customer needs and will vary typically with time of day, week and season. Such a bilateral deal will be difficult for wind generation to meet, as it is variable. Therefore wind generation will need a balancing market where it can buy electricity to fulfil its side of the bilateral contract.

7.3 Balancing, capacity and ancillary services

Regardless of which mechanism is used in the electricity market to trade energy there will always be some form of central market for balancing. The balancing market is necessary as the quantities of energy required by the load for different times in the future are not known exactly, and there is a need to balance supply and demand in real time (Section 5.2). There is also a need to consider technical constraints that are difficult to deal with in bilateral arrangements and may not be dealt with in centralised arrangements. For example, the SEM market being proposed for the island of Ireland is a gross pool arrangement, but is unconstrained. Fossil-fuel and hydro generators are controllable and can generally set their outputs to desired levels as required. Wind power lacks this ability and therefore contributes to the need for balancing.

While the balancing market will have similarities to a gross pool, it will differ in a number of important respects. The balancing market will only cover very short periods of time and is sometimes referred to as a real-time market. The volume of energy that is traded in the balancing market is small, but the prices are extremely important to the overall market as they will reflect short-term supply/demand imbalances. During times of low load and high generator availability, prices should be low. In contrast, during times of high load and/or low generator availability, the prices should rise. Persistent high prices indicate the need to invest in generating capacity.

In some markets there are explicit payments for capacity, made on the basis of the ability to generate. Markets that do not have capacity payments are referred to as *energy only*. SEM has an explicit capacity mechanism while BETTA does not. The impact of these capacity payments should be to encourage long-term investment, as the investor has some level of guaranteed income, as opposed to the energy only market where income will be based purely on production of energy. Capacity payments should also moderate the energy price received by the generators, particularly during times of shortage. If there are no explicit capacity payments then the value of capacity is reflected in the energy price, either within the pool or through bilateral trades. With no explicit capacity mechanism, and assuming a competitive market, the argument can be made that there should be no limit to the price of energy, that is, whatever the market will bear. It is during times of shortage when the price is high that capacity is being paid for implicitly. With an explicit capacity mechanism in place, it can be argued that the energy price should be capped by the regulator to avoid participants paying for capacity twice. The capacity payments are based largely on the concept of capacity credit. The capacity credit for a generator is the additional load that can be served while maintaining system reliability. Wind does have a capacity credit that decreases with wind penetration (Section 5.3.5) and should receive some payments under a capacity mechanism.

Operation of an electricity supply system in a reliable manner requires the provision of what are termed ancillary services. These technical services include reserves, reactive power, congestion management and black-start capability (Section 5.3.6). Ancillary services are being provided increasingly through market mechanisms, and wind generation has the potential to earn revenue from them. They can be contracted on a long-term basis in a competitive manner. They can be provided through centralised markets, for example, locational pricing for congestion management, and co-optimised energy and reserve markets (Baldick *et al.*, 2005). A wind turbine can provide reserve by not operating at its maximum power output for the wind available, that is *spilling wind*. This may make economic sense in certain circumstances.

The more modern wind turbines/farms can control reactive power and can participate in voltage control. Black-start capability is not possible for a wind turbine as it depends on wind availability.

In a market of the bilateral type there is a perceived tendency to under-value wind power as operators find it difficult, economically and technically, to make firm bilateral arrangements. Intermittent sources such as wind are penalised in the balancing market because they are seen to be placing a burden on the system in terms of extra provision of ancillary services such as regulation, reserve, etc. The whole ancillary services market concept is poorly developed in comparison with the energy market, and this is leading to concerns over reliability and lack of incentives for power plant such as wind.

There is a very strong coupling between energy markets, balancing, capacity mechanisms and ancillary services. If bilateral trading and/or a gross pool were practical up to real time there would be no need for a balancing mechanism. Reserves and capacity are closely related. If, for example, the TSO contracts for reserve in advance over long periods of time, then this is effectively a capacity mechanism. Reserves that are dispatched by the balancing mechanism are in the form of energy. Therefore great care must be taken when these terms are being interpreted.

7.4 Support mechanisms

For wind to compete effectively with other sources of electricity, it is necessary that its advantages, that is, low emissions and security of supply, are rewarded and recognised. This can be achieved by supporting wind generation directly or by penalising conventional generation for harmful emissions. Direct support consists generally of specific targets for renewable energy penetration. Indirect support aims to capture the true costs of other forms of electricity production, for example the external cost of emissions being represented in the cost of electricity generation from fossil-fuel plant. Emission taxes, for example a carbon tax that penalises fossil fuels, would make electricity more expensive and would give wind power and other renewable sources a competitive advantage.

The United Kingdom has a system of renewable obligations certificates (ROCs). ROCs are separate products from the electrical energy but are related directly to the quantity of renewable electricity produced. Suppliers are obliged to purchase a certain quantity of ROCs or face a penalty charge if they are short. The penalty was originally set at £30/MWh and is index-linked. The quantity was 5.5 per cent of energy volume for 2005/2006. The penalty charge places a cap on the value of the certificates. As the number of certificates increases with more renewable generation, a point will be reached when the supply equals demand, after which the price of the certificates will collapse. Such a possibility makes these types of mechanisms somewhat risky for investors. In order to address this, the target increases with time and is set at 6.7 per cent for 2006/2007, rising to 10.4 per cent by 2010. The penalty revenue that is collected from suppliers who do not meet their targets is recycled pro rata to all certificate holders. The value of the certificates is therefore enhanced, and they are currently valued at about £45/MWh. The European emissions trading system has similar characteristics to the UK ROC system, but requires all conventional generators to have a certificate for every tonne of CO_2 emitted, with penalties for non-compliance. The emissions trading mechanism internalises the external costs of conventional generation and helps make wind generation more competitive.

One of the most popular mechanisms to support renewable generation is the feed-in tariff. Essentially the wind farm is guaranteed a price per MWh of electricity for a period of time. The feed-in tariff is not a market-based mechanism, and limits are placed on the volumes, with different prices for different technologies. For example, offshore wind will typically attract a higher price than on-shore wind as the technology is less mature and more expensive. A similar scheme is a competitive tender where wind developers bid in a price they are willing to accept per MWh over a defined period (e.g. 15 years). The lowest bids are accepted up to the volume required and the contracts are issued. The Alternative Energy Requirement (AER) in Ireland is an example of this type of mechanism (DCMNR, 2005).

While the various support mechanisms do encourage and support wind power in the market, they all have their strengths and weaknesses. From a market perspective they are all less than ideal. The best approach is to have wind power compete in a fair and open market where all costs are included.

7.5 Costs

The concept of *causer pays* is a phrase commonly coined in discussions concerning electricity markets. The costs fall into a number of categories. There are the costs of the physical networks (transmission and/or distribution) that are required to deliver the energy from the various sources. The network assets need to be paid for, and individual generators are charged transmission use of system charges and distribution use of system charges depending on where in the system they are connected. Network charges can include a locational element that encourages generation and/or load to locate or not locate in certain places. Such locational signals will be driven by transmission losses and congestion. Losses can also be included in locational prices (Keane and O'Malley, 2006). The other cost category is associated with the provision of ancillary services such as reactive power and reserve. Some of the wind turbine technologies create a need for reactive power, but these can be self-provided and this is the approach that is being pursued through grid codes (Table 4.2). Additional reserves will be needed for balancing for large penetrations of wind power due to the prediction error. Self-provision of reserve is generally not very efficient and is best done centrally. There will be a financial overhead associated with the market, comprising transaction costs, market system costs and the cost of regulation.

In some markets the system costs are socialised, that is added to the price of electricity and passed through to the consumer. However, this is not good practice, as costs should be allocated to those who can control them, thus encouraging them to invest in solutions to reduce them. Certain supply system costs are best managed by the wind generators themselves – for example, the need for reserves to balance wind can be reduced by better wind forecasting. Other costs are best dealt with by the network operator (e.g. losses) and others by a combination of generators and network operators (e.g. reactive power).

Proper allocation of the costs of wind power will optimise the amount of wind generation. Many in the wind power industry are motivated to obtain as much wind generation connected as soon as possible. However, if this is allowed to occur without proper allocation of the costs to the wind industry, wind generating capacity may saturate at a lower level of penetration than would otherwise be feasible. Therefore, it is in the long-term interest of the wind power industry to accept and pay the costs that it imposes on the system. Quantification of these costs is non-trivial and is often the subject of debate.

Recently, many reports have been published on the impact of wind on electricity systems (ESBNG, 2004a; Gardner *et al.*, 2003). The studies attempt to quantify the impacts of wind generation on the system in terms of additional reserves (Doherty and O'Malley, 2005), impact on emissions (Denny and O'Malley, 2006) and system inertia (Lalor *et al.*, 2005), starts and stops of other units, ramping, load following, etc. The technical impacts are then converted into costs. These studies can be controversial for a number of reasons. First, other generation technologies impose costs on the system, yet they do not attract as much attention. Second, the studies are very difficult to perform because they are looking into the future, and there is no real experience of operating power systems with substantial amounts of wind power. There is a consensus, however, that wind power does have some adverse impact on power system operation, in particular the associated need for balancing and ancillary services, and that this adds to the cost of wind power. The incremental cost increases with wind penetration, and it is this aspect that leads commentators to suggest that there is a maximum level of wind generation that should be allowed. However, it is also clear that these costs are different for different systems due to the nature of the network, other generating plant and system operating practices.

There is no doubt that there are plant portfolios that are particularly suited to wind power and there are those that are not. For example, flexible open-cycle gas turbines may be deemed to be beneficial to wind power, while in contrast inflexible base-loaded nuclear generation is not complementary. In all electricity systems there are plant and load portfolios that are more beneficial to wind power than others. The best mixes are different for different regions, climatic conditions, natural resources, industrial characteristics, etc. Large-scale wind generation with its particular characteristic is altering the load/generation mix and if it is to thrive the market needs to encourage investment in plant that will optimise the mix.

The reliability and security of a power system requires an instantaneous power balance. Therefore the variability of wind power over different time scales – minutes, hours and days – requires that there is a plan/schedule in place to commit/de-commit other plant (typically thermal) to maintain the balance. The greater the amount of wind power installed, the larger the error in the expected value. Also, the longer the time horizon of the wind forecast, the greater the error. Shorter time frames for forecasting will reduce the error to very small amounts, but this will require flexible plant to operate the system. Load forecasting also introduces a degree of unpredictability but is not as sensitive to time horizon, and for large wind penetrations the wind forecast error will predominate over the longer time horizons. Clearly the matching thermal generators need to have characteristics that make them suitable to follow this variability and/or unpredictability. Under the *causer pays* principle, any additional costs incurred in providing these balancing services need to be borne by the wind power generators. Costs of increased wind generation include: capital cost of flexibility in matching thermal plant, extra start-ups and shut-downs and ramping of conventional plant (Denny *et al*., 2006).

Operation of a power system with significant amounts of wind power is a unique challenge that will require technical and market innovation. The design and development of these new electricity markets are heavily influenced by the operational characteristics of the electricity system. Wind power has characteristics that make it significantly different from the more traditional generation technologies, in particular its variability and unpredictability. Variability of the wind power is the expected change in output over a time horizon; the error in this expectation is a measure of the unpredictability. In order to extract maximum benefit from installed wind power it may be necessary to modify the existing operational methods (Chapter 5), and these modifications need to be reflected in the corresponding market structure. For example, unit commitment has traditionally been formulated as a deterministic optimisation problem. This was based on the assumption that the forecasts for load and unit availability were generally accurate. With wind power it may be appropriate to formulate the unit commitment problem as a stochastic optimisation problem as the wind forecast error can be large. Failure to modify and/or adapt operational methods may hinder the development of wind power. Through the market the costs of these modifications and additional services need to be allocated effectively.

System operations and electricity market arrangements need to be closely aligned, that is the economic and engineering principles need to be coordinated in order to optimise the electricity industry. There are some who believe that certain operational and market practices are not favourable to wind power and have put forward arguments in favour of altering market rules and/or operational practices. For example it can be argued that a long gate closure time, such as 12 hours, is detrimental to wind power, as the forecast error can be very large over this time frame (30 per cent – Chapter 6). This exposes wind operators to major buy and sell volumes in the balancing mechanism. The price in the balancing mechanism will typically be high when wind is short and low when wind is long. Shorter gate closure times will reduce the volumes that are exposed to the balancing mechanism. This argument was successfully used to shorten the gate closure time in the predecessor to the BETTA market from 4 hours to 1 hour (OFGEM, 2002). However, the fundamental issue remains – wind power is difficult to predict well in advance, and long-term bilateral contracting for wind power will rely on a balancing mechanism or generators that will enter into short-term bilateral arrangements a few hours in advance of real time. The best option for wind power – the balancing mechanism or a bilateral arrangement – will depend on the details of the market.

7.6 Investment and risk

Investors in wind power or any other form of generation will expect a return on their investment through various market mechanisms. Investors will expect higher returns on their investment if the risks are higher. There are significant risks in electricity markets. It is a capital-intensive business and the assets are difficult to move; for example, it is not viable to move wind turbines once they are installed. The volatility in the energy price can be a significant deterrent to investment. It can be hedged between the generation and supply businesses, and this has encouraged a level of vertical integration and/or the use of financial instruments. The simplest and most useful financial instrument is a contract for difference, where generators and suppliers agree a long-term price for energy regardless of the underlying short-term price (Lowrey, 1997). One of the biggest risks in an electricity market is regulatory risk. Regulators can and do have immense influence over the market, and any perceived weaknesses in the regulatory framework or regulatory ability will discourage investment.

Fostering true competition in an electricity market is not a trivial task. However, trying to achieve this with a dominant player is nearly impossible. Market power exists when a single market player can set the price. This can easily occur in an electricity market and is a common problem in generation. In particular, if the old monopolies are not properly dealt with by forcing them to divest generation assets prior to setting up a market, it is inevitable that they will have some level of market power, at least initially. Market power can be used to drive up the price and make abnormal profits, or to drop the price below cost to deter market entry. Market power in electricity markets can also occur without dominance – this type of market power occurs when individuals engage in gaming in the electricity market.

As wind penetration levels rise, operational and network constraints may require the curtailment of wind power (Section 5.3). At low load levels and at times of high wind production it may be necessary for system security reasons to curtail the amount of wind generation. The market needs to be able to accommodate this. A simple example of this is in a low load situation where for frequency control and/or voltage-control reasons there is a need to keep a minimum number of flexible units on-line. Such units will have minimum running levels, and in order to maintain supply/demand balance it may be necessary to curtail the wind generation. Under certain network, load and generation patterns there will be congestion on the network and there will be a need to decrease generation in one location and increase it in another to relieve congestion. Such curtailment occurs for other generation technologies and is a potential source of risk if no financial compensation is received for lost revenue. Compensation will be paid in circumstances where a generator has firm access. Firm access means that the generator has the right to export its energy onto the grid in all reasonable circumstances. For example, if a generator with firm access is constrained down to relieve a transmission constraint, then it will be compensated for the lost opportunity cost, typically the difference between its generation cost and the price times the volume of curtailment. Non-firm access means that no such compensation is paid and is a risk that generators must manage. To encourage renewable generation, legislation is sometimes enacted to give it priority access, which is a concept similar to firm access.

7.7 The future

The characteristics of the electricity industry detailed above have made the creation of truly competitive electricity markets problematic. Some would argue that the real-time nature, limited transmission capacity and scale of investment make the industry virtually impossible to run in a competitive framework. The debate continues and the competitive market structures being introduced are continually undergoing changes and are in many ways experimental. It is interesting to note that in Ireland there is a transition from a market of the bilateral type to a market of the pool type, while in the United Kingdom the transition from The Pool to NETA/BETTA was in the opposite direction. Only time will tell if the restructuring experiments were successful.

The success of an electricity market can be assessed by a number of quantifiable criteria. The market must ensure adequate capital investment to maintain the reliability and security of supply. With large amounts of wind power being installed there is a requirement that this investment is in plant and technology that complement wind power characteristics. The price signals to the market must encourage and reward correct behaviour among participants. The market should result in an optimal operation of the power system such that the cost/benefit to society (social welfare) is maximised. This is not an easy matter to quantify as it involves optimal operation in short, medium and long time frames. It involves solving the optimal dispatch, unit commitment and optimal planning problems, which are still the subject of research. The market needs to reward innovation and needs to minimise barriers to entry. The market should promote energy efficiency and minimise harmful emissions. The cost to society of these emissions needs to be built into the market and hence the prices. Some will also argue that the price to consumers should be low. This type of argument is naïve, as the price in a competitive market will reflect the costs. A sustainable, reliable, efficient, and low-emission power supply system is a costly proposition that needs to be paid for.

Wind power is not the only recent addition in electricity markets. The load in an electricity system was assumed traditionally to be largely inflexible and hence was treated as a price taker. With the demand side of the market being largely inflexible, market power on the generation side is harder to avoid (particularly in small markets at times of high demand). This has led to serious market power issues which have, in many cases, undermined the entire concept of electricity markets. However, with technology developments, particularly in telecommunications, the load is becoming much more flexible and can participate more fully in the emerging electricity markets. Not only will this load flexibility enable the markets to function more efficiently and effectively, but it should also allow more wind power to be connected to the grid as there will be the opportunity to match the wind variability and unpredictability with flexible load. Energy storage may also have a future in electricity markets as a complementary technology to wind power (Section 5.4). However, most market studies would indicate that storage is only viable in niche areas.

It could be argued that the recent well-publicised blackouts in North America and Europe highlight the failure to restructure the industry correctly. In this environment there are many critics of wind power who suggest that its variability will threaten the reliability of the electricity system. This argument will have merit if the market fails to encourage behaviour leading to correct investment in the appropriate plant mix and technologies that will complement wind power. Some have argued that the introduction of competitive electricity markets has led to under-investment in generation and transmission. They claim that the assets are being sweated for short-term gain with little or no long-term planning. The success of wind power penetration is heavily dependent on investment in the transmission system, and on the appearance of thermal plant that will serve to replace wind power during lulls.

Appendix

FACTS technology

Flexible alternating current transmission system (FACTS) devices are expensive, and it is necessary to ensure that their functionality is required before specifying them. They perform four basic functions, which can be combined in different devices:

1. Power transfer between electrically separated systems
2. Active power management
3. Reactive power management
4. Waveform quality management.

Active power management devices (for example phase shifters, static synchronous series compensators and unified power flow controllers) that are less relevant to wind technology, either internally or in connection terms, will not be discussed here. Current source converters for high-voltage direct current (HVDC) are dealt with for completeness, although it is likely that most large wind farms would be connected by voltage source converter technology, and doubly fed induction generators (DFIGs) apply this technology in their rotor circuits.

Synchronous connection by lines and cables makes two electrical systems behave as one. This clearly is inappropriate if the frequencies are different or one system has a stability or fault level problem that would be exacerbated by connection with another source. In these circumstances the systems may be maintained as separate entities through a converter/inverter DC path. In each of the other applications, traditional technology exists but performs its function slowly or within a narrow range. It is the need for rapid action in controlling active and reactive power and the need to manage a wide range of waveform distortion problems that requires the application of FACTS devices.

Two fundamental converter technologies are used in separating electrical systems:

The current source converter. This is usually a line-commutated device, that is, the thyristors are switched on by a gate control but remain on until the current reaches zero. Self-commutated current source converters have been constructed using gate turn-off (GTO) or insulated-gate bipolar transistor (IGBT) technology. In line-commutated

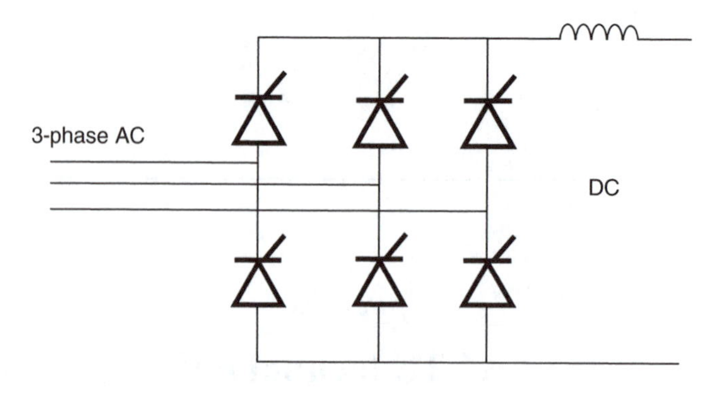

Figure A.1 6-pulse converter bridge

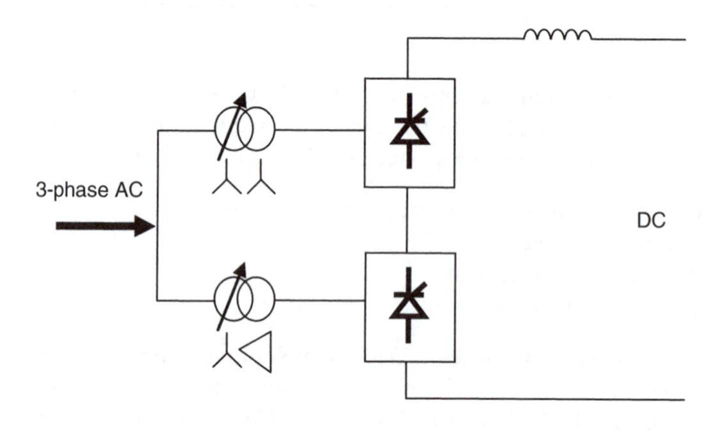

Figure A.2 12-pulse converter bridge

devices the current must lag the voltage, whereas in self-commutated devices either lead or lag is possible, giving power factor control. The converters are normally arranged in a 6- or 12-pulse bridge. A 6-pulse bridge is shown in Figure A.1.

Twelve-pulse bridges (Fig. A.2) produce smoother AC and DC waveforms. They are normally created from two 6-pulse bridges: one connected to a star-configured part of the grid transformer and the other to a delta-configured part. This ensures that commutation takes place every 30° rather than every 60°, as for a 6-pulse bridge.

In order to operate successfully, such converters have a commutation overlap period and hence have short periods in each cycle when two phases are effectively short circuited. The DC-side leakage inductances of the two phase contributions to the DC link determine the magnitude of the short-circuit current. The larger the current, the longer will be the commutation time. The effects are more significant in inverter operation. If the system is weak, the system voltage will collapse during these periods and the inverter control will lose its commutation reference. For this reason, current source converters can only be used where both systems are relatively strong in an electrical sense, as measured by fault level. Manufacturers indicate

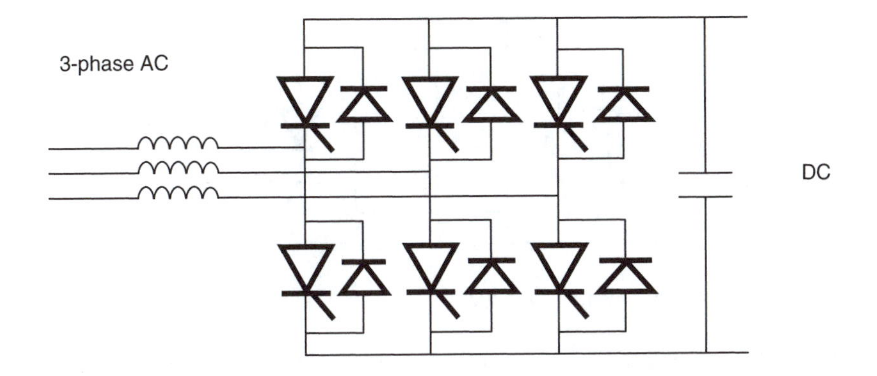

Figure A.3 Voltage source converter

that commutation failure starts to become an issue when the fault level is less than six times the transferred power and is not possible below three times the transferred power. Clearly the commutation gives rise to ripple on the DC side, which is smoothed through an inductor, and to imperfect sine waves on the AC-side. These imperfections result in DC harmonics at $m \times n$, where m is an integer and n is the number of pulses in the converter, so 12, 24, 36, 48 and so forth for a 12-pulse bridge. The AC side experiences harmonics of $m \times n \pm 1$, so 11, 13, 23, 25, 35, 37 and so forth.

The voltage source converter. This converter can produce an AC voltage that is controllable in magnitude and phase, similar to a synchronous generator or synchronous compensator. The device commutates independently of the AC-side voltage and therefore the voltage source converter can be used on a load-only system, that is, a system with zero fault level. This makes it useful for DFIG rotor connection, wind farm connection, connection of oil platforms and so forth. The device configuration is shown in Figure A.3.

At present, economics suggest that it is viable for transfers up to 200–300 MW, above which it becomes too expensive. The efficiency of voltage source converters is lower than current source devices which can achieve 98 per cent or greater. Voltage source converters operate by switching the devices at frequencies higher than line frequency using pulse width modulation (PWM). By varying the pattern from the requirement for a standard sinewave, harmonics can be negated at source. Thus the converter allows control over both the amplitude of the voltage and the waveform quality. A penalty for higher frequency switching is increased losses. It must therefore be part of the objective of any selective harmonic elimination (SHE) scheme employed in PWM control to minimise switching frequency. Magnetic circuits can be applied with coils suitably wired to cancel some harmonics and minimise others, hence reducing the SHE duty. If PWM is not used, as when SHE is applied to a fixed and restricted pattern of harmonics, then the output voltage control is achieved by varying the DC bus capacitor voltage. This is achieved by charging or discharging the DC bus capacitor through variation of the firing angle. Clearly this process takes time and does not give the fast response of PWM. Current loop control can be linked

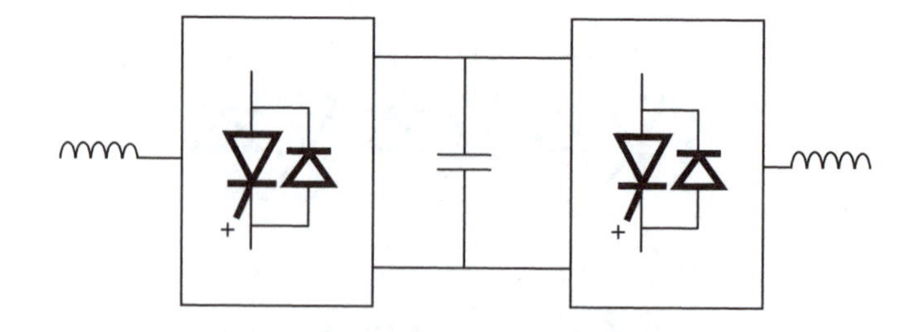

Figure A.4 Converter system for DFIG

with the voltage control achieved in PWM, resulting in a device with a remarkably fast and flexible response. Figure A.4 shows one phase of a converter system which can be applied to the rotor of a DFIG device.

Wind farms often require reactive compensation for voltage regulation under normal conditions, and to assist ride-through under fault conditions. The relevant devices are:

Static VAr compensator (SVC). This device, which has been in use since the 1970s, is a combination of thyristor-switched capacitor banks and thyristor-controlled inductor banks. At low system voltage conditions the device generates reactive power, that is, delivers capacitive current, whereas during high voltage conditions it absorbs reactive power, behaving like an inductor. Commonly the devices have a steady-state rating and a transient rating (in the absorbing direction only) with the aim of *rescuing* a dangerously high voltage condition and allowing time for other system action. It has two basic uses: to regulate bus voltage at a node, and, placed in the mid-point of a line, to allow the line impedance to be compensated, hence extending its rating. A problem with the device is that its capacitive contribution is most effective at higher voltage. Below 0.9 pu voltage, it falls off linearly to 0 at zero voltage, as shown in Figure A.5.

This is because the reactive power capability of the capacitors reduces with voltage. It therefore must be seen as a device that is useful within a *normal* voltage range. Obviously where the problem is related to fault ride-through of wind generators, the voltage will fall much below this range and, to be really useful, the capacitors must be charged until the disturbance is cleared. Taken alone, the device has no way of discriminating between supplying reactive power to the system disturbance and to the wind generator.

Static compensator (STATCOM). This device, formerly known as the advanced static VAr compensator (ASVC), is based on a voltage source converter rather than a thyristor-controlled capacitance. The voltage source converter DC terminals are connected to a small capacitor which is maintained at some voltage level (see Fig. A.6).

The output voltage of the device lags the system voltage by a small angle to maintain the capacitor charged. The angle is varied to adjust the voltage of the capacitance, which determines the reactive power (MVAr) injection into the system. The output is achieved by creating the current wave with a 90° phase shift to the output voltage wave. The advantage of this device is that it can maintain constant reactive power

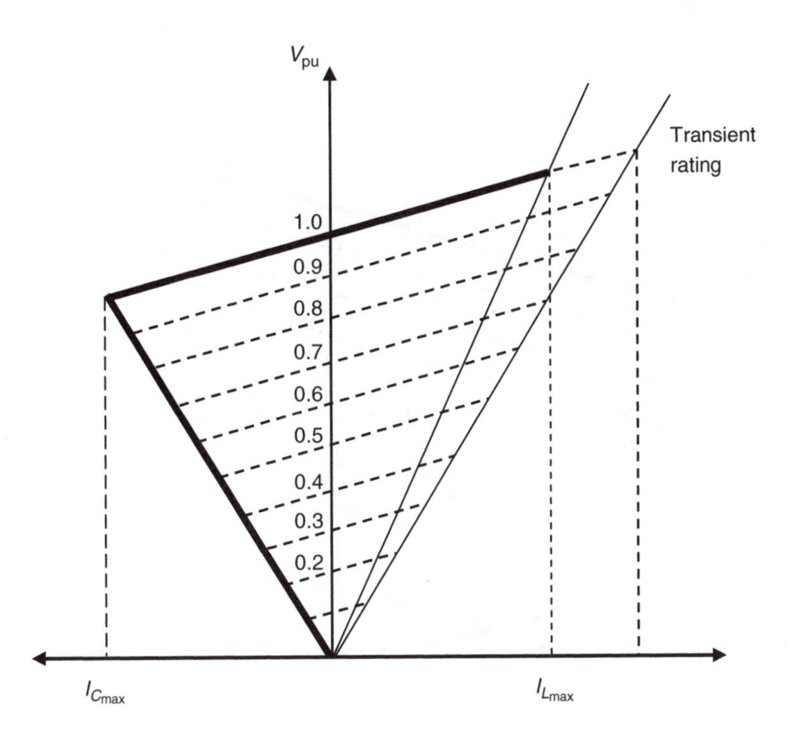

Figure A.5 SVC operating chart

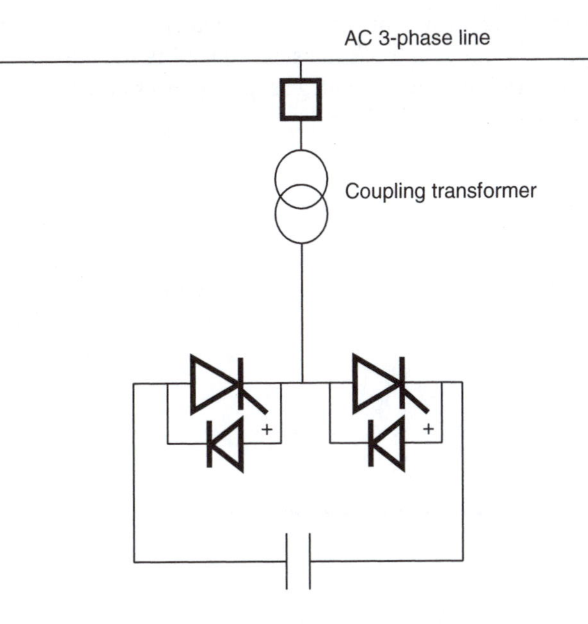

Figure A.6 STATCOM configuration

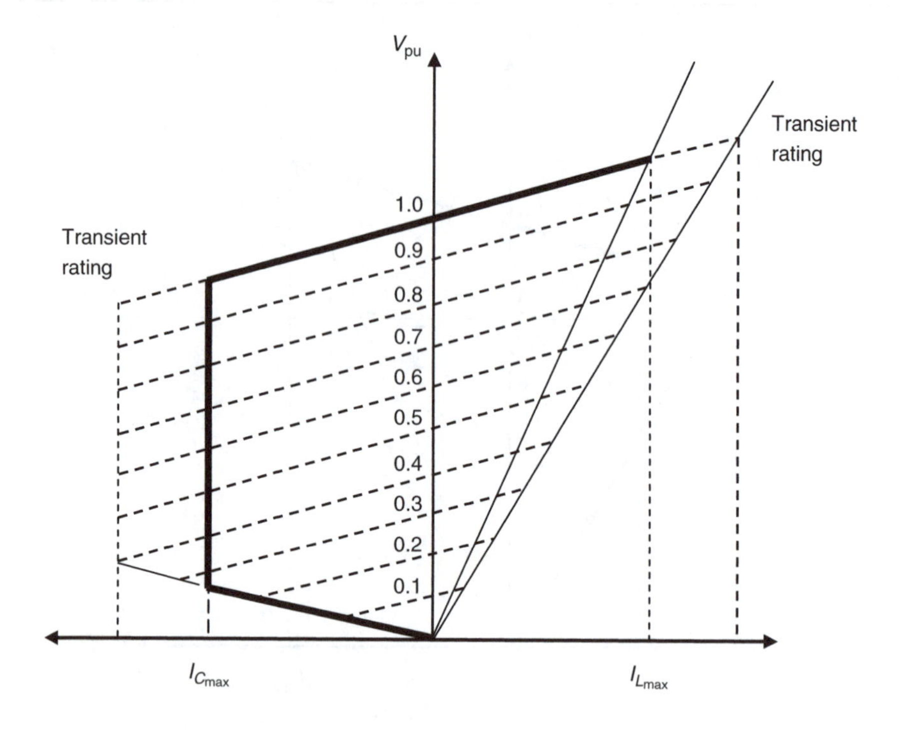

Figure A.7 STATCOM operating chart

output down to about 0.2 pu system voltage, after which it falls off proportionally to 0 at zero voltage (see Fig. A.7). It has transient capability in both the capacitive and inductive quadrants. Being based on voltage source converter technology, the device can also act as an active harmonic filter.

References

Akhmatov, V., Abildgaard, H., Pedersen, J., and Eriksen, P. B.: 'Integration of offshore wind power into the western Danish power system', *Offshore Wind Conference*, Copenhagen, Denmark, October 2005

Akhmatov, V., Nielsen, A. H., Pedersen, J. K., and Nymann, O.: 'Variable-speed wind turbines with multi-pole synchronous permanent magnet generators. Part I: modelling in dynamic simulation tools', *Wind Engineering*, 2003;**27**:531–48

Anderson, D., and Leach, M.: 'Harvesting and redistributing renewable energy: on the role of gas and electricity grids to overcome intermittency through the generation and storage of hydrogen', *Energy Policy*, 2004;**32**:1603–14

Andersson, G., Donalek, P., Farmer, R., Hatziargyriou, N., Kamwa, I., Kundur, P., *et al.*: 'Causes of the 2003 major grid blackouts in North America and Europe, and recommended means to improve system dynamic performance', *IEEE Transactions on Power Systems*, 2005;**20** (4):1922–8

Armor, A. F.: 'Management and integration of power plant operations' in Flynn, D. (ed.), *Thermal power plant simulation and control* (IEE Power and Energy Series, London, 2003) pp. 395–416

Árnason, B., and Sigfússon, T. I.: 'Iceland – a future hydrogen economy', *International Journal of Hydrogen Energy*, 2000;**25** (5):389–94

Asmus, P.: 'How California hopes to manage the intermittency of wind power', *The Electricity Journal*, July 2003;**16** (6):48–53

Atkinson, B. W. (ed.): *Dynamical Meteorology* (Methuen, New York, 1981)

Australian Greenhouse Office (AGO): *National wind power study: an estimate of readily accepted wind energy in the national electricity market* (AGO, 2003)

Bach, P. F.: 'A power system with high penetration of renewables', *CIGRE Colloquium on Grid Integration of Renewables – Achievements and Challenges*, Dublin, Ireland, September 2005

Baldick, R., Helman, U., Hobbs, B. F., and O'Neill, R. P.: 'Design of efficient generation markets', *Proceedings of the IEEE*, 2005;**93** (11):1998–2012

Barry, D., and Smith, P.: 'Analysis of the integration of wind generation into the Irish system', *IEEE PowerTech Conference*, St Petersburg, Russia, June 2005

BTM Consult ApS: 'International wind energy development', *World Market Update*, Ringkobing, Denmark, 2003

Bueno, C., and Carta, J. A.: 'Wind powered pumped hydro storage systems, a means of increasing the penetration of renewable energy in the Canary Islands', *Renewable and Sustainable Energy Reviews*, 2006; **10** (4): 312–40

Burton, A., Sharpe, D., Jenkins, N., and Bossanyi, E.: *Wind Energy Handbook* (John Wiley & Sons, Chichester, 2001)

Castronuovo, E. D., and Peças Lopes, J. A.: 'Optimal operation and hydro storage sizing of a wind-hydro power plant', *Electrical Power and Energy Systems*, 2004;**26**:771–8

Christiansen, P.: 'A sea of turbines', *IEE Power Engineer*, February 2003;**17** (1):22–4

Christopoulos, C.: *An introduction to applied electromagnetism* (John Wiley & Sons, Chichester, 1990)

CIGRE: *Modelling of gas turbines and steam turbines in combined-cycle power plants*, CIGRE Task Force 38.02.25, 2003

Commission for Energy Regulation (CER), Ireland (2005) *The Single Electricity Market – proposed high-level design* [online], AIP/SEM/06/05, available from: http://www.cer.ie [30 June 2006]

Crabtree, G., Dresselhaus, M. S., and Buchanan, M. V.: 'The hydrogen economy', *Physics Today*, 2004;**57** (12):39–44

Cregan, M., and Flynn, D.: 'Advances in power plant technology' in Flynn, D. (ed.) *Thermal power plant simulation and control* (IEE Power and Energy Series, London, 2003) pp. 1–14

Dale, L. (2002) *NETA and wind* [online], EPSRC Blowing Workshop, Manchester, available from: http://www.ee.qub.ac.uk/blowing/ [30 June 2006]

Dany, G.: 'Power reserve in interconnected systems with high wind power production', *IEEE PowerTech Conference*, Porto, Portugal, September 2001

Dawber, K. R., and Drinkwater, G. M.: 'Wind farm prospects in Central Otago, New Zealand', *Renewable Energy*, 1996;**9** (1–4):802–5

Denny, E., Bryans, G., Fitzgerald, J., and O'Malley, M.: 'A quantitative analysis of the net benefits of grid integrated wind', *IEEE Power Engineering Society General Meeting*, Paper 1-4244-0493-2/06, Montreal, Canada, 2006

Denny, E., and O'Malley, M.: 'Wind generation, power system operation and emissions reduction', *IEEE Transactions on Power Systems*, 2006;**21** (1): 341–7

Department of Communications, Marine and Natural Resources (DCMNR), Ireland (2005) *AER programme and profile of AER I – VI competitions* [online], available from: http://www.dcmnr.gov.ie [30 June 2005]

Department of Trade and Industry (DTI): *Digest of UK Energy Statistics, 2005* (DTI, UK, 2005)

Deutsche Energie-Agentur (DENA): *Planning of the grid integration of wind energy in Germany onshore and offshore up to the year 2020* (DENA, Germany, 2005)

Dinic, N., Fox, B., Flynn, D., Xu, L., and Kennedy, A.: 'Increasing wind farm capacity', *IEE Proceedings Part C*, 2006;**153** (4):493–8

Doherty, R., and O'Malley, M.: 'New approach to quantify reserve demand in systems with significant installed wind capacity', *IEEE Transactions on Power Systems*, 2005;**20** (2):587–95

Doherty, R., Outhred, H., and O'Malley, M.: 'Establishing the role that wind generation may have in future generation portfolios', *IEEE Transactions on Power Systems*, 2006;**21** (3):1415–22

Donalek, P.: 'Advances in pumped storage', *Electricity Storage Association Spring Meeting*, Chicago, Illinois, May 2003

DTI: *Report into network access issues*, Embedded Generation Working Group, England, 2000

Econnect Ltd: *Wind turbines and load management on weak networks*, ETSU Report, W/33/00421/REP, 1996

Electricity Supply Board International (ESBI) and Energy Technology Support Unit (ETSU): *Total renewable energy source in Ireland – final report*, European Commission Altener programme, Contract XVIII/4.1030/t4/95/IRL, 1997

Electricity Supply Board National Grid (ESBNG): 'Impact of wind power generation in Ireland on the operation of conventional plant and the economic implications', ESBNG, Ireland, 2004a

Electricity Supply Board National Grid (ESBNG): *Review of frequency issues for wind turbine generators* (ESBNG, Ireland, 2004b)

Ender, C.: 'Wind energy use in Germany – Status 30.6.2005', *DEWI Magazine*, 27, 2005

Energy Technology Support Unit (ETSU): *New and renewable energy: prospects in the UK for the 21st century; supporting analysis*, ETSU R-122, 1999

Ensslin, C., Ernst, B., Rohrig, K., and Schlögl, F.: 'On-line monitoring and prediction of wind power in German transmission system operation centres', *European Wind Energy Conference*, Madrid, Spain, June 2003

Environmental Change Institute (ECI): *Wind power and the UK wind resource* (ECI, England, 2005)

European Commission, DG XII: *ExternE – externalities of energy* (Office for Official Publications, Luxembourg, 1995)

European Commission: *The promotion of electricity produced from renewable energy sources in the internal electricity market*, Directive 2001/77/EC, 2001

European Wind Energy Association (EWEA): *Large scale integration of wind energy in the European power supply: analysis, issues and recommendations* (EWEA, Belgium, 2005)

Federal Energy Regulatory Commission (FERC): *Standardised market design and structure* (FERC, USA, 2003)

Ford, R.: 'Blackout scare underlines industry dilemma', *IEE Power Engineer*, June/July 2005, 16

Gardner, P., Snodin, H., Higgins, A., and McGoldrick, S.: *The impacts of increased levels of wind penetration on the electricity systems of the Republic of Ireland and Northern Ireland*, Report for the Commission for Energy Regulation/OFREG NI, 2003

Giebel, G.: 'The capacity credit of wind energy in Europe, estimated from reanalysis data', *Shaping the Future, World Exposition 2000*, Hanover, Germany, July 2000

Giebel, G. (ed.): *Wind power prediction using ensembles*, Risø-R-1527(EN), Risø National Laboratory, 2005

Giebel, G., Landberg, L., Kariniotakis, G., and Brownsword, R.: 'State-of-the-art on methods and software tools for short-term prediction of wind energy production', *European Wind Energy Conference*, Madrid, Spain, June 2003

Golding, E. W.: *The generation of electricity by wind power* (E & F N Spon, London, 1955 – reprinted by Halstead Press, New York, 1976)

Grainger, J. J., and Stevenson, W. D.: *Power system analysis* (McGraw-Hill, New York, 1994)

Gyuk, I., Kulkarni, P., Sayer, J. H., Boyes, J. D., Corey, G. P., and Peek, G. H.: 'The United States of storage', *IEEE Power and Energy Magazine*, 2005;**3** (March/April):31–9

Hamoud, G., and Bradley, I.: 'Assessment of transmission congestion cost and locational marginal pricing in a competitive electricity market', *IEEE Transactions on Power Systems*, 2004;**19** (2):769–75

Hartnell, G., and Milborrow, D.: *Prospects for offshore wind energy* [online], EU Altener contract XVII/4.1030/Z/98-395, 2000, available from: http://www.bwea.com [30 June 2006]

Heffner, G. C., Goldman, C. A., and Moezzi, M. M.: 'Innovative approaches to verifying demand response of water heater load control', *IEEE Transactions on Power Delivery*, 2006;**21** (1):388–97

Heier, S.: *Grid integration of wind energy conversion systems* (John Wiley & Sons, Chichester, 1998)

Holdsworth, L., Ekanayake, J. B., and Jenkins, N.: 'Power system frequency response from fixed-speed and doubly-fed induction generator based wind turbines', *Wind Energy*, 2004;**7**:21–35

Holdsworth, L., Wu, X. G., Ekanayake, J. B., and Jenkins, N.: 'Comparison of fixed-speed and doubly-fed induction wind turbines during power system disturbances', *IEE Proceedings Part C*, 2003;**150** (3):343–52

Holttinen, H.: 'Hourly wind power variations in the Nordic countries', *Wind Energy*, 2005a;**8** (2):173–95

Holttinen, H.: 'Impact of hourly wind power variations on the system operation in the Nordic countries', *Wind Energy*, 2005b;**8** (2):197–218

Holttinen, H.: 'Optimal electricity market for wind power', *Energy Policy*, 2005c;**33** (16):2052–63.

Hor, C. L., Watson, S. J., and Majithia, S.: 'Analyzing the impact of weather variables on monthly electricity demand', *IEEE Transactions on Power Systems*, 2005;**20** (4):2078–85

Hudson, R., Kirby, B., and Wan, Y. H.: 'The impact of wind generation on system regulation requirements', *AWEA Wind Power Conference*, Washington, DC, 2001

Hughes, E.: *Electrical and electronic technology* (Pearson Prentice-Hall, Harlow, 2005)

ILEX Energy Consulting and Strbac, G.: *Quantifying the system costs of additional renewables in 2020* (Department of Trade and Industry, UK, 2002)

Institut fur Solar Energieversorgungstechnik (ISET): *Wind energy report, Germany: annual evaluation*, Kassel, Germany, 1999/2000

International Energy Agency (IEA): *Key world energy statistics* (IEA, Paris, 2005)

Jauch, C., Matevosyan, J., Ackermann, T., and Bolik, S.: 'International comparison of requirements for connection of wind turbines to power systems', *Wind Energy*, 2005;**8** (3):295–306

Jenkins, N., Allan, R., Crossley, P., Kirschen, D., and Strbac, G.: *Embedded Generation* (IEE Power and Energy Series, London, 2000)

Johnson, A., and Tleis, N.: 'The development of grid code requirements for new and renewable forms of generation in Great Britain', *Wind Engineering*, May 2005;**29** (3):201–15

Johnson, G. L.: *Wind energy systems* (Prentice Hall, New Jersey, 1985)

Kahn, E. P.: 'Effective load carrying capability of wind generation: initial results with public data', *The Electricity Journal*, December 2004, pp. 85–95

Kaldellis, J. K., Kavadias, K. A., Filios, A. E., and Garofallakis, S.: 'Income loss due to wind energy rejected by the Crete island electricity network – the present situation', *Applied Energy*, 2004;**79**:127–44

Kalnay, E., Lord, S. J., and McPherson, R. D.: 'Maturity of numerical weather prediction: medium range', *Bulletin of the American Meteorological Society*, 1998;**79**:2753–69

Kariniotakis, G., Halliday, J., Brownsword, R., Palomares, M. A., Ciemat, I. C., Madsen, H., *et al.*: 'Next generation short-term forecasting of wind power – overview of the ANEMOS Project', *European Wind Energy Conference*, Athens, Greece, February 2006a

Kariniotakis, G., Pinson, P., Siebert, N., Giebel, G., and Barthelmie, R.: 'The state of the art in short-term prediction of wind power – from an offshore perspective', *European Wind Energy Conference*, Athens, Greece, February 2006b

Keane, A., and O'Malley, M.: 'Optimal allocation of embedded generation on distribution networks', *IEEE Transactions on Power Systems*, 2005;**20** (3):1640–6

Keane, A., and O'Malley, M.: 'Optimal distributed generation plant using novel loss adjustment factors', *IEEE Power Engineering Society General Meeting*, Montreal, Canada, 2006

Keith, G., and Leighty, W. C.: 'Transmitting 4000 MW of new wind power from North Dakota to Chicago: new HVDC electric lines or hydrogen pipeline', *Synapse Energy Economics*, Vol. 11, 2002

Kirby, B.: *Spinning reserve from responsive loads*, Oak Ridge National Laboratory, ORNL/TM-2003/19, 2003

Kirby, N. M., Xu, L., Luckett, M., and Siepman, W.: 'HVDC transmission for large offshore wind farms', *IEE Power Engineering Journal*, 2002;**16** (3):135–141

Kishinevsky, Y., and Zelingher, S.: 'Coming clean with fuel cells', *IEEE Power and Energy Magazine*, 2003;**1** (6):20–5

Koch, F., Erlich, I., Shewarega, F., and Bachmann, U.: 'The effect of large offshore and onshore wind farms on the frequency of a national power supply system: simulation modelled on Germany', *Wind Engineering*, 2003;**27** (5):393–404

Kondoh, J., Ishii, I., Yamaguchi, H., Murata, A., Otani, K., Sakuta, K., *et al.*: 'Electrical energy storage systems for energy networks', *Energy Conservation and Management*, 2000;**41** (17):1863–74

Krause, P. C., Wasynczuk, O., and Sudhoff, S. D.: *Analysis of electric machinery and drive systems* (IEEE Series on Power Engineering, Wiley, 2002)

Kristoffersen, J. R., and Christiansen, P.: 'Horns Rev offshore wind farm: its main controller and remote control system', *Wind Engineering*, 2003;**27** (5):351–60

Kundur, P.: *Power systems stability and control* (McGraw-Hill, New York, 1994)

Lalor, G., Mullane, A., and O'Malley, M.: 'Frequency control and wind turbine technologies', *IEEE Transactions on Power Systems*, 2005;**20**:1905–13

Lalor, L., Ritchie, J., Rourke, S., Flynn, D., and O'Malley, M.: 'Dynamic frequency control with increasing wind generation', *IEEE Power Engineering Society General Meeting*, Denver, Colorado, June 2004

Landberg, L.: *Short-term prediction of local wind conditions*, Risø-R-702(EN), Risø National Laboratory, 1994

Landberg, L., Hansen, M. A., Versterager, K., and Bergstrøm, W.: *Implementing wind forecasting at a utility*, Risø-R-929(EN), Risø National Laboratory, 1997

Lang, S., Möhrlen, C., Jørgensen, J., O'Gallchóir, B., and McKeogh, E.: 'Application of a multi-scheme ensemble prediction system of wind power forecasting in Ireland and comparison with validation results from Denmark and Germany', *European Wind Energy Conference*, Athens, Greece, February 2006a

Lang, S., Möhrlen, C., Jørgensen, J., O'Gallchóir, B., and McKeogh, E.: 'Aggregate forecasting of wind generation on the Irish grid using a multi-scheme ensemble prediction system', *Renewable Energy in Maritime Island Climates*, Dublin, Ireland, April 2006b

Larsson, A.: 'Flicker emissions of wind turbines during continuous operation', *IEEE Transactions on Energy Conversion*, 2002;**17** (1):114–23

Laughton, M. A.: 'Implications of renewable energy in electricity supply', IMechE Seminar Publication *Power Generation by Renewables*, London, UK, 2000

Le Gourieres, D.: *Wind power plants theory and design* (Pergamon Press, Oxford, 1982)

Leonhard, W., and Müller, K.: 'Balancing fluctuating wind energy with fossil power stations', *Electra*, October 2002, pp. 12–18

Littler, T. L., Fox, B., and Flynn, D.: 'Measurement-based estimation of wind farm inertia', *IEEE PowerTech Conference*, St Petersburg, Russia, June 2005

Lowrey, C.: 'The Pool and forward contracts in the UK electricity supply industry', *Energy Policy*, 1997;**25** (4):413–23

Lund, H.: 'Large-scale integration of wind power into different energy systems', *Energy*, 2005;**30**:2402–12

Madsen, H., Kariniotakis, G., Nielsen, H. A., Nielsen, T. S., and Pinson, P. (2004) *A protocol for standardising the performance of a short-term wind power prediction* [online], available from: http://anemos.cma.fr/download/ANEMOS_D2.3_EvaluationProtocol.pdf [30 June 2006]

Manwell, J. F., McGowan, J. G., and Rogers, A. L.: *Wind energy explained* (John Wiley & Sons, Chichester, 2002)

Matevosyan, J., and Söder, L.: 'Evaluation of wind energy storage in hydro reservoirs in areas with limited transmission capacity', *4th International Workshop on Large-scale Integration of Wind Power and Transmission Networks for Offshore Wind Farms*, Billund, Denmark, October 2003

McCartney, A. I.: 'Load management using radio teleswitches within NIE,' *IEE Power Engineering Journal*, 1993;**7** (4): 163–9

McIlveen, R.: *Fundamentals of weather and climate* (Chapman & Hall, 1992)

Meteorological Office: *Meteorological Glossary* (HMSO, 1991)

Milborrow, D.: 'Towards lighter wind turbines', *8th British Wind Energy Association Conference*, Cambridge, 1986 (Mechanical Engineering Publications Ltd, London, 1986)

Milborrow, D.: '2000 review shows improving output', *Windstats*, 2001a;**14**(1)

Milborrow, D.: *Penalties for intermittent sources of energy*, Working Paper for PIU Energy Review, 2001b

Milborrow, D.: 'Assimilation of wind energy into the Irish electricity network' in *Perspectives from Abroad* (Sustainable Energy Ireland, Dublin, May 2004)

Milborrow, D.: 'Nuclear suddenly the competitor to beat', *Windpower Monthly*, 2006;**21**:43–7

Milligan, M., and Porter, K.: 'The capacity value of wind in the United States: methods and implementation', *The Electricity Journal*, March 2006, pp. 91–9

Morimoto, S., Sanada, M., and Takeda, Y.: 'Wide-speed operation of interior permanent magnet synchronous motors with high-performance current regulator', *IEEE Transactions on Industry Applications*, 1994; **30** (4):920–6

Müller, S., Deicke, M., and De Doncker, R. W.: 'Doubly fed induction generator systems for wind turbines', *IEEE Industry Applications Magazine*, 2002, pp. 26–33

National Grid Company: *Evidence to House of Lords' Select Committee on 'Electricity from Renewables'*, The Stationery Office, HL78-II, 1999

Nielsen, T. S., Joensen, A., Madsen, H., Landberg, L., and Giebel, G.: 'A new reference for predicting wind power', *Wind Energy*, 1998;**1**:29–34

Nielsen, T. S., Madsen, H., Nielsen, H. A., Landberg, L., and Giebel, G.: 'Prediction of regional wind power', *Global Wind Power Conference*, Paris, France, April 2002

Nielsen, T. S., Madsen, H., Nielsen, H. A., Pinson, P., Kariniotakis, G., Siebert, N., Marti, I., *et al.*: 'Short-term wind power forecasting using advanced statistical methods', *European Wind Energy Conference*, Athens, Greece, February 2006

Nourai, A., Martin, B. R., and Fitchett, D. R.: 'Testing the limits', *IEEE Power and Energy Magazine*, 2005;**3** (2): 40–6

Office of Gas and Electricity Markets (OFGEM): *Initial proposals for NGC's system operator incentive scheme under NETA*, A Consultation Document and Proposed Licence Modifications (OFGEM, London, 2000)

OFGEM (2002) *The review of the first year of NETA* [online], Vol. 2, available from: http://www.ofgem.gov.uk [30 June 2006]

OFGEM (2004) *BETTA at a glance* [online], available from: http://www.ofgem.gov.uk [30 June 2006]

O'Kane, P. J., Fox, B., and Morrow, D. J.: 'Impact of embedded generation on emergency reserve', *IEE Proceedings Part C*, 1999;**146** (2):159–63

P28: *Planning limits for voltage fluctuations caused by domestic, industrial and commercial equipment in the UK* (Energy Networks Association, UK, 1989)

Palutikof, J., Cook, H., and Davies, T.: 'Effects of geographical dispersion on wind turbine performance in England: a simulation', *Atmospheric Environment*, 1990;**24**:(1):213–227

Pantaleo, A., Pellerano, A., and Trovato, M.: 'Technical issues for wind energy integration in power systems: projections in Italy', *Wind Engineering*, 2003; **27** (6):473–93

Papazoglou, T. M. (2002) *Sustaining high penetration of wind generation – the case of the Cretan electric power system* [online], EPSRC BLOWING Workshop, Belfast, available from: http://www.ee.qub.ac.uk/blowing/ [30 June 2006]

Parsons, B., Milligan, M., Zavadil, R., Brooks, D., Kirby, B., Dragoon, K., *et al.*: 'Grid impacts of wind power: a summary of recent studies in the United States', *Wind Energy*, 2004;**7** (2):87–108

Peças Lopes, J. A.: 'Technical and commercial impacts of the integration of wind power in the Portugese system having in mind the Iberian electricity market', *IEEE PowerTech Conference*, St Petersburg, Russia, June 2005

Persaud, S., Fox, B., and Flynn, D.: 'Modelling the impact of wind power fluctuations on the load following capability of an isolated thermal power system', *Wind Engineering*, 2000;**24** (6):399–415

Petersen, E. L., Mortensen, N. G., Landberg, L., Højstrup, J., and Frank, H. P.: *Wind power meteorology*, Risø-I-1206(EN), Risø National Laboratory, 1997

Pinson, P., Ranchin, T., and Kariniotakis, G.: 'Short-term wind power prediction for offshore wind farms – evaluation of fuzzy-neural network based models', *Global Windpower Conference*, Chicago, Illinois, March 2004

Piwko, R., Osborn, D., Gramlich, R., Jordan, G., Hawkins, D., and Porter, K.: 'Wind energy delivery issues: transmission planning and competitive electricity market operation', *IEEE Power and Energy Magazine*, 2005;**3** (6):47–56

Pritchard, G.: 'The must-run dispatch auction in the electricity market', *Energy Economics*, 2002;**24** (3):196–216

Putnam, P. C.: *Power from the wind* (Van Nostrand, New York, 1948)

Rodríguez, J. M., Martos, F. S., Baeza, D. A., and Bañares, S.: 'The Spanish experience of the grid integration of wind energy sources', *IEEE PowerTech Conference*, St Petersburg, Russia, June 2005

Romanowitz, H., Muljadi, E., Butterfield, C. P., and Yinger, R.: 'VAr support from distributed wind energy resources', *World Renewable Energy Congress VIII*, Denver, Colorado, August 2004

Schaber, S., Mazza, P., and Hammerschlag, R.: 'Utility-scale storage of renewable energy', *The Electricity Journal*, 2004;**17** (6):21–9

Schwartz, M.: 'Wind forecasting activities at the U.S. National Renewable Energy Laboratory', *IEA Topical Expert Meeting on Wind Forecasting Techniques*, Boulder, Colorado, April 2000

Sherif, S. A., Barbir, F., and Veziroglu, T. N.: 'Towards a hydrogen economy', *The Electricity Journal*, 2005;**18** (6):62–76

Smith, J. C.: 'Winds of change: issues in utility wind integration', *IEEE Power and Energy Magazine*, 2005;**3** (6):20–5.

Sorensen, B., Petersen, A. H., Juhl, C., Ravn, H., Sondergren, C., Simonsen, P., *et al.*: 'Hydrogen as an energy carrier: scenarios for future use of hydrogen in the Danish energy system', *International Journal of Hydrogen Energy*, 2004;**29** (1):23–32

South Western Electricity Plc: *Interaction of Delabole wind farm and South Western Electricity's distribution system*, ETSU Report, W/33/00266/REP, 1994

Spera, D. A. (ed.): *Wind turbine technology, fundamental concepts of wind turbine engineering* (ASME Press, New York, 1994)

Steinberger-Wilckens, R.: 'Hydrogen as a means of transporting and balancing wind power production' in Ackermann, T. (ed.), *Wind Power in Power Systems* (John Wiley & Sons, Chichester, 2005)

Sustainable Energy Ireland (SEI): *Operating reserve requirements as wind power penetration increases in the Irish electricity system* (Sustainable Energy Ireland, Dublin, August 2004)

Sweet, W.: 'Get rich quick scheme', *IEEE Spectrum*, January 2006, pp. 8–10

Tan, K., and Islam, S.: 'Optimum control strategies in energy conversion of PMSG wind turbine system without mechanical sensors', *IEEE Transactions on Energy Conversion*, 2004;**19** (2):392–9

Tande, J. O. G., and Vogstad, K. O.: 'Operation implications of wind power in a hydro-based power system', *European Wind Energy Conference*, Nice, France, August 1999, pp. 425–8

Troen, I., and Petersen, E. L.: *European wind atlas* (Risø National Laboratory, Roskilde, 1989)

Union for the Co-ordination of Transmission of Electricity (UCTE): *Final report of the investigation committee on the 28 September 2003 blackout in Italy*, UCTE, Brussels, April 2004

van der Linden, S.: 'The potential for bulk energy storage in the USA, current developments and future prospects', *World Renewable Energy Congress VIII*, Denver, Colorado, August 2004

van Zuylen, E. G., Ramackers, L. A. M., van Wijk, A. J. M., and Verschelling, J. A.: 'Wind power fluctuations on a national scale', *European Wind Energy Conference*, Gothenburg, Sweden, 1996, pp. 986–9

Vince, G.: 'Smart fridges could ease burden on energy supply', *New Scientist*, July 2005, p. 24

Voorspools, K. R., and D'Haeseleer, W. D.: 'An analytical formula for the capacity credit of wind power', *Renewable Energy*, 2006;**31**:45–54

Warren, J., Hannah, P., Hoskin, R., Lindley, D., and Musgrove, P.: 'Performance of wind farms in complex terrain', *17th British Wind Energy Association Conference*, Warwick, 1995 (Mechanical Engineering Publications Ltd, London, 1995)

Watson, J.: 'The technology that drove the "dash for gas"', *IEE Power Engineering Journal*, 1997;**11** (1):11–19

Watson, J.: *Advanced fossil-fuel technologies for the UK power industry*, Submission to the UK Government Review of Energy Sources for Power Generation, Science Policy Research Unit, University of Sussex, UK, 1998

Watson, R., and Landberg, L.: 'Evaluation of the Prediktor wind power forecasting system in Ireland', *European Wind Energy Conference*, Madrid, Spain, 2003

Weedy, B. M., and Cory, B. J.: *Electric power systems* (John Wiley & Sons, New York, 1998)

Windpower Monthly: 'The Windicator', 1997–2005 (April issues)

Wood, A. J., and Wollenberg, B. F.: *Power generation operation and control*, 2nd edn (John Wiley & Sons, New York, 1996)

Index